Dinosaurs
A Concise Natural History
THIRD EDITION

The ideal textbook for non-science majors, this lively and engaging introduction encourages students to ask questions, assess data critically and think like a scientist. Building on the success of the previous editions, *Dinosaurs* has been reorganized and extensively rewritten in response to instructor and student feedback. It continues to make science accessible and relevant through its clear explanations and extensive illustrations. Updated to reflect recent fossil discoveries and to include new taxa, the text guides students through the dinosaur groups, emphasizing scientific concepts rather than presenting endless facts. It is grounded in the common language of modern evolutionary biology – phylogenetic systematics – so that students examine dinosaurs as professional paleontologists do. The key emerging theme of feathered dinosaurs, and the many implications of feathers, have been integrated throughout the book, highlighted by the inclusion of stunning new photographs in this beautifully illustrated text, now in full color throughout.

DAVID FASTOVSKY is Professor and Chair of the Department of Geosciences at the University of Rhode Island. His interest in dinosaurs started as a child when he read about a paleontologist's adventures in the Gobi Desert early in the twentieth century. Dinosaurs won out years later when he had the tough decision of choosing between a career in music (he takes his viola on his many field trips) or paleontology, and he has carried out fieldwork all over the world. He's known as a dynamic teacher as well as a respected researcher on the environments in which dinosaurs roamed, as well as their extinction.

DAVID B. WEISHAMPEL is Professor in the Center for Functional Anatomy and Evolution at The Johns Hopkins University School of Medicine. His research focuses on dinosaur evolution and how dinosaurs functioned, and he is particularly interested in herbivorous dinosaurs and the dinosaur record of Europe. Among his many publications he is senior editor of *The Dinosauria*, and has contributed to a number of popular publications, including acting as consultant to Michael Crichton in the writing of *The Lost World*, the inspiration for Steven Spielberg's film *Jurassic Park*. He was recently honored in an International Symposium on duck-billled dinosaurs, dedicated to him and his research.

JOHN SIBBICK has been creating illustrations of extinct life forms and their environments for well over 30 years, producing numerous books on dinosaurs, as well as pterosaurs and general books on prehistoric life. His work has appeared in scientific magazines, television documentaries, and museums, plus a set of stamps depicting dinosaurs and other prehistoric reptiles for the United Kingdom's Royal Mail.

Frontispiece. Gideon Mantell (1790–1852), the "father" of modern dinosaur paleontology.

Dinosaurs
A Concise Natural History

THIRD EDITION

DAVID E. FASTOVSKY
University of Rhode Island

DAVID B. WEISHAMPEL
The Johns Hopkins University

With illustrations by **JOHN SIBBICK**

CAMBRIDGE
UNIVERSITY PRESS

CAMBRIDGE
UNIVERSITY PRESS

University Printing House, Cambridge CB2 8BS, United Kingdom

Cambridge University Press is part of the University of Cambridge.

It furthers the University's mission by disseminating knowledge in the pursuit of education, learning and research at the highest international levels of excellence.

www.cambridge.org
Information on this title: www.cambridge.org/9781316501153

First published 2009
Second edition 2012
Third edition 2016

Printed in the United States of America by Sheridan Books, Inc.

A catalog record for this publication is available from the British Library

Library of Congress Cataloging-in-Publication data
Names: Fastovsky, David E. | Weishampel, David B., 1952-
Title: Dinosaurs : a concise natural history / David E. Fastovsky, University of Rhode Island, and David B. Weishampel, The Johns Hopkins University ; with illustrations by John Sibbick.
Description: Cambridge : Cambridge University Press, 2016. | Includes indexes.
Identifiers: LCCN 2016000278| ISBN 9781107135376 (Hardback : alk. paper) | ISBN 9781316501153 (Paperback : alk. paper)
Subjects: LCSH: Dinosaurs. | Dinosaurs–Research–History. | Paleontology–History. | Fossils–Collection and preservation.
Classification: LCC QE861.4 .F27 2016 | DDC 567.9–dc23 LC record available at http://lccn.loc.gov/2016000278

ISBN 978-1-107-13537-6 Hardback
ISBN 978-1-316-50115-3 Paperback

Additional resources for this publication at www.cambridge.org/dinosaurs3

CONTENTS

PREFACE TO THE THIRD EDITION

Non-bird dinosaurs ("dinosaurs") are very much in the news these days: dinosaurs had feathers; dinosaurs were warm-blooded; dinosaurs were snuffed out in mere seconds; we've just discovered the largest dinosaur ever; we've just discovered the smallest dinosaur ever; here is *Jurassic World*, there is *The Good Dinosaur...* it's a media blizzard. And why not? Ideas about dinosaurs have moved conceptually far beyond the twentieth century: a few weirdo reptiles that lived on this planet, but were too stupid to survive. Today dinosaurs are recognized as centerpieces of vertebrate evolution; they're just a whole lot more *interesting* than they used to be!

On the one hand, our book is written to help sort out who's who and what's what in this barrage of things dinosaurian; on the other hand, and more significantly, our book is written as an introduction to how scientists in general, and natural historians in particular, think about scientific problems.

So what has changed in this third edition? The answer is, a lot; most of it driven by the many exciting discoveries and ideas that have shaped an understanding of dinosaurs since the second edition was published. Even the very number of dinosaurs known has almost trebled just since we began our careers in paleontology, and with things moving forward so quickly, we must keep up.

This new edition:

- contains the newest discoveries as well as new scientific hypotheses; these include feathers and feather *colors* in ornithodirans, new insights into the origin of dinosaurs, lots of new dinosaurian faces, many new (and updated) cladograms, and a gorgeous bestiary of feathered Chinese dinosaurs that has rewritten the origin of birds;
- has over 50 new illustrations, including many photographs shot especially for this book;
- includes newly modified versions of existing illustrations, for increased clarity;
- is for the first time printed in full color, to show the photographs and hand-drawn artwork to best advantage;
- is heavily rewritten, reorganized, and expanded to encompass improved treatments of dinosaur warm-bloodedness and the significance of feathers, a revised look at dinosaur–plant co-evolution, enlarged discussion of the insights gotten through geochemistry and CT-scanning, and an introduction to some stand-out members of the younger generation of paleontologists. In fact, every chapter has been thoroughly revised and, in some cases, completely rewritten; and
- is completely reorganized, reflecting the newest ideas about the origin and subsequent evolution of dinosaurs.

For all that, however, the book retains its essential qualities:

- it is unique among dinosaur textbooks designed for non-majors, because it introduces and exclusively uses the language and methods of professional paleontologists;

- multiple perspectives and hypotheses are presented; science, of course, doesn't consist of immutable answers;
- the beautiful J. Sibbick illustrations continue to grace its pages; and
- it still contains the same lively style, aimed at clear, understandable explanations of dinosaurs and their world.

To the student:

We've written DCNH to introduce you to dinosaurs, amazing creatures that last lived 66 million years ago! We'd also like to use these magnificent beasts to give you insights into natural history, evolution, and the ways that scientists study Earth history.

There are all kinds of questions you can ask about dinosaurs. Simple examples include: Were they really that stupid? Did they die all at once overnight? What were the horns for? Did the mothers take care of their babies? Was *T. rex* really nasty? Were all dinosaurs covered with feathers? What color were they? Could *Brontosaurus* run? How fast? Along with getting answers to these and many other questions, you'll also meet legendary and charismatic dinosaur hunters, younger and older (including the original models for *Raiders of the Lost Ark's* Indiana Jones and *Jurassic Park's* Alan Grant). DCNH will help you think like a scientist, while your knowledge of dinosaurs, natural history, and science grows with each chapter you read.

The authors, D. E. Fastovsky and D. B. Weishampel, are active dinosaur researchers, with a combined experience of almost 80 years of studying dinosaurs in the field and at universities. The book is enriched by the gorgeous original artwork of John Sibbick, one of the world's best and most famous dinosaur illustrators.

DAVID E. FASTOVSKY is Professor and Chair of Geosciences at the University of Rhode Island; he has also taught about dinosaurs and Earth history at the University of California (Berkeley), the National Autonomous University of Mexico and at the University of Vienna. His interest in dinosaurs started as a child when he read about Roy Chapman Andrews in the Gobi Desert (whose story you'll find in the pages of the book you are holding). Fastovsky has had many of his own adventures in far-flung parts of the world, including France, Argentina, Mexico, Venezuela, the western USA and Canada, and Mongolia. He is known as a dynamic teacher and a respected researcher with a focus on the extinction of the dinosaurs, as well as the environments in which they roamed. He has made several television documentary appearances, and was a recipient of the Distinguished Service Award from, and is a Fellow of, the Geological Society of America.

DAVID B. WEISHAMPEL is Professor in the Center for Functional Anatomy and Evolution at The Johns Hopkins University. Recipient of two teaching awards, Weishampel teaches human anatomy, evolutionary biology, cladistics, and, of course, a course on dinosaurs. His research focuses on dinosaur evolution and how dinosaurs function, and he is particularly interested in herbivorous dinosaurs and the dinosaur record of eastern Europe and Mongolia. He is the senior editor of the immensely well-received *The Dinosauria*, and has written or co-written, not including the DCNH series, four books and a many scholarly articles. Weishampel has contributed to a number of popular publications as well, including acting as consultant to Michael Crichton in the writing of *The Lost World*, the inspiration for Steven Spielberg's film *Jurassic Park*. In 2011, a special symposium on duck-billed dinosaurs was convened by the Royal Ontario and Royal Tyrrell Museums, Canada, to honor Dr. Weishampel's contributions to dinosaur paleontology.

JOHN SIBBICK has over 35 years of illustration experience working on subjects ranging from mythology to natural history and is probably best known for his depictions of prehistoric scenes and dinosaurs. In the first stage of any commission he takes the fossil evidence and consults with specialists in their field and works out a number of sketches to build up an overall picture of structure, surface detail, and behavior. From his base in the United Kingdom, he has provided images for books, popular magazines such as the *National Geographic*, and television documentaries, as well as museum exhibits and one-man shows of original artwork. For this book he has provided around 250 pieces of original art.

To the instructor:

Dinosaurs: A Concise Natural History is designed to introduce first- and second-year university students, many commonly seeking to fulfill general science requirements, to the logic of scientific inquiry and to concepts in natural history and evolutionary biology. The perspectives and methods introduced through dinosaurs have a relevance that extends far beyond the dinosaurs, engendering in students scientific logic and critical thinking. The approach has been successful, and new discoveries and interpretations now merit this third edition.

In its preparation, Cambridge University Press devoted considerable energy to obtaining extensive feedback from many instructors who had had experience teaching from previous editions. Their thoughtful, detailed, and, in many cases, comprehensive, answers were particularly useful in determining the ways in which this edition could be strengthened as a teaching tool; indeed, we responded, whenever possible, to all suggestions and recommendations. The care that this group of veteran instructors put into their answers has surely enriched our book.

A unique conceptual approach:

Names, dates, places, and features are available everywhere these days. But litanies of names, dates, and places is not science; the *creative* synthesis of these data is far more important and, fortunately, far more interesting. The goal of this book is to help students achieve that synthesis.

Uniquely among dinosaur textbooks, phylogenetic systematics is immanent in DCNH. This approach, however, follows current practice in evolutionary biology, and allows students to understand dinosaurs as professional paleontologists do. To have had an entire class in dinosaurs, and yet be insensible to the underlying phylogenetic connections among these (and all) organisms strikes us as indefensible; it would be akin to studying biology without evolution. The cladograms used in this book are thus drawn in a way that highlights the evolutionary relationships they depict, ensuring that both the methods and conclusions of phylogenetic systematics remain accessible.

Part I introduces the fundamental intellectual tools of the trade. Chapters 1 and 2 treat geology, the geological time scale, fossils, collecting, and what happens after the bones leave the field. The third chapter, a carefully crafted introduction to the logic of phylogenetic systematics, uses familiar and common examples to acquaint students with the method. Chapter 4 takes students, phylogenetically, from basal Vertebrata to Dinosauria.

Parts II and III cover, respectively, Saurischia and Ornithischia. The chapters within Parts II and III cover the major groups within Dinosauria, treating them in terms of phylogeny and evolution, behavior, and lifestyle. For the first time in the DCNH series, this edition puts Saurischia ahead of Ornithischia: this organization allows students to move smoothly from basal (non-dinosaur) dinosauromorphs to the earliest dinosaurs, whose affinities are clearly saurischian, although perhaps murky within that group (Chapter 5). Birds (Chapter 8) follow from the theropod chapters (6 and 7), because birds are living saurischians. In this edition, the normally prominent status accorded to the venerable *Archaeopteryx* has been diminished, since the astounding Liaoning fossil discoveries in the past 20 years have diminished the uniqueness of *Archaeopteryx* as a transition to birds. In recognition of the close relationship between dromaeosaurs and avialans (including birds), we introduce the group Paraves for the first time in this new edition. Ornithischians are treated in Chapters 10–12, culminating in Ornithopoda, a group that, with the new millennium, has been phylogenetically somewhat fraught. By the time students reach this chapter, however, they will be in a position to understand, appreciate, and assimilate some of the uncertainty.

Part IV covers the aspects of the paleobiology of Dinosauria, from their metabolism (Chapter 13), to the great rhythms that drove their evolution (and co-evolution; Chapter 14), to a fully updated exposition on their extinction (Chapter 16). Chapter 15, the penultimate chapter, is devoted to the history of dinosaur paleontology. Although commonly introduced at the beginning of dinosaur books via a scaffolding of names, dates, and discoveries,[1] our history chapter – a history of *ideas* – is placed toward the end, so that the thinking that currently drives the field can be understood in context. We believe that the history of dinosaur paleontology is much more resonant when students already know something about the fossils being hunted and the ideas being developed. Finally, the book ends, like the dinosaurs themselves, with a discussion of the great Cretaceous–Paleogene mass extinction. Here we might say, as so many have, that Earth then entered the Age of Mammals, but, paradoxically, we'll try to persuade readers that we're still in the "Age of Dinosaurs".

We would cheat our readers if we left out accounts of the dinosaur hunters, whose colorful personalities and legendary exploits make up the lore of dinosaur paleontology; so we've included many of their stories as well (Chapter 15). In the first and second editions of DCNH, we transparently introduced our own generation as "Young Turks;" forgetting that, like everybody else, we've aged too! The third edition, therefore, introduces its readers to a younger (thirties and forties) generation of paleontologists whose striking accomplishments bode well for their future achievements.

Finally, as in all previous editions, any errors that appear in this work are entirely Dave's fault.

Features:

DCNH is designed to help instructors to teach and to help students learn.

- The book is richly illustrated with new, especially commissioned, art by John Sibbick, one of the world's foremost illustrators of dinosaurs. These images effectively highlight and reinforce the concepts in the text. Many pages are also graced by photographs, generously contributed by professional paleontologists or taken especially for this edition.
- The chapters are arranged so that they present the material in order of increasing complexity and sophistication, building the confidence of the student early on, and extending the sophistication of their learning gradually through the book.
- The tone of the text is light, lively, and readable, engaging the student in the science, and dispelling the apprehension many students experience when they pick up a science textbook.
- "Objectives" at the beginning of each chapter help students to grasp chapter goals.
- Boxes scattered throughout the book present a range of ancillary topics, from dinosaur poetry, to extinction cartoons, to how bird lungs work, to colorful accounts of unconventional, outlandish, and extraordinary people, places, and stories.
- A comprehensive series of "Topic questions," to be used as study guides, are located at the end of each chapter. The questions probe successively deeper levels of understanding, and students

---

[1] Indeed, many of our respondents requested this.

who can answer all of the "Topic questions" will have a good grasp of the material. Variants of these questions can serve as excellent templates for examination questions.

- A Glossary ties definitions of key terms into the page numbers where the term is used.
- There are two indices: an Index of subjects and an Index of genera that includes English translations of all dinosaur names.
- Appendices are included in certain chapters to introduce material that students may need in order to understand chapter concepts, such as the chemistry necessary to understand radioactive decay, plate tectonics, and the basic principles of evolution by natural selection (Darwinian evolution).

Online resources to help you deliver your dinosaur course include:

- electronic files of the figures and images within the book;
- lecture slides in PowerPoint with text and figures to help you to structure your course; and
- solutions to the questions in the text for instructors.

All resources are available to instructors at www.cambridge.org/dinosaurs3.

DEDICATION

To **Lesley, Naomi,** and **Marieke,**
my family.
To poor **Robert,** because. . .

To **Sarah** and **Amy.**
Thanks for continuing to remind your dad
that there are things other than dinosaurs!

PART I
REMEMBRANCE OF THINGS PAST

Dinosaur tales

This book is a tale of dinosaurs; who they were, what they did, and how they did it. But, more significantly, it is also a tale of natural history. Dinosaurs enrich our concept of the **biosphere**, the three-dimensional layer of life that encircles the Earth. Our biosphere has a 3.8 *billion*-year history; we, and all other life around us, are merely Earth's latest tenants. To be unaware of the history of life is to be unaware of our organic connections to the biosphere and ultimately to our Earth. Dinosaurs have significant lessons to impart in this regard, because as we learn who dinosaurs really are, we can better understand who *we* really are.

Since the first dinosaur bones were identified in the early nineteenth century, both the total number of dinosaurs known, and the rate of finding them, has spiraled (Figure I.1).

Not only that; it is clear that they are no longer your parents' cast-iron clunkers. Their relationships to us are not what we thought; their lifestyles are not what we thought; they look very different from what we'd imagined; indeed, almost everything about them has changed since the late 1960s and, thankfully, they're a whole lot more interesting now. It seems as if we truly are in a Golden Age of Dinosaur Discovery!

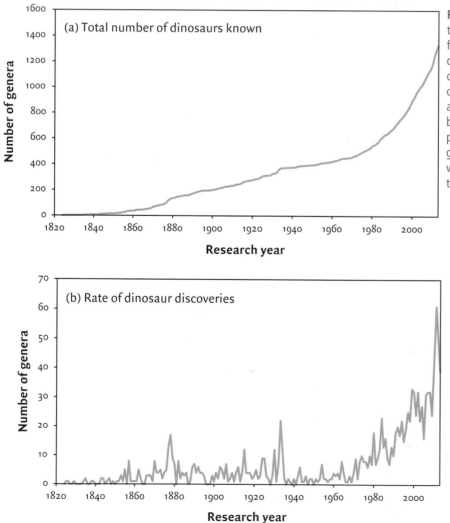

Figure I.1 Paleontologist Michael Benton's compilation of dinosaur discoveries from 1822 to 2014. (a) Total number of dinosaurs known; (b) rate of dinosaur discoveries. There were spikes in the rate of dinosaur discoveries in the late 1880s and in the early to mid twentieth century, but all of that pales in comparison to the present. As recently as the previous generation of paleontologists, dinosaurs were not so well understood as was thought!

The word "dinosaur" in this book

The term "dinosaur" (*deinos* – terrible; *sauros* – lizard) was invented in 1842 by the English naturalist Sir Richard Owen (see Box 15.2) to describe a few fossil bones of large, extinct "reptiles." With modifications (for example, not all dinosaurs are "large"), the name has proven resilient. It has become clear in the past 25 years, however, that not all dinosaurs are extinct; in fact, almost all specialists now agree that *birds are living dinosaurs*. The technically correct term **non-avian dinosaurs** to specify all dinosaurs except birds is a mouthful, so we prefer to use the term "dinosaurs" as shorthand for "non-avian dinosaurs." The distinction between non-avian dinosaurs and all dinosaurs will be most relevant only when we discuss the origin of birds and their early evolution in Chapters 8 and 9; there, we'll take care to avoid confusing terminology.

Science

Ours is also a tale of science, which is in turn all about imagination and creativity, with a little help from data. Creativity is the currency of science; the value of the science is surely dependent upon the creativity of the scientist. In the following pages, we hope to build a sense of the intellectual richness of science; a quality that philosopher of science Karl Popper called the "logic of scientific discovery." And so if this is a story about science, we'd better articulate what we mean by the word "science."

Science = testing hypotheses

Science is the business of constructing testable hypotheses and then testing them. An example of a simple scientific hypothesis is: "The Sun will rise tomorrow." This hypothesis makes specific predictions. Most importantly, the hypothesis that the Sun will rise tomorrow is **testable**; that is, it makes a prediction that can be assessed. The test is relatively straightforward: we wait until tomorrow morning and either the Sun rises or it doesn't.

So what's *not* science? Important questions that you can't get at using science might be "Is there a God?", "Does she love me?", and "Why don't I like hairy men?" In *Music Man*, Marian "the librarian" Paroo asks "What makes Beethoven great?" She won't learn it from science. Some questions are far better suited to science than others.

The "proof" is in... the test!

In our example of a scientific hypothesis (above), if the Sun does not rise, the statement has been *falsified*, or demonstrated to be not correct, and the hypothesis can be rejected. On the other hand, if the Sun rises, the statement has *failed falsification*, and the hypothesis cannot be rejected. For a variety of relatively sophisticated philosophical reasons, scientists do not usually claim that they have *proven* a statement to be true; rather, the statement has simply been tested and not falsified. It turns out that, in a philosophical sense, it is very difficult to "prove" anything. For this reason, scientists rarely use words like "prove" or "true" when discussing their work.

But we *can* test hypotheses, and one of the basic tenets of science is that it consists of hypotheses that have predictions which can be tested. We will see many examples of hypotheses in the coming chapters; to be valid, all must involve testable predictions. Without testability, it may be very interesting; it may be very exciting, but it is not science.

Science in the popular media

"Science" is everywhere in the popular media, from Animal Planet to *National Geographic*, to the Weather Channel. That's good; our lives are increasingly intersected by science, and the more savvy we are scientifically, the better the decisions that we can make. But the problem is that not all of the "science" that one sees in the popular media is very scientific!

Among scientists, work is recognized and accepted only after it is **peer-reviewed**; that is, when it has been vetted and carefully scrutinized by other scientists qualified to evaluate the work. Very commonly, the work goes through multiple cycles of review and revision by the authors in response to the critiques of professionals in the field. At that time – and only at that time – it can be published. This is the time-honored – and time-proven – way of producing the very highest quality scientific research.

In the popular media, the goals are somewhat different. In movies, on television, and on the radio, the goals are generally entertainment as well as immediacy: publicizing the newest discoveries as quickly as possible. Both of these have very little to do with high-quality science. In the case of immediacy, the rush to publicize can mean that half-baked theories and interpretations are popularized long before they have seen anything like peer review. And when peer review finally comes, the results can be (and have been) devastating.

The goal of entertainment can be equally destructive, in terms of the science. For example, television and radio reports commonly present "the other side" of a particular scientific issue, even when the vast preponderance of scientists take a particular position and the "other side" has very little legitimacy. People can always be found who will voice heterodox opinions, even when their views are plainly way outside of the mainstream of peer-reviewed science. It's always *entertaining* to present controversy.

In paleontology, the quest for entertainment brings a variety of excesses, beyond mere controversy. Size sells; teeth and claws sell; asteroids sell; extinction sells. But ultimately, we hope in this book to show that the real riches – and entertainment – lie in a balanced, undistorted treatment of dinosaurs: a more magnificent, fascinating group of creatures could hardly have been invented, and the ongoing scientific process of revealing who they were and what they did is compelling drama.

The web, a popular and extremely important source of information, is driven by a variety of sometimes-contradictory imperatives (knowledge; entertainment; retail), and our relationship to it is more nuanced. Accessability, ease, and breadth are its great strengths; reliability and depth can be, but as always, care about sources is required. Much really good, peer-reviewed science (and lots of other information!) is available on the web; indeed, most major peer-reviewed journals have a significant web presence; some can be found *only* on the web. Information from the websites of professional societies (for example, the Society of Vertebrate Paleontology) is reliable in the way that the peer-reviewed journals published by such societies are reliable. Academic institutions tend to have reliable web pages; although personal web pages, even in an academic context, generally reflect the opinions of their authors, idiosyncratic or not. Wikipedia – a stupendous tool for research and, in our view, a stunning contribution to the availability of human knowledge, reflects the web: it is often very accurate, but it is not without idiosyncratic entries. In short, the web is a powerful tool, to be sure; but *caveat emptor*: Not all sources of information are equivalently authoritative.

To catch a dinosaur

1

CHAPTER OBJECTIVES

- Understanding fossils and fossilization
- Collecting dinosaur fossils
- Preparing dinosaur specimens

Preservation and fossils

That we even know there ever were such creatures as dinosaurs is near-miraculous: throughout their time on Earth, a few individuals of the innumerable dinosaurs that must have once lived, happened to be preserved as **fossils**, the buried remains of organic life. There are many types of fossils: most familiar are **body fossils**, which commonly involve some part of the animal (e.g., bones). **Trace fossils**, impressions such as footprints, are also familiar. Finally, there are a variety of **other** kinds of fossils, including things as disparate as molecules and skin impressions.

Dinosaurs last romped on Earth ~66 million years ago, and until the very latest part of the twentieth century, paleontologists mostly assumed that dinosaur **soft tissues** – muscles, blood vessels, organs, skin, fatty layers, etc. – were long gone, with the only observable vestiges being the **hard parts**: generally, bones and teeth. Hard parts were presumably not as easily degraded over time as the soft tissues that constitute most of the body. Yet as it turns out, new techniques and determined study are revealing all kinds of preservation in unexpected places, including tissues, cells, and molecules (for example, the discovery of actual red blood cells and connective tissues from *Tyrannosaurus*; see Box 6.3). Fair warning: it's no longer just about old bones, and we'll visit some of these newer discoveries throughout this book.

Making body fossils

Before burial

Consider what might happen to a dinosaur – or any land-dwelling vertebrate – after it dies (Figure 1.1).

Carcasses are commonly **disarticulated** (dismembered), first by predators and then by scavengers ranging from mammals and birds to beetles and bacteria. Some bones might be stripped clean of meat and left to bleach in the sun. Others might get carried off and gnawed. Sometimes the disarticulated remains are trampled by herds of animals, breaking and separating them further. Plants produce acids that dissolve bone. But ultimately, the nose knows that most of the heavy lifting in the world of decomposition is done by bacteria that feast on rotting flesh. The sum total of all the earthly remains of the animal will end up lying there: a few disarticulated bleached bones in the vegetation (Figure 1.1).

Figure 1.1. Bones. A wildebeest carcass, partly submerged in mud and water and on its way to becoming permanently buried and fossilized. If the bones are not protected from scavengers, air, and sunlight, they decompose rapidly and are gone in 10–15 years. Bones destined to become high-quality fossils must be buried soon after the death of the animal.

Replacement and/or permineralization

Nearly complete specimen exposed

Quick burial

(a)

Dinosaur dies

(b)

Dismemberment before burial –
scavenging and other natural
processes

Isolated bones buried and mineralized

Isolated bones exposed

Figure 1.2. Two endpoint processes of fossilization. In both cases, the first step is the death of the animal. Some decomposition occurs at the surface. In the upper sequence (a), the animal dies, the carcass undergoes quick burial, followed by bacterial decomposition underground, and permineralization and/or replacement. Finally, perhaps millions of years later, there is exposure. Under these conditions, when the fossil is exhumed, it is largely complete and the bones articulated (connected). This kind of preservation yields bones in the best condition. In the lower sequence (b), the carcass is dismembered on the surface by scavengers and perhaps trampled and distributed over the region by these organisms. The remains may then be carried or washed into a river channel and buried, replaced, and/or permineralized, eventually to be finally exposed perhaps millions of years later. Under these conditions, when the fossil is exhumed, it is disarticulated, fragmented, and the fossil bones may show water wear and/or the gnaw marks of ancient scavengers. Different conditions of fossil preservation tell us something about what happened to the animals after death.

If the animal isn't disarticulated right away, it is not uncommon for a carcass to bloat, as feasting bacteria produce gases that inflate the dead body. After a bit, the carcass will likely deflate (sometimes explosively), and then dry out, leaving bones, tissues, ligaments, tendons, and skin crusty and inflexible.

Burial

Sooner or later bones are either destroyed or buried. If they aren't gnawed and digested as somebody's lunch, their destruction can come from **weathering**, which means that the minerals in the bones break down and the bones wash away. But the game gets interesting for paleontologists when weathering is stopped by rapid burial. At this point, they (the bones, not the paleontologists) become fossils. Figure 1.2 shows two of the many paths bones might take toward fossilization.

Figure 1.3. Permineralized bone from the Jurassic-aged Morrison Formation, Utah, USA. The fossilized bone is now a solid piece of rock.

After burial

The burial environment is not chemically static, and the *bad* news is that bone is made out of calcium–sodium hydroxyapatite, a mineral that is not terribly stable on the Earth's surface. As a result, the original bone minerals are easily **replaced** by other minerals, commonly retaining the details and form of the bone. This is especially likely if the bone comes into contact with fluids rich in dissolved minerals, such as commonly occurs after burial. The *good* news is that, by few choice cation substitutions while buried, calcium–sodium hydroxyapatite commonly alters to more stable forms of apatite, a familiar one being fluorapatite.

If no fluids are present throughout the history of burial (a time interval that could be measured in millions of years), the bone may be unreplaced, which is to say that 100% of the original bone mineralogy remains. On the other end of the spectrum, a completely replaced bone is a magnificent natural forgery: chemically and texturally not bone, but retaining the exact shape and delicate features of the original bone. Given burial stays measured in tens of millions of years, however, most dinosaur bones are altered (replaced) to a greater or lesser degree. Alteration/replacement tends to be progressively greater in the case of older and older fossils; degrees of alteration are generally the norm.

Since bones are porous, the spaces once occupied by blood vessels, connective tissue, and nerves easily fill up with minerals. This is called **permineralization** (Figure 1.3).

Most fossil bones undergo a combination of some degree of replacement and permineralization.

The preceding description explains how isolated fossil bones, or perhaps most of a single animal, might come to be preserved. But along with these isolated finds, on rare and lucky days, we can sometimes come upon **bonebeds,** rich finds with hundreds – even thousands – of bones preserved. Sometimes, these bonebeds are **monospecific,** that is, they contain fossils of mostly one species, and one can't help but wonder if these perhaps reflect some kind of **gregarious,** or herding behavior captured in the fossil record.

Trace and other kinds of fossils

Trace fossils

Still the single most important type of dinosaur fossil, other than the bones themselves, is trace fossils. Dinosaur trace fossils (sometimes also called **ichnofossils** [*ichnos* – track or trace]), come as **footprints** or as complete **trackways.** Figure 1.4 shows a **mold,** or impression, of a dinosaur footprint. We also find **casts,** which are made up of material filling up the mold. Thus a cast of a dinosaur footprint is a three-dimensional object that formed inside the impression (or mold).

Figure 1.4. Theropod dinosaur footprint from the Early Jurassic Moenave Formation, northeastern Arizona, USA. Human foot for scale.

In the past 20 years the importance of ichnofossils has been recognized. Ichnofossils have been used to show that dinosaurs walked erect, to reveal the position of the foot, and to reconstruct the speeds at which dinosaurs traveled. Trackways tell remarkable stories, such as that fateful day 150 or so million years ago when a large predatory theropod stalked a herd of smaller animals (Figure 1.5).

Other fossils

For want of a more imaginative name, we'll lump the various other kinds of fossils under "Other." Sometimes the fossilized feces of dinosaurs and other vertebrates are found. Called **coprolites**, these occasionally impressive relics can give an intestine's-eye view of dinosaurian diets. Likewise, as we shall see later in this book, fossilized eggs and also skin impressions have been found. Nests are also known (Figure 1.6). Molecules; cells; stomach stones (**gastroliths**; see Figure 6.20); soft tissue; the list is really as long as there are parts of a dinosaur!

Finding fossils

So, if the fossils are buried, how is it that we find them? The answer is really in the luck of geology: if fossil-bearing **sedimentary rocks** happen to be eroded, and a **paleontologist** happens to be looking for fossils at the moment that one is actively eroding from a rock, the fossil *may* be observed and *may* be collected. Who knows how many times, throughout their 160+ million-year existence on Earth, dinosaurs stepped on exposed, weathering fossils of earlier dinosaurs, now surely lost to eternity (Figure 1.7)?

It's a tenuous connection: the vagaries of fossil preservation, the chance of geological exposure, and their discovery by an ambitious paleontologist. And yet they're out there – the fossils and the paleontologists!

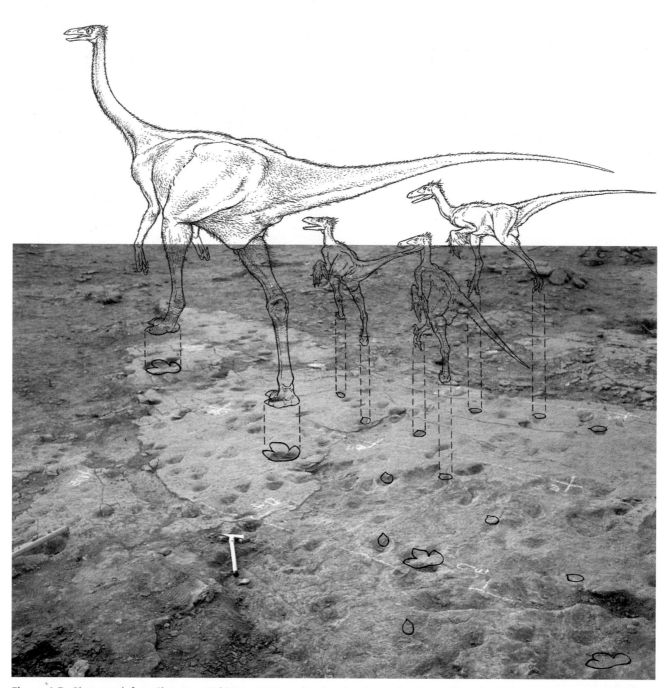

Figure 1.5. Photograph from Shar-tsav, Gobi Desert, Mongolia, showing the tracks of a medium-sized theropod dinosaur among those of a pack of smaller theropods. Our drawing suggests an interpretation consistent with the evidence: a *National Geographic* moment in the Late Cretaceous when a pack of *Velociraptor* attacked a single *Gallimimus*.

Collecting

The romance of dinosaurs is bound up with collecting: exotic and remote locales, heroic field conditions, and the manly extraction of gargantuan beasts (see Chapter 15). But ultimately dinosaur collecting is a process that draws upon good planning, a strong geological background, and a bit of luck. The steps are:

(1) **planning**;

(2) **prospecting**; that is, hunting for fossils;

(3) **collecting**, which means getting the fossils out of whichever (usually remote) locale they are situated; and

(4) **preparing** and **curating** them; that is, getting them ready for viewing, and incorporating them into museum collections.

These steps involve different skills and sometimes different specialists.

Planning

Collecting dinosaur fossils is not to be undertaken lightly. Dinosaur bones are – even in the richest sites – quite rare, and the moment they are disturbed the loss of important information becomes a concern. For this reason, most professional paleontologists have advanced degrees – often a PhD in the geological or biological sciences – but before actually leading an expedition themselves, *all* have acquired many years of experience both in the logistical as well as the scientific ends of fieldwork.

Running an expedition

The logistical end of an expedition involves keeping one's team fed, watered, healthy, and happy in remote places where, in many cases, these commodities don't come easily. Relentless sun, extreme heat, dust, lack of amenities, subsistence on a limited diet, and isolation from the "real world," all conspire to wear down even the most robust of people. It's all happening in the Great

Figure 1.6. Fossil burrow of the dinosaur *Oryctodromeus*. Careful study of the sedimentary context of this dinosaur revealed the burrow.

Outdoors, true, but it's nothing like a camping catalog! Add to these, language problems when you are working in other countries and limited access to medical facilities in the event of an accident involving either you or one of your crew, and the potential for disaster increases dramatically.

Many expeditions have to carry everything with them – fuel, water, food, all gear for the maintenance of daily life – as well as all the maps and equipment necessary to successfully carry out the science and safely retrieve heavy, yet delicate, dinosaur bones. This takes some serious planning and experience; you and your crew's lives may depend upon it (Figure 1.8). You have to know what you are doing.

Fossils generally, and dinosaurs in particular, are not **renewable resources**, which means that collecting a dinosaur is a one-shot deal: it must be done right, because we will never be afforded another chance to do it again. Any information that is lost – any piece of it that is damaged – may be lost or damaged forever. For this reason, there are many regulations associated with collecting vertebrate fossils.

The most basic are the collection permits required for work on public lands. Obtaining the permits requires advanced planning because the agencies in charge of issuing the permits reasonably require detailed accounts of your plans before the process can go forward.

One important part of the permit-obtaining process, especially in the case of dinosaur fossils (which tend to be large and heavy), is the eventual location of the fossils. Who gets them? Does

Figure 1.7. A pair of *Parasaurolophus* walking over some exposed fossilized bones of an earlier dinosaur that are weathering out of the cliff. Fragments of the fossilized bone have fallen at the dinosaurs' feet.

that person or place have the proper resources – or even the space – to store, preserve, and make them accessible to scientists and the general public? How is all this to be accomplished? Most of the truly great collections and many of the most important dinosaur fossils are housed in major museums, such as the American Museum of Natural History (New York), the Field Museum (Chicago), the Yale Peabody Museum (New Haven, CT), Tyrrell Museum (Alberta), the Smithsonian (Washington, DC), the Natural History Museum (London), the Musée National d'Histoire Naturelle (Paris) and many others around the world. These institutions have the resources required for the care of important specimens and the data associated with them.

Work overseas – and paleontology generally involves a lot of foreign travel, no matter where you live – involves a whole new level of administrative preparation. All of the problems described above are compounded by language barriers, by the necessity to obtain visas along with permits, by the logistics of preparing a field expedition in a foreign country, and by the necessity of arranging for the eventual disposition of the fossils. What country, after all, would gladly see its fossil resources dug up and exported elsewhere? It's a delicate balance, sometimes requiring the skills of a diplomat.

Box 1.1   A dino named "Sue"

"Sue" is a spectacular fossil of a very large *T. rex* – and a notorious object lesson in the kinds of misunderstandings and problems that can arise when collecting dinosaurs. Paleontologist Peter Larson, co-founder and President of the Black Hills Institute of Geological Research (BHIGR; a company that collects, prepares, and sells fossils and casts), believed that he had obtained permission from South Dakota rancher Maurice Williams, for the rights to collect (and then, presumably, prepare and sell) fossils he and his crews might find on Mr. Williams' land. They agreed that BHIGR would pay Mr. Williams for any fossils that he found. In July, 1990, Sue Hendrickson, a volunteer crew member, discovered the dinosaur (hence the name). It was *very* rare – a *T. rex*: somewhat jumbled, but clearly large, semi-articulated, well-preserved, and obviously extremely valuable. Larson and his crews devoted considerable resources (ultimately, ~$209 000) and time to collecting the many parts of the specimen (the skull alone was almost 2 m long!), starting with shaving almost 10 m off of the top of the hill in which it had been found so that they could reach the bones. By the end of the field season, they had brought a very large chunk of the specimen back to their laboratory in South Dakota, where preparation began. In accordance with Mr. Larson's understanding of the original agreement, a check for $5000 was issued to Mr. Williams for rights to the specimen. Talks were presented at national scientific meetings; publications were planned; preparation continued apace; so far, it seemed as if it was exciting specimen, but not a socio-political phenomenon.

Then the patient went septic. Mr. Williams claimed that he had not given BHIGR rights to the specimen; only that that they were allowed to remove the fossil and prepare it. Complicating the issue, because Mr. Williams is a member of the Sioux Tribe, the Sioux Nation claimed that the fossil belonged to it. And as if that weren't difficult enough to sort out, the US Federal Government claimed that Mr. William's land was held in trust by the United States, and therefore the fossil belonged to the United States as represented by the Department of the Interior!

In 1992, without warning, the feds moved in: guns, FBI agents, police.[a] Somewhat unceremoniously (by all accounts), the FBI searched the place, and the US National Guard forcibly removed Sue's remains from the BHIGR, storing them at the University of South Dakota School of Mines and Technology, a well-known facility able to handle a specimen of this magnitude – and equally importantly, presumably, neutral ground.

Then the courts took over. The dinosaur languished. After a rather lengthy trial, the court returned the specimen to Mr. Williams, who sought to auction it through the famous auction house of Sotheby's. In the mean time, Mr. Larson served jail time for permit infractions, unrelated to Sue; however, there are those who claim that the charges and subsequent sentencing were in effect political.

The value of the specimen was undoubted, and the concern was that it would leave the United States and end up in a private collection somewhere where it could not be appreciated. So, an unholy alliance formed between the Field Museum in Chicago, the California State University System, Disney Parks, McDonalds, the Ronald McDonald House, and private donors, who all pitched in to see the fossil displayed at the Field Museum, which won the bid for a total output of about $8.3 million.

Preparation was carried out both at the Field Museum and at Disneyworld in Orlando, Florida, where thousands of visitors to both venues saw the bones being freed of matrix, cast, and readied for mounting. Sue is now the star attraction of the Field Museum. All's well that ends well, right?

The "Sue" story is perhaps the most egregious example of the kinds of things that can go awry in paleontological collecting. By most accounts it ended well: the dinosaur is safe, intact – estimates are that it is an almost unbelievable 80% complete – and on display where it is properly maintained and optimally appreciated. On the other hand, maybe things didn't end so well: Sue changed the landscape of dinosaur paleontology, because now everybody recognizes that dinosaur fossils, formerly thought to be valueless, can potentially be sold for millions of dollars. Collecting dinosaurs has become an expensive business, and many scientists no longer have the resources to stay in the game.

[a] Larson quotes an FBI agent as saying, "This can be real easy, or real hard, depending on whether or not you are willing to cooperate." We favor Robert De Niro for the role.

Figure 1.8. Supplies for one of the American Museum of Natural History's 1920s expeditions to the Gobi Desert. In the intervening 90 years, nobody has found a way to get around hauling the basic necessities into the field.

Science

All of that care expended upon all those logistics is meaningless unless our planning extends to the science as well. Paleontologists don't just go to weird places and grab bones. If they did, they'd lose, forever, essential information bearing upon four major problems.

(1) What kind of environment was it in which the dinosaur was preserved (because it might have lived somewhere else)?
(2) Where did it live (geographically)?
(3) When it did it live?
(4) How did it die?

Oryctodromeus, a small burrowing, herbivorous dinosaur (Figure 1.6), is a perfect example of the importance of geological context. Here was an animal found fossilized in its own burrow. Had the important geological context not been properly interpreted, the burrow would not have been recognized and this animal's unusual behavior (for a dinosaur, at least) would have gone unappreciated.

So before even collecting the fossil, the **locality** – the area in which the fossil or fossils occur – must be mapped geologically, in a way that records the most information possible about the setting in which the fossil was found. This kind of information requires specialized geological study of the **paleoenvironments**, that is the ancient environments represented by the rocks in which the fossils are found, as well as the geological context above and below the fossil. Usually this is accomplished by geological mapping and by detailed study of the sedimentary geology of the locality. Interpreting the ancient environment in which the bones came to rest commonly involves teaming up with **sedimentologists** – geoscientists with specialized knowledge of sedimentary rocks and the ancient environments that they preserve. This kind of teamwork allows paleontologists to develop the most complete picture of the fossils and the conditions in which they lived and died.[1]

---

[1] The study of all that happens to an organism after its death is called "taphonomy," and is a specialized field combining sedimentology and paleontology. Understanding the taphonomy of a fossil is the best way to know whether the animal actually lived in the environment in which its fossils were found, or whether its carcass was just deposited there after death.

A question commonly asked, is "How do you know where the dinosaur fossils are?". The simplest answer is, "We don't." There is no secret, magic formula for finding dinosaurs, unless it is long, hard hours of pre-expedition library time and careful assessment of potentially **productive** – that is, fossil-bearing – regions. On the other hand, a well-educated guess rooted in knowing something about the kinds of environment in which dinosaurs lived can greatly increase the odds of finding fossils.

Some basic criteria help the success of the search. These are:

(1) Right rocks: the rocks must be sedimentary.
(2) Right time: the rocks must be of the right age.
(3) Living on land: the rocks must be terrestrial.

Right rocks

Sedimentary rocks have the best potential to preserve dinosaur fossils. Indeed, sedimentary rocks form in, and represent, sedimentary environments, many of them places where dinosaurs lived and died. Dinosaurs are known from other types of rocks, but their fossils are most likely found in sedimentary rocks.

Right time

If the rocks you search were *not* deposited sometime between the Late Triassic and the Late Cretaceous, you won't find dinosaurs. Older and younger rocks will yield amazing fossil creatures, but not dinosaurs.

Living on land

Dinosaurs were **terrestrial,** that is, non-marine beasts through and through, which means that their bones will generally be found in rocks that preserve the remnants of ancient river systems, deserts, and deltas. However, dinosaur remains are also known from lake deposits and from near-shore marine deposits.

Many of the richest fossil localities in the world are in areas with considerable rock exposure, such as **badlands**. Fossil localities are common in deserts: plant cover on the rocks is low, and the dry air slows down the rates of weathering so that, once a fossil is exposed, it isn't chemically destroyed or washed away. Paleontologists, therefore, don't often find themselves in the jungle looking for fossils; the weathering rates are too high and the rocks are covered by vegetation. The chances of finding fossils are best in deserts, or at least fairly dry regions. Still, not all dinosaur material has been found in deserts. As long as the three criteria above are met, there is a possibility of finding dinosaur fossils, and that's usually reason enough for going in and taking a look.

Because fossils are a non-renewable resource, collecting them should be treated with the utmost circumspection. Poor planning, indifference to the significance of fossils, lack of training, and ignorance when retrieving them from the ground, will at best lose important data, and at worst place you and your team's lives in jeopardy.

Prospecting

Once we've planned properly, and we've got ourselves and our equipment to outcrops that match the criteria above, and once we have developed a concept of their geological context, what then? Simply put, we drop our eyes and start searching for bone weathering out of the rock. So, when people talk about going on a dinosaur "dig," it is a misnomer. Nobody digs into sediment to find bones (although

Figure 1.9. Field paleontologist extraordinaire René Hernandez-Rivera bags another dinosaur: duck-billed dinosaur vertebrae from the Late Cretaceous of Baja California, México. Photo courtesy of Maria Luisa Chavarría.

you might dig a bit to excavate them; bones are found because they are spotted weathering out of sedimentary rocks. If we're lucky and/or have a good eye, we'll spot something. Some collectors fare better than others, and a "feel" borne of experience is surely involved in finding bone, as well as an experienced eye and often luck (Figure 1.9). Ace Los Angeles County Museum fossil collector Harley Garbani used to say that the fossils "whistled" to him.

Collecting

Collecting is the arena in paleontology in which finesse meets brute force. Delicacy is required in preparing the fossils for transport; raw power is required for lifting blocks of bone and **matrix** (the rock which surrounds the bone) – commonly weighing many hundreds of pounds – out of the ground and back to civilization. Because of their size and delicacy, most dinosaur fossils are encased in a rigid **jacket**, or protective covering. Figure 1.10 shows how this is done.

Transport out of the field can be difficult, depending upon the size and weight of the jackets. A small jacket (soccer-ball size) can be carried out easily enough; but large jackets can require braces, hoists, winches, cranes, flatbed trucks, front-end loaders, helicopters, and even freight cars.

Back at the ranch

Once the fossil dinosaur bone is out of the field and back where it can be studied, the jacket must be cut open, and the fossil **prepared**, or freed from the matrix. This runs from simple brushing, to scraping with dental needles, to sophisticated treatments such as acid removal of the rock

Figure 1.10. Jacketing. (a) A fossil is found sticking out of the ground; now it needs to be cleaned off so that its extent can be assessed. Exposing bone can be done with a variety of tools, from small shovels, to dental picks, to fine brushes. As the bone is exposed, it is "glued"; that is, impregnated with a fluid hardener that soaks into the fossil and then hardens. (b) The pedestal. When the surface of the bone is exposed, the rock around it is then scraped away. For small fossils, this can be quite painless; however, for large fossils, this can mean taking off the face of a small hill. Anyway, this process continues until the bone (or bones) is sitting on a pedestal, a pillar of matrix underneath the fossil. (c) Toilet paper cushion. Padding is placed around the fossil to cushion it. The most cost-effective cushions are made from wet toilet paper patted onto the fossil. It takes a lot of toilet paper: for example, a 1 m thigh bone (femur) could take upward of one roll. On the other hand, this is not a step where we should cut corners, because returning from the field with a shattered specimen, or one in which the plaster jacket is stuck firmly to the fossil bone, is not so good. (d) Plaster jacket. Jackets are made of strips of burlap cloth soaked in plaster, and then applied to the toilet-paper-covered specimen. A bowl of plaster is made up, and then pre-cut, rolled strips of burlap are soaked in it and then unrolled onto the specimen and the pedestal. (e) Turning the specimen. After the plaster jacket is hardened, the bottom of the pedestal is undercut, and the specimen is turned; that is, separated at the base of the pedestal from the surrounding rock and turned over. This is a delicate step in which the quality of the jacket is tested. (f) The top jacket. More plaster and burlap are then applied to the open (former) bottom of the jacket, now its top. At this point, the fossil is fully encased in the plaster-and-burlap jacket, and is ready for transport from the field.

Figure 1.11. Scenes from a prep lab, in this case, that of the Museum of Northern Arizona in Flagstaff, USA. (a) Foreground: a large plaster jacket containing the theropod dinosaur *Coelophysis*. Background: sand tables for stabilizing specimen fragments (to be glued), open jackets, storage shelves, and grinding equipment. (b) The jacket shown in (a). The *Coelophysis* specimen is visible in the foreground. The arms and hands are to the left; the pelvis, legs, and tail are visible to the right. (c) A dinosaur skull (*Pentaceratops*) laid out for study. In the background are the large bays in which specimens are stored.

matrix. These techniques are generally carried out in a **preparation laboratory** (or prep lab; Figure 1.11) by **preparators**, people whose job it is to free the fossil from its rock matrix and stabilize it for study and display. The work is very long (and thus, costly), difficult, requires great skill, considerable knowledge, and real patience. Every paleontologist knows it: preparators are the unsung heroes of paleontology.

We often expect that the final result will be a free-standing display in a museum. Mounts of real fossil bone are attractive, but also time-consuming and costly, and the metal frames that support the bones can be destructive to the fossil. Moreover, mounted specimens commonly undergo damage over time; slight shifts in the mounts because of the extraordinary weights of the fossil bones, or vibrations in the buildings in which the bones are housed, or museum patrons lifting apparently "insignificant" bits all diminish the quality of mounted specimens. In addition, when the specimens are assembled and mounted, they can be hard to examine for study.

Many museums, therefore, cast the bones in fiberglass and other resins, and display the casts. Such displays are virtually indistinguishable from the originals. With their light weight, and the possibility of internal frames, they can be spectacular and dynamic (Figure 1.12).

Leaving the bones disarticulated, properly curated, and available for study maximizes returns on the very substantial investments that are involved in collecting dinosaur remains. Paleontology is carried out in large part by public support, and mounted casts give the public the best value for money.

Figure 1.12. A spectacular mount of the sauropod *Barosaurus* and the theropod *Allosaurus*. This mount is made of fiberglass and epoxy resin, cast from the bones of the original specimens. A dynamic pose like this would not have been possible using the original fossil bones.

SUMMARY

Fossils, the buried remains of organic life, are divided into body fossils, trace fossils, and a grab bag of other types of fossils. The body fossils include bones, shells, and other organic remains; trace fossils consist of tracks, trackways, and other impressions in the form of molds and casts. "Other" includes everything (and anything) else.

Fossilization is a process that occurs after the organism dies. It consists of burial, and commonly involves a variety of types of partial or complete replacement, in which the original organic and mineral material of the once-living organism is naturally replaced by other minerals while buried.

Obtaining fossils, particularly dinosaur fossils, requires rigorous training and preparation, along with perhaps a bit of educated guessing. Four steps are involved: planning, prospecting, collecting, and laboratory preparation and curation. The planning ranges from choosing where to look, to getting the legal permission to carry out the study, to outfitting an expedition properly to safely meet its goals. The prospecting requires a well-trained, experienced eye. Collecting is a process designed to bring delicate fossils safely back to where they can be prepared, which involves cleaning, reconstruction, and protection. Curation makes it possible for fossils to be safely stored on the long term, and for them to be accessible to researchers and to an interested public alike.

SELECTED READINGS

Behrensmeyer, A. K. and Hill, A. P. (eds.) 1980. *Fossils in the Making*. University of Chicago Press, Chicago, IL, 338 pp.

Cvancara, A. M. 1990. *Sleuthing Fossils: The Art of Investigating Past Life*. John Wiley and Sons, New York, 203 pp.

Farlow, J. O., Chapman, R. E., Breithaupt, B., and Matthews, N. 2012. The scientific study of dinosaur footprints. In Brett-Surman, M. K., Holtz, T. R. Jr., Farlow, J. O., and Walters, B. (eds.) *The Complete Dinosaur*. Indiana University Press, Bloomington, pp. 713–759.

Gillette, D. D. and Lockley, M. G. (eds.) 1989. *Dinosaur Tracks and Traces*. Cambridge University Press, New York, 454 pp.

Horner, J. R. and Gorman, J. 1988. *Digging Dinosaurs*. Workman Publishing, New York, 210 pp.

Kielan-Jaworowska, S. 1969. *Hunting for Dinosaurs*. Maple Press Co. Inc., York, PA, 177 pp.

Larson, P. and Donnan, K. 2004. *Rex Appeal: The Amazing Story of Sue, the Dinosaur That Changed Science, the Law, and My Life*. Invisible Cities Press Llc, Monpelier, Vermont, 424 pp.

Martin, R. E. 1999. *Taphonomy: A Process Approach*. Cambridge University Press, Cambridge, 524 pp.

Rogers, R. and Eberth, D. A. 2008. *Bonebeds: Genesis, Analysis, and Paleobiological Significance*. University of Chicago Press, Chicago, 512 pp.

Sternberg, C. H. 1985 (originally published 1917). *Hunting Dinosaurs in the Badlands of the Red Deer River, Alberta, Canada*. NeWest Press, Edmonton, Alberta, 235 pp.

TOPIC QUESTIONS

1. Define: fossil, dinosaur, replacement, preparation, coprolite, biosphere, body fossil, molds, casts, and ichnofossils.
2. What kind of training does it take to be a paleontologist?
3. What steps are involved in the collection of dinosaur bones?
4. What criteria maximize the likelihood of finding dinosaur fossils?
5. What kinds of paleoenvironments are most likely to preserve dinosaur bones? Why?
6. Why is it that, generally, the older the fossil, the less likely it is that original bone material will be preserved?
7. Why is understanding the geological context so important when collecting dinosaurs?
8. Why do dinosaur fossils tend to be preserved better in deserts than elsewhere?
9. What kinds of conditions might be required to find a fully articulated dinosaur fossil?
10. What kinds of geological activity are important to carry out before a fossil is extracted from the ground? Why?

Dinosaur days

CHAPTER OBJECTIVES

- Introducing geological time and stratigraphy
- Learning about continental drift during the time of the dinosaurs
- Learning about ancient climates during the time of the dinosaurs

When did dinosaurs live (and how do we know)?

Fossils, including dinosaur remains, are found in layers of rock, commonly called **strata.** The field of **stratigraphy** is the geological specialty that tells us how old or young particular strata are; thus, stratigraphy is a means of learning the age of dinosaur fossils. Stratigraphy is divided into **geochronology** or Earth time stratigraphy (*geo* – Earth; *chronos* – time), **lithostratigraphy** or rock stratigraphy (*lithos* – rock), and **biostratigraphy** or stratigraphy based upon the presence of fossils (*bios* – organisms).

Geochronology

Geologists generally signify time in two ways: in numbers of years before present, and by reference to blocks of time with special names (like "Triassic" and "Mesozoic"). For example, we say that the Earth was formed 4.6 billion *years before present*, meaning that it was formed 4.6 billion *years ago* and is thus 4.6 billion years old. Unfortunately, determining the precise age in years of a particular rock or fossil is not always easy, or even possible. For this reason, geologists have divided time into blocks of varying lengths, and rocks and fossils can be referred to these blocks of time, depending upon how exactly the age of the rock or fossil can be estimated. For example, you might not know that a fossil was 92.3 million years old, but you might be able to determine that it was within the interval of time known as the Late Cretaceous, meaning that its age is somewhere between 100.5 and 66.1 million years old, dates about which you have more information.

We start our discussion intuitively, by talking about the age in years, or the numerical age. Later we will address the division of time into blocks of varying lengths.

Geochronology: the ages of the ages

Geoscientists are happiest when they can learn the numerical age of a rock or fossil; that is, its age in years before present. Ages in years before present are reckoned from the decay of unstable **isotopes** (see Appendix 2.1 at the back of this chapter) found in certain minerals. Unstable isotopes spontaneously decay from an energy state that is not stable (that is, that "wants" to change) to one that is more stable (that is, that will not change, but rather remain in its present form). The change from unstable to stable gives off a bit of energy, and occurs over short or long amounts of time, depending upon the isotope. The basic decay reaction runs as follows:

> unstable "parent" isotope $\rightarrow$ stable "daughter" isotope
> $+$ nuclear products $+$ energy (generally, heat)

Here is an example: The element potassium (K; atomic number 19; atomic weight 39). Potassium has a relatively rare unstable isotope, ^{40}K, which has 19 protons and 21 neutrons. When ^{40}K decays, a proton in the nucleus of the atom captures an electron (the process is called, unsurprisingly, "electron capture"), and in so doing, it makes a neutron, which decreases the atomic number from 19 to 18 (a proton is lost when it became a neutron). So now we're no longer dealing with the element potassium (atomic number 19); we're dealing with a different element: argon (Ar; atomic number 18). However, the atomic weight stays at 40, since no mass has been lost (a proton was simply converted to an electron). Table 2.1 summarizes these quantities.

Following the general pattern of decay reactions (above), this reaction can be written as follows:

$$^{40}K \xrightarrow{\hspace{4cm}} {}^{40}Ar \quad + \text{ energy}$$

[unstable "parent" potassium isotope $\rightarrow$ stable "daughter" argon isotope $+$ energy]

with potassium being the unstable "parent" and argon being the stable "daughter." (Please see Appendix 2.1 for a quick review of the chemistry underlying these concepts.)

Table 2.1 **Protons, neutrons, atomic number, and atomic weights of isotopes of potassium and argon.**			
	Stable K isotope	**Unstable K Isotope**	**Stable Ar Isotope**
	^{39}K	^{40}K	^{40}Ar
Protons	19	19	18
Neutrons	20	21	22
Atomic weight	39	40	40
Atomic number	19	19	18

The *rate* of the decay reaction is the key to obtaining a numerical age. If we know:

(1) the original amount of parent isotope at the moment that the rock was formed or the animal died (before becoming a fossil);
(2) how much of the parent isotope is left; and
(3) the rate of the decay of that isotope,

we can estimate the amount of time that has elapsed. For example, suppose we know that 100% of an unstable isotope was present when a rock was new, but now only 50% remains. If we know the rate at which the element decayed, we can estimate the amount of time that has elapsed since the rock was formed; that is, the *age of the rock*. This is shown in Figure 2.1.

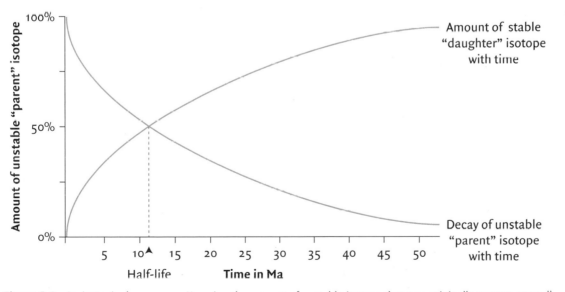

Figure 2.1. An isotopic decay curve. Knowing the amount of unstable isotope that was originally present, as well as the amount of unstable isotope now present and the rate of decay of the unstable isotope, it is possible to determine the age of a rock with that isotope in it. Suppose we found a rock with a ratio of 25 per cent unstable parent : 75 per cent stable daughter of a particular isotope. That would mean *two* half-lives had elapsed (half-life no. 1 = 50 per cent of 100 per cent parent (50 per cent parent : 50 per cent daughter); half-life no. 2 = 50 per cent of 50 per cent parent (25 per cent parent : 75 per cent daughter)). The amount of time represented by two half-lives can be read on the axis marked "Time in Ma"; in this case, about 25 million years. The rock would thus be about 25 million years old.

Choosing the right isotope

When people think of geological dating, they commonly think "carbon-14" (^{14}C). But when it comes to dinosaurs, ^{14}C is definitely *not* the way to go! Simply put, it decays too quickly. But different unstable isotopes have different constant rates of decay,[1] so the idea is to pick an isotope with a rate of decay that is appropriate to the amount of time you are interested in.

It is convenient to summarize the rate of decay by a single number. That number is called the **half-life**, which is the amount of time that it takes for 50% of the atoms of an unstable isotope to decay (leaving half as much parent as was originally present). The half-life, then, provides guidance about which isotope is appropriate for which amount of time. For example, to date human remains, not likely more than several thousand years old, the rubidium/strontium isotopic system ($^{87}Rb/^{87}Sr$), with a half-life of 48.8 billion years, would hardly be the ideal isotopic system. This would be a bit like timing a 100 m dash with a sundial. Likewise, dating dinosaur bones using ^{14}C, which has a half-life of 5730 years, would be like giving your own age in milliseconds. Dinosaurs lived tens to hundreds of millions of years ago (abbreviated Ma),[2] and after so many half-lives had elapsed, there wouldn't be nearly enough isotope left to measure. Isotopes with appropriate half-lives are the way to go.

Unstable isotopes are powerful dating tools, but they cannot be used directly on dinosaur bone. There has to be a source of unstable isotopes, which occur more commonly (although they are still extremely rare) in certain minerals, some of which form as lava cools and the minerals crystallize. The decay process begins when the unstable isotope is first formed (that is, when it

---

[1] In fact the rate at which individual molecules decay fluctuates in the short term but is statistically constant over long periods of time.

[2] We use the expression Ma, from *mille annos* – a million years. Thus, 66 Ma is 66 million years ago.

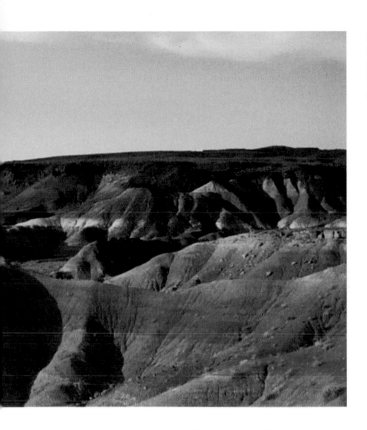

Figure 2.2. Superposition of strata in Petrified Forest National Park, Arizona, USA. Thick stacks of red mudstones and grey sandstones were deposited by rivers 210 million years ago, to produce the succession of layers visible in this photograph.

crystallizes in the lava), so, age of the rock is obtained from the moment the crystal formed as the lava cooled.

No dinosaur ever lived *within* hot lava – at least not for very long – so how can we get the age of the dinosaur bone when all we have is a date from some lava? This – the relationship between one body of rock (in this case, containing dinosaur bone) to another (here, the cooled lava) – is the province of lithostratigraphy.

Lithostratigraphy

Sedimentation and sedimentary rocks

Sediments – sand, silt, mud, dust, and other less-familiar materials – are deposited over geographic scales of m^2 to hundreds of km^2, and are the direct result of sedimentation such as flowing water, wind, or explosion of a volcano, to name a few more or less common processes. Virtually every geographical location we can think of – a river, a desert, a lake, an estuary, a mountain, the bottom of the ocean, the *pampas* – has sedimentary processes peculiar to it that will produce distinctive sediments and, with time and burial, distinctive sedimentary rocks. These geological processes – most commonly, wind and water – deposit sediments in strata. The strata pile up on one another, stacking into thick sequences of sedimentary rocks (Figure 2.2).

Relative dating

It is a fact that younger sediments are deposited upon older sediments (exemplified in Figure 2.2), and yet this apparently self-evident insight is the fundamental basis of all correlations of sedimentary strata in time. Ascertaining the relative ages of two strata is termed **relative dating** and, while not providing the age in years before present, provides the age of one stratum *relative* to another stratum. Here, then, is part of the solution to dating dinosaur bone. Suppose that a

Figure 2.3. Bone between two dated horizons. As we know the ages of the two horizons, the age of the bone can be interpolated between them (see the text).

stratum containing a dinosaur bone is sandwiched between two layers of volcanic ash. Ideally, a numerical age date could be obtained from each of the ash layers. We would know that the bone was younger than the lower layer but older than the upper layer. Depending upon how much time separates the two layers, the bone between them can be dated with greater or lesser accuracy (Figure 2.3).

But how can one tell that two geographically separated deposits were deposited at the same time if numerical ages are unknown? In this, fortunately, stratigraphers are aided by one last, extremely important tool: biostratigraphy.

Biostratigraphy

Biostratigraphy is a method of relative dating that utilizes the presence of fossil organisms. It is based upon the idea that a particular time interval can be characterized by a distinctive **assemblage**, or group, of organisms. For example, if one knows that dinosaurs lived from ~230 to 66 Ma, then any rock containing a dinosaur fragment must fall within that age range. Although biostratigraphy cannot provide ages in years before present, the fact that many species of organisms have existed on Earth for 1–2 million-year intervals enables them to be used as powerful dating tools. For example, *Tyrannosaurus rex* lived for only ~1 million years, from 67 to 66 Ma. Therefore when Sue Hendrickson found the *T. rex* "Sue" in South Dakota (Box 1.1), she knew that it could be correlated with *T. rex*-bearing sediments in Montana that have been well dated at 67–66 Ma, or any other location where *T. rex* has been found.

Eras and Periods and Epochs, oh my!

Geological time is hierarchical, much as our time is divided into years, months, weeks, days, hours, minutes, and seconds. We'll begin with large blocks of time called Eras. The Eras are, from oldest to youngest, the Paleozoic (*paleo* – ancient; *zoo* – animal), the Mesozoic (*meso* – middle), and the Cenozoic (*cenos* – new). Within each Era are smaller subdivisions (still consisting of tens of millions of years each) called Periods, and within these are yet smaller subdivisions of time called Epochs (consisting of several millions of years each). Figure 2.4 shows the part of the geological time scale during most of which dinosaurs roamed the Earth.

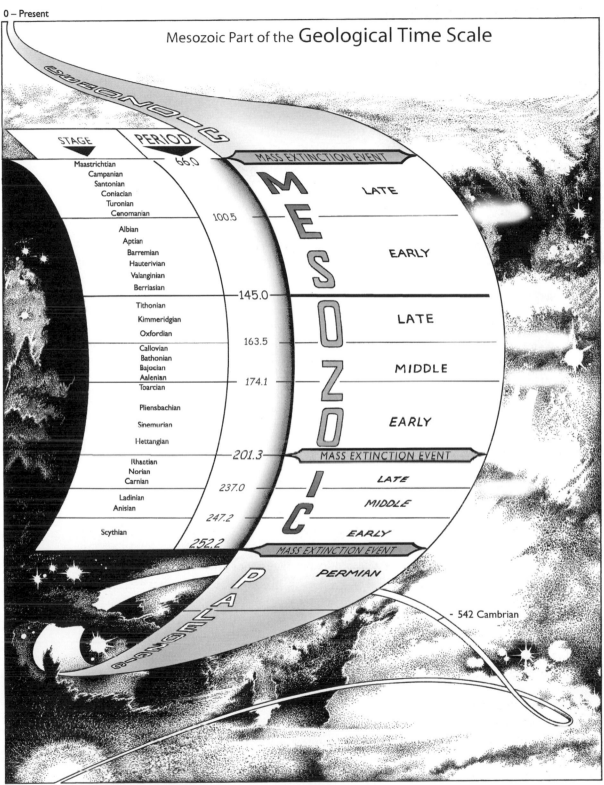

Figure 2.4. The Mesozoic part of the geological time scale. The Mesozoic constitutes only a rather tiny fraction of the expanse of Earth time. If you compacted Earth time into a single year, from January 1 (the formation of the Earth) to December 31, then dinosaurs were on Earth from about December 11 to December 25 (dates from International Commission on Stratigraphy (ICS); www.stratigraphy.org; v. 2016/04).

Continents and climates

Although the idea is just over 100 years old, most people are aware that, throughout its history, somehow the continents have moved around the Earth's surface; the Earth has been anything but static. The process is commonly called "**continental drift**" but is in fact part of a much more significant concept: **plate tectonics** (see Appendix 2.2). Plate tectonics, when it was first conceived in the late 1960s, revolutionized our understanding of the Earth, because it ultimately provided explanations for phenomena as different as the rise of mountains, oceanic circulation, and even the distribution of organisms.

Here, because we are interested in dinosaurs, we take the luxury of bypassing a mere *3.7 billion* years of continental evolution and zip right up to the Late Triassic, when all the continents coalesced into a single landmass, now known as **Pangaea** (Figure 2.5).

Late Triassic

Pangaea was the dominant land feature of the Late Triassic Earth. Like any large landmass, Pangaea had great mountain ranges; however, it was at least theoretically possible to walk on land from any place to any other. Unlike today, where the faunas and floras on different continents differ, Late Triassic faunas and floras all around the world were somewhat similar.

Early and Middle Jurassic

The initial break-up of Pangaea took place in the Early Jurassic. The effect was like unzipping the great supercontinent from south to north. Sediments in the eastern seaboard and Gulf Coast regions of North America, and in Venezuela and western Africa, record the opening and widening of a seaway through the Middle Jurassic. The single continent of Pangaea was beginning to reorganize into northern and southern supercontinents (see below), although even as late as Middle Jurassic time, a land bridge is thought to have existed between the two.

Figure 2.5. The positions of the present-day continents during the Late Triassic (237–201 Ma). Earth was dominated by the unified landmass Pangaea.

Figure 2.6. The positions of the continents during the Late Jurassic (163.5–145.0 Ma). Pangaea has begun its dismemberment while the southern continent of Gondwana remains together.

Also at this time, some of the earliest **epicontinental** (or "epeiric") seas of the Mesozoic Era first made their appearances. Epicontinental seas are shallow marine waters that cover parts of continents. In the past, epicontinental seas were considerably more widespread than they are today, because then **eustatic** (or global) sea levels were higher than they are now. Epicontinental seas are a very familiar part of Earth's geologic past and, with global climate change, there is good reason to think that they are also part of its future.

Late Jurassic

In the Late Jurassic (Figure 2.6) and Early Cretaceous, continental separation was well underway. A broad seaway, the *Tethyan Seaway* (after the Greek titan goddess *Tethys*, a goddess of the Sea), ran between two supercontinents, one in the north known as **Laurasia** and one in the south called **Gondwana**.

Early Cretaceous

The first half of the Cretaceous was a time of active mountain-building, sea-floor spreading, high eustatic sea levels, and broad epeiric seas. The **Tethys Ocean**, a sea that eventually became the Atlantic Ocean, remained a dominant geographical feature, as Europe continued to separate from North America.

In Gondwana, a stable continental marriage dating back to the Early Paleozoic Era finally underwent rifting involving two of its largest constituents – Africa and South America – as well as two smaller constituents, India and Madagascar. While India and Madagascar were in the first bloom of unconfinement, Australia and Antarctica remained together (a union that would not end until about 50 Ma), and a land connection remained, as it *almost* does today, between South America and Antarctica.

Late Cretaceous

The global positions of continents during the Late Cretaceous would be almost familiar to us (Figure 2.7). North America became nearly isolated, connected only by a newly emergent land

Figure 2.7. The positions of the continents during the Late Cretaceous (100.5–66.0 Ma). The positions of the continents did not differ greatly from their present-day distributions. Note the land bridge between Asia and North America, as well as the European archipelago. By this point in time, both of the supercontinents, Gondwana and Laurasia, had disintegrated.

connection across the modern Bering Straits to the eastern Asiatic continent. Although best known from the last Ice Age (100 000 years ago), this land bridge has come and gone several times since the Cretaceous. Africa and South America were fully separated, the former retaining its satellite, Madagascar, and the latter retaining a land bridge to the Antarctica/Australia continent. India was by now well on its way towards colliding with southern Asia.

Climates during the time of the dinosaurs?

Earth has recorded traces that allow us to infer at least *aspects* of past climates, and indeed the flavor of the Mesozoic would be lost without some general sense of Mesozoic **paleoclimates** (ancient climates). Distributions of the landmasses as well as the amount and distribution of the oceans on the globe drastically modify temperatures, humidity, and precipitation patterns. In the following, we explore this, comparing the extreme case of the continents coalesced into a single landmass (see Figure 2.5) with the equally extreme (but more familiar) case of the continents widely distributed around the globe (see Figure 2.7).

Heat retention in continents and oceans

Continents (land) and oceans (bodies of water) respond very differently to heat from the Sun: you can jump into a swimming pool and be comfortable long after the air and land have gotten chilly! This is due to differences in respective **heat capacity**[3]: how easily the temperature of a given material can be changed. Water has a higher heat capacity than continental crust, which means that it takes more

---

[3] Defined as the amount of heat required to change the temperature of a mole of material 1 °C.

energy to change its temperature than it takes to change the temperature of continents. The result is that the water takes more time to heat up, and more time to cool down. This is also why temperatures at the coast are almost always more mild than those found inland.

Consider how these properties of continents and oceans might modify climates at the dawn of the Mesozoic, when the continents were united into the single landmass Pangaea (see Figure 2.5). Here, "continental effects" – more rapid warming and cooling of continents than oceans – would have been more intense than today. Pangaea must have experienced wide temperature extremes. It would have heated up quickly and got hotter, and then cooled off more rapidly and got colder faster than modern continents, whose continental effects are mitigated by the broad, temperature-stabilizing expanses of oceans between them.

The post Late Triassic break-up of Pangaea weakened the strong continental effects. With the rise in eustatic sea level and supercontinental dismemberment, the effects of the large epeiric seas were superimposed upon the diminishing continental effects. These large bodies of water would have stabilized global temperatures, decreasing the magnitude and rapidity of the temperature fluctuations experienced on the continents during times of lower sea level.

Climates through the Mesozoic

The Late Triassic and Early Jurassic were times of heat and aridity. In general, throughout the Triassic, the evidence is that heat and aridity increased, with a possible episode of increased humidity during the early part of the Late Triassic. They also were times of marked **seasonality**; that is, well-defined seasons, strongly affected by the Pangaea continental mass. By the latter two-thirds of the Jurassic, however, as well as most of the Cretaceous, Earth is thought to have been *without* polar ice or glaciers on the northern parts of the continents. This is quite beyond our own experience; now, glaciers occur at high latitudes at both poles, and the poles themselves are covered in ice. The conclusion that there were no ice or glaciers above the Arctic and Antarctic Circles (latitudes 66° N and 66° S) is based largely upon the presence of warm climate indicators such as the fossils of warmth-loving plants and certain fish at high latitudes, and upon the absence of any evidence of continental glaciation from this time.

The absence of polar ice had an important consequence for climates: water that would have been bound up in ice and glaciers was instead in ocean basins. This in turn meant higher global sea levels, which led to extensive epeiric seas. The increased abundance of water on the continents as well as in the ocean basins had a stabilizing effect on temperatures (because it decreased continental effects), and decreased the amount of seasonality experienced on the continents.

Continental climates are enormously variable, however, and in North America Upper Jurassic terrestrial deposits, features preserved in the rocks, such as oxidized sediments and calcium carbonate deposits, suggest that in the Late Jurassic there was marked by seasonally arid conditions. So much for dinosaurs in steamy, swampy jungles!

Global paleoclimates in the Cretaceous are somewhat better understood than those of the preceding periods. During the first half of the Cretaceous at least, global temperatures remained warm and equable. The poles continued to be ice-free, and the first half of the Cretaceous saw far less seasonality than we see today. This means that, although equatorial temperatures were approximately equivalent to those we experience today, the temperatures at the poles were somewhat warmer. Temperatures at the Cretaceous poles have been estimated at 0–15 °C, which means that the temperature difference between the poles and the equator was only between 17 and 26 °C, considerably less than the ±41 °C of the modern Earth.

The first half of the Cretaceous was synergistic: **tectonic** activity, such as mountain building and sea-floor spreading, caused an increase in atmospheric CO_2 and a decrease in the volume of the ocean basins, which in turn increased the area of epeiric seas. The seas thus stabilized climates already warmed by enhanced absorption of heat in the atmosphere.

Sound familiar? The Early to mid-Cretaceous experienced the notorious "greenhouse" conditions that are currently of such concern today. Because several times in its past history, including in the Early Cretaceous, the Earth has "experimented" with greenhouse conditions, Earth history has a lot to offer to the dialog about global warming.

The last 30 million years of the Cretaceous produced a mild deterioration in the equable conditions of the mid-Cretaceous. A pronounced withdrawal of the seas took place, and evidence exists of more pronounced seasonality.

SUMMARY

Determining the ages of dinosaurs is accomplished by a mixture of biostratigraphy, lithostratigraphy, and geochronology. These allow paleontologists to date rocks and fossils in relation to each other, as well as to obtain estimates of their ages in years before present. Using these techniques, geoscientists have constructed and refined a geological time scale for the entire history of the Earth. The time scale is hierarchically divided into successively more refined time intervals: Eras, Periods, and Epochs.

The earliest dinosaurs appeared during the Late Triassic, a time in which the Earth's continents were united into a single supercontinent called Pangaea. Since then, the continents have separated, moving to their present positions.

The presence of a single supercontinent had major implications for climates, which were strongly seasonal. These mitigated throughout the Jurassic, although the Late Jurassic appears to have had strong seasonality, at least in North America. By mid-Cretaceous time, high sea levels, melted polar ice, and high levels of atmospheric CO_2 appear to have acted synergistically to produce global warming and greenhouse conditions.

SELECTED READINGS

Barron, E. J. 1983. A warm, equable Cretaceous: the nature of the problem. *Earth Science Reviews*, 19, 305–338.

Berry, W. B. N. 1987. *Growth of a Prehistoric Time Scale Based on Organic Evolution*. Blackwell Scientific Publications, Boston, MA, 202 pp.

Crowley, T. J. and North, G. R. 1992. *Paleoclimatology*. Oxford Monographs in Geology and Geophysics, no. 18. Oxford University Press, New York, 339 pp.

Gradstein, F., Ogg, J., and Schmitz, M. (eds.) 2012. *A Geological Time Scale 2012*. Elsevier, Amsterdam, 1176 pp.

Lewis, C. 2000. *The Dating Game: One Man's Search for the Age of the Earth*. Cambridge University Press, Cambridge, 253 pp.

Robinson, P. L. 1973. Palaeoclimatology and continental drift. In Tarling, D. H. and Runcorn, S. K. (eds.) *Implications of Continental Drift to the Earth Sciences*, vol. I. NATO Advanced Study Institute, Academic Press, New York, pp. 449–474.

Ross, M. I. 1992. *Paleogeographic Information System/Mac Version 1.3: Paleomap Project Progress Report no. 9*, University of Texas at Arlington, 32 pp.

TOPIC QUESTIONS

1. Define: eustatic, biostratigraphy, stratigraphy, geochronology, half-life.
2. What is the difference between "numerical" dates and relative ages?
3. If a dinosaur bone is found between two layers dated at 90 million and 70 million years, respectively, what is the age of the dinosaur bone?
4. Referring to question no. 3, how precisely can that bone be dated?
5. What is the numerical value for the half-life shown in Figure 2.1?
6. If there were dinosaur-bearing rocks in North America and Africa, and the African dinosaur remains could be dated biostratigraphically, how could you correlate the North American deposits with the African ones?
7. Compare the Late Triassic Earth with the Late Cretaceous Earth.

APPENDIX 2.1 CHEMISTRY QUICK 'N DIRTY

Earth is made up of elements, as seen[4] on a Periodic Table. Many of these, such as hydrogen, oxygen, nitrogen, carbon, and iron, are familiar, while others, such as berkelium, iridium, and thorium, are less so. All elements are made up of **atoms**, an atom being the smallest particle of any element that still retains the properties of that element. Atoms, in turn, are made up of **protons, neutrons,** and yet smaller **electrons**, which are collectively termed **subatomic** ("smaller-than-atomic") particles. Protons and neutrons reside in the central core, or **nucleus,** of the atom. The electrons are located in a cloud surrounding the nucleus. Some electrons are more tightly bound around the nucleus and others are less tightly bound. Those that are less tightly bound are, as one might expect, more easily removed than those that are more tightly bound (Figure A2.1).

Keeping that in mind, let us further consider the subatomic particles. Protons and electrons are electrically charged; electrons have a charge of -1 and protons have a charge of $+1$. Neutrons, as their name implies, are electrically neutral and have no charge. To keep a charge balance in the atom, the number of protons (positively charged) must equal the number of electrons (negatively charged). This number – which is the same as the number of protons – is called the **atomic number** of the element, and is displayed to the lower left of the elemental symbol. In our example, the element potassium is identified by the letter K (its Latin name is *kalium*), and it has 19 protons. Its atomic number is thus 19, and it may be written $_{19}$K.

Along with having an electrical charge, some subatomic particles also have mass. Rather than force us to work with the extremely small mass of a proton (one of them weighs about 1.67×10^{-24} grams!), it is assigned a mass of 1. Neutrons have a mass of 1 as well. Because relative to protons and neutrons the masses of electrons are negligible, the mass number of an element is composed of the total number of neutrons *plus* the total number of protons. In the case of the element potassium, for example, the mass number equals the total number of neutrons (20) plus the total number of protons (19); that is, 39. This is usually written ^{39}K. Because the atomic number is always the same for a particular element, it is commonly not included when the isotope is discussed. Thus $^{39}_{19}$K is usually abbreviated to ^{39}K.

---

[4] And heard, sung by asapSCIENCE (www.asapscience.com) to J. Offenbach's "Can Can" from "Tales of Hoffman," and by Tom Lehrer to A. S. Sullivan's "Major General's Song" from the album *An Evening Wasted with Tom Lehrer.* Both can be heard on YouTube.

Figure A2.1. Diagram of a potassium atom. In the nucleus are 19 protons and 20 neutrons. In a cloud around the nucleus are 19 electrons, whose position relative to the nucleus is governed by their energy state.

Variations in elements that have the same atomic number but different mass numbers are called isotopes. For example, an important (for our purposes) isotope of ^{39}K is ^{40}K. Since ^{40}K is an isotope of potassium, it has the same atomic number as ^{39}K (because it has 19 electrons and 19 protons). The change in *mass* number results from additional neutrons. ^{40}K has 21 neutrons, which, with the 19 protons, increase its atomic mass to 40. Because it is potassium, of course, its atomic number remains 19.

APPENDIX 2.2 PLATE TECTONICS

THE MOVEMENT OF THE PLATES IS REFERRED TO AS THE THEORY OF PLATE TECTONICS AND IT IS AS IMPORTANT FOR UNDERSTANDING THE WAY EARTH WORKS AS THE THEORY OF EVOLUTION HAS BEEN FOR UNDERSTANDING LIFE!

WE'LL RETURN TO THE BOILING HOT CHOCOLATE ANALOGY TO UNDERSTAND THE MECHANICS OF PLATE TECTONICS.

BOILING ON A STOVE, HOT CHOCOLATE DEVELOPS A SCUM, WHICH CONTINUOUSLY MOVES AS HEAT CIRCULATES THROUGH THE POT OF LIQUID. NEW HOT SCUM IS FORMED, AND AS THIS OCCURS, OLDER, COOLER SCUM PUCKERS AND WRINKLES TO ACCOMMODATE THE NEW MATERIAL.

SO IT IS WITH THE LITHOSPHERE: NEW ADDITIONS OF LITHOSPHERE OCCUR ALONG ELONGATE GEOGRAPHIC FEATURES – SPREADING CENTERS – PLACES WHERE HOT MOLTEN MATERIAL COMES UP FROM THE ASTHENOSPHERE TO COOL AND MAKE MORE LITHOSPHERE. AS NEW HOT ASTHENOSPHERE IS ADDED, OLDER, COOLER LITHOSPHERE IS PUSHED AWAY.

OBVIOUSLY THE LITHOSPHERE ALREADY NEXT TO THE SPREADING CENTER HAS TO GO SOMEWHERE, AND WHEN IT DOES, IT COLLIDES WITH OTHER LITHOSPHERE, LIKE TWO CARS CRASHING HEAD-ON. THE CARS CAN:

SLIDE OVER THE OTHER – THIS CAN OCCUR WHEN OCEANIC LITHOSPHERE MEETS OCEANIC LITHOSPHERE ~ BUT BECAUSE OCEANIC LITHOSPHERE IS DENSER THAN CONTINENTAL LITHOSPHERE, OCEANIC LITHOSPHERE RARELY OVERRIDES CONTINENTAL LITHOSPHERE WHEN THERE IS A OCEANIC-CONTINENTAL LITHOSPHERE COLLISION.

SLIDE UNDER THE OTHER – ON THE OTHER HAND, THIS IS THE CLASSIC RESULT OF AN OCEANIC-CONTINENTAL LITHOSPHERIC COLLISION: THE OCEANIC LITHOSPHERE SLIDES UNDER THE CONTINENT. WE SAY THAT THE OCEANIC LITHOSPHERE HAS BEEN **SUBDUCTED**.

CRUMPLE AGAINST EACH OTHER – IN THIS CASE, CONTINENTAL LITHOSPHERE CRUMPLING AGAINST CONTINENTAL LITHOSPHERE – IS HOW BIG MOUNTAINS ARE MADE. THE CONTINENTAL LITHOSPHERE OF INDIA COLLIDED WITH THE CONTINENTAL LITHOSPHERE OF ASIA TO PRODUCE THE HIMALAYAS, THE WORLD'S HIGHEST MOUNTAIN RANGE.

SO HERE'S THE COMPLETE GEOLOGICAL PACKAGE: HOT MANTLE MOVES UP, FORMING AN OCEANIC SPREADING CENTER; FAR AWAY, AT THE EDGE OF A CONTINENT, THIS CAUSES THE OLDER OCEANIC CRUST THAT FORMED EARLIER TO BE SUBDUCTED UNDER THE CONTINENT. BECAUSE PIECES OF THE SUBDUCTING SLAB ARE HOT, THEY TEND TO MELT AND RISE THROUGH THE OVERRIDING CONTINTENTAL LITHOSPHERE. WHEN THEY FINALLY REACH THE SURFACE, ALL THAT MANTLE-DERIVED HEAT ENERGY, AND GAS WANTS TO ESCAPE, AND SO VOLCANIC MOUNTAIN RANGES ARE BUILT; FOR EXAMPLE, THE ANDES IN SOUTH AMERICA AND IN NORTHWESTERN UNITED STATES, THE CASCADES.

PLATE TECTONICS REALLY EXPLAINS HOW THE WORLD *WORKS* ~ FROM WHY THE CONTINENTS HAVE MOVED THROUGH TIME, TO WHY VOLCANOES ARE WHERE THEY ARE, TO THE ORIGIN OF MOUNTAIN RANGES (TERMED "OROGENY"), TO THE REASONS FOR AND LOCATIONS OF EARTHQUAKES (MOVEMENTS IN THE CRUST).

THIS IS WHY GEOLOGISTS LIKE TO SAY, "SUBDUCTION LEADS TO OROGENY!"

Who's related to whom – and how do we know?

Who are you? Who? Who? PETE TOWNSEND, The Who

CHAPTER OBJECTIVES

The goal of this chapter is to get comfortable with the following subjects, because we'll revisit them again and again throughout this book:

- Phylogeny
- Evolution
- Phylogenetic systematics
- Cladograms

Who are you?

Identity is *the* fundamental question in all animals – including humans. Knowing who we are provides the essential context for us, and, for this reason, a favorite dramatic technique (think soap operas) is to have a key character lose some aspect of his or her identity. If you think about it, who we are governs every aspect of our behavior.

"Who are you?" is really the same question as "To whom are you related?" because *relationship*, as in human families, is the key to identity. So to understand who dinosaurs are, we need to know their relationships – to each other and to other animals.

Those relationships – who we are and where we come from – are the special province of **phylogeny**: the history of the descent of organisms. And here is where evolution comes in: evolution is the cause of the fundamental genealogical connection among organisms.

Evolution

Evolution refers to *descent with modification*: the concept that organisms have changed and modified their **morphology** (*morph* – shape; *ology* – the study of) through each succeeding generation (see Appendix 3.1 for a brief refresher on the theory of evolution). That makes each new generation the most recent bearer of the unbroken genetic thread that connects life. Each new generation is forward-looking in that its members potentially contain new features that might be useful for whatever the organism encounters in its life; but it is also connected to its ancestors by features that it has inherited. That connection is **relationship**.

Relationship and the classification

Organisms are commonly classified according to the classification system devised by the Swedish naturalist Carolus Linnaeus (1707–1778). His hierarchical system is the very famous (or infamous!) ranking of organisms in groups of decreasing size: kingdom, phylum, class, order, family, genus, and species. Individuals are generally referred to by italicized **generic** (genus) and **specific** (species) names, for example, in the case of a famous large dinosaur, *Tyrannosaurus rex*. Any name in the hierarchy – representing a group of organisms – is considered a **taxon** (plural **taxa**).

All classifications serve a purpose, for example, our movies are classified by both subject (Drama, Horror, Comedy, etc.) as well as by suitability for viewing (PG-13, R, etc.). The Linnaean classification has come to reflect degree of relatedness. Thus, all members of a taxon – at any level in the hierarchy – are said to be more closely related to each other than to anything else. So, for example, in the case of the taxon "Mammalia," all members (mammals) are thought to be more closely related to each other (other mammals) than to anything else.

For all its ubiquity, the Linnaean classification represents a kind of historical sleight-of-hand. Linnaeus developed his classification system well before evolution was properly understood. Linnaeus himself did not think that evolution was responsible for the Earth's diversity (he was in fact what we would now call a "creationist"), and so his classification was developed solely on the basis of overall similarities among organisms; it had nothing to do with relationship as we understand it today. The idea of relationship, of descent with modification, quietly slipped in later, after Darwin had published *On the Origin of Species* (1859). It is reasonable, of course: things that are more closely related tend to look more similar, but as we shall see in Chapter 4, overall similarity is not always the best indicator of evolutionary relationship. For this reason, although the Linnaean classification is still widely in use, many evolutionary biologists have observed that the Linnaean classification poorly accommodates a world where biotic diversity is generated by evolution. In Chapter 4, we will see some of the problems that arise.

Determining relationships

Homology

If there are genetic relationships among organisms, then there must be genetic relationships among their parts. For example, the five "fingers" in the human "hand" and the five "toes" in the front "foot" of a lizard didn't just occur independently. They're all digits of the forelimb, a particular feature that happens to have been evolutionarily retained in these two lineages (humans and lizards). In principle and in fact, the digits on the forelimbs of lizards and humans can be traced back in time to digits in the forelimb of the common ancestor of humans and lizards. We call two anatomical structures **homologous** when they can, at least in theory, be traced back to a single original structure in a common ancestor (Figure 3.1). Thus we infer that the digits in the forelimbs of all mammals are homologous with those of, for example, dinosaurs. The wings of a fly, however, are not homologous with those of a bird, since they cannot be traced to a single structure on a common ancestor. Because the wings of a fly and the wings of a bird do the same thing (enable flight), they are said to be **analogous** (Figure 3.2).

Here is a subtle, but important distinction: the *limbs* of the pterosaur, bird, and mammal shown in Figure 3.1 are surely homologous; however, their function as *wings* is equally surely not. If wings were homologous, it would imply that wings evolved once – in the common ancestor – and all these wings are variants on the original wing. In fact, we have good reason to conclude that each of the ancestors of each of these three animals was non-flying, and that in each lineage, wings (and flight) evolved separately. Thus, wingedness in these vertebrates is analogous. Evolution teaches us that there are many ways to make a wing.

Chopping down the "tree of life"

This brings us to the "tree of life." We've all seen "trees of life"; they begin with a pulsating, primordial slime-blob that eventually evolves into everything else as you trace the branches outward (Figure 3.3 is an example). Such trees show who came from whom, and when that occurred. They are common in textbooks and museum displays, and deeply influence our ideas about evolution. But how do we know who gave rise to whom? After all, no human witnessed the zillions of speciation events that constitute the "tree of life."

The problem is worse than that. We learned in Chapter 1 that fossil preservation is a rare event indeed. Is it likely that some fossil that happens to be preserved and that we happen to find, turns out to be the actual ancestor of some other that happens to be preserved and that we happen to find? The chances of this occurring are vanishingly small. Thus the oldest fossil in the human family is very unlikely to be the great great great granddaddy of all subsequent humanity.

Still, that fossil *is* likely to have many features that the real great great great granddaddy possessed. And therein lies the key to recognizing who is related to whom: while we'll never find the *actual* ancestor, we have a really good chance of finding out more or less how that ancestor looked. Because trees of life specify actual ancestors, and because, as we have seen, we're not likely to actually have the real ancestor in hand, we'll avoid trees of life, and instead use a revolutionary method to understand who is related to whom or, thinking of it another way, the course of evolution. That method is called **phylogenetic systematics**.

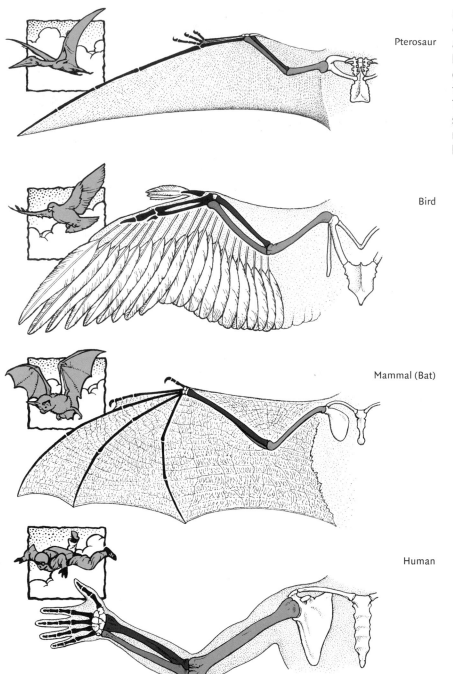

Pterosaur

Bird

Mammal (Bat)

Human

Figure 3.1. Homologs. The front limbs of humans, bats, birds, and pterosaurs are all homologous, and retain the same basic structure and bone relationships even though the appearance of these forelimbs may be outwardly different. The remarkable similarities in structure suggest shared ancestry, and for this reason, homology forms the basis for hypotheses of evolutionary relationships.

Phylogenetic systematics – the reconstruction of phylogeny

Phylogenetic systematics was developed in the mid 1900s specifically for reconstructing the course of evolution. As we shall see later in this chapter, it is the only *scientific* means of determining relationship, and for this reason it is used by virtually all evolutionary biologists and paleontologists.

Human   Grasshopper

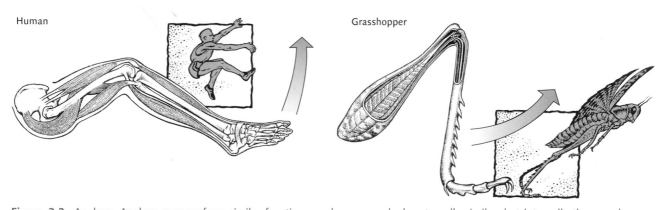

Figure 3.2. Analogs. Analogs may perform similar functions, and may even look outwardly similar, but internally they can be very different. Here a human leg is contrasted with that of a grasshopper. Although both have "legs," the two structures have little in common. For starters, in humans, the skeleton is internal, whereas in grasshoppers, the skeleton is external. These and a zillion other differences suggest that these two "legs" are analogs, rather than homologs.

And although it hasn't yet trickled down into the popular literature, the only way to really understand dinosaurs is through the lens of phylogenetic systematics. And so that's what we do in this book.

To reconstruct phylogeny, we need a way to recognize how closely two creatures are related. Superficially this is very simple: *things that are more closely related tend to share specific features.* We know this intuitively because we can see that organisms that we believe are closely related (for example, a dog and a coyote) share many similarities. Breeders have taken advantage of this for thousands of years. They've depended upon the fact that offspring look, and sometimes act, very much like their parents to obtain plants and animals with the qualities for which they select. Phylogenetic systematics is the technique by which relationships between organisms can be inferred using unique features of organisms. It depends first and foremost upon the hierarchical distribution of features in the natural world.

Hierarchy

One of Linnaeus' great insights was his recognition that all features in the natural world are organized in a **hierarchy**, which can be understood to be a successive ranking of subsets within sets. A familiar hierarchy, for example, is ranks within the military. Or, to choose a biological example, all creatures possessing fur[1] are a subset of all animals possessing a backbone, which are in turn a subset of all living organisms (Figure 3.4); these features are distributed hierarchically. *In fact, all features of living organisms are hierarchically distributed in nature,* from the possession of DNA – which is almost ubiquitous – to highly restricted features such as the possession of a brain capable of producing a written record of culture.

Always, however, unmodified or slightly modified vestiges of the original plan remain, and these provide the keys to the fundamental hierarchical relationships that reveal who's related to whom.

---

[1] A number of organisms in the world are fur-covered besides mammals; for example, bees and some spiders (like tarantulas) have a fur-like covering. But the fur on mammals is unique: it is made of unique proteins, grows in a unique way, and thus is different from these other furry or hairy coatings. When we refer to "fur" in this chapter, it is the mammalian type of fur to which we refer.

Figure 3.3. A tree of life. This particular one is a satire by Matt Groening. The image of evolution as a tree, however, is completely familiar. From the *Big Book of Hell* © Matt Groening. All Rights Reserved. Reprinted by permission of Pantheon Books, a division of Random House, Inc., NY.

Figure 3.4. The natural hierarchy illustrated as a wooden jigsaw puzzle. The different organisms represent the larger groups to which they belong. For example, the mouse, representing Mammalia, and the lizard, representing Reptilia, together fit within the puzzle to represent Vertebrata, itself a subset of bilaterally symmetrical organisms (Bilateria), which would include invertebrates such as a lobster or a mosquito. Bilateria and other groups constitute those organisms we call Animalia (animals). Animalia, then, is a subset of all living organisms.

Characters

Identifying the features themselves is a prerequisite to establishing life's hierarchy, so we need to look more closely at what we mean by "features." Observable features of anatomy are termed **characters**. Unique bones or unusual morphologies would all be characters. On the other hand, "observable features" would not include what something does – or how it does it. So, for example, "bites hard" is not a character, but perhaps "big jaw muscles" would be.

Characters acquire their meaning not as a single feature on a particular organism but when their *distribution* among a selected group of organisms is considered. For example, modern birds are generally linked on the basis of having feathers. All living feathered animals are birds and all birds have feathers. Thus, not only is a penguin a bird but so are eagles, ostriches, and kiwis: they all have feathers. And if someone told us that something is a bird, we could confidently predict that it too has feathers.

Because characters are distributed hierarchically, their *position* in the hierarchy is obviously dependent upon the groups they are being used to identify. Consider again the simple example of mammalian fur. Because mammals uniquely have this type of fur, it follows that, if you wanted to tell a mammal from a non-mammal (that is, any other organism), you need only observe that the mammal is the one that has that type of fur. On the other hand, the character "possession of fur" is not useful for distinguishing a bear from a dog; both have mammalian fur. To distinguish a bear from a dog, you'd need some character other than fur.

These distinctions are extremely important in establishing the hierarchy, and for this reason, characters function in two ways: as **diagnostic** characters and as non-diagnostic characters. The word "diagnostic" here has the same meaning as in medicine. Just as a doctor diagnoses a malady by distinctive and unique properties, so a group of organisms is diagnosed by distinctive and unique characters.

The same character may be diagnostic in one group, but non-diagnostic in a smaller subset of that group (because it is now being applied at a different position in the hierarchy). We saw that fur allowed us to tell a mammal from a non-mammal, but it can't distinguish one mammal from another: it wouldn't tell a bear from a dog.

Cladograms

Cladograms (*clados* – branch; *gramma* – letter) are simply branching diagrams that show hierarchies of diagnostic characters. But, as we'll see, they're not just visual aids, they're the keys to understanding who's related to whom.

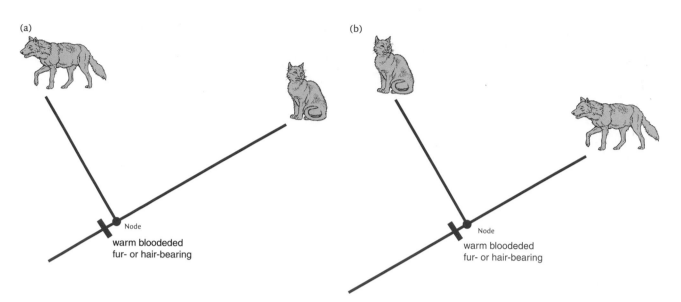

Figure 3.5. (a) A cladogram. The cat and dog are linked by the characters listed at the hatch mark (or bar), just below the node. The node itself defines the things to be united; commonly a name is attached to the node that designates the group. Here, such a name might be "mammalian carnivores." (b) The same cladogram with the dogs and cats reversed. Reversing the order of two taxa around a node does not change the cladogram's meaning, because the cladogram is only stating that both taxa share the same characters.

To understand how a cladogram works, we begin with two familiar animals; say, a cat and a dog. A cladogram of a cat and a dog is shown in Figure 3.5.

So we're looking for diagnostic characters for these animals. Here, we choose:

(1) possession of fur;
(2) possession of a backbone; and
(3) possession of carnivorous teeth of a unique design.

The cladogram links two separate objects – the cat and the dog – based upon the characters that they *share*. The features are listed on the cladogram adjacent to the **node**, which is a split point (bifurcation) in the diagram (see Figure 3.5).

Cladograms, of course, are not evolutionary trees; they are simple branching diagrams, showing two organisms uniquely sharing character(s). So it follows that there is absolutely no difference in the order in which the two organisms sharing the characters are presented. It could go, from left to right, dog, cat; or cat, dog. Either way, both organisms share the character listed at the node. This is shown in Figure 3.5 as well.

The issue becomes more complicated (and more interesting) when a third animal is added to the group (Figure 3.6), in this case a monkey. Now, for the first time, because none of the three animals is identical, two of the three will have more in common with each other than either does with the third. It is in this step that the hierarchy is established. The group that contains all three animals is diagnosed by certain features shared by all three (fur and posession of a backbone). Notice that a subset containing two animals (the cat and the dog) has also been established, linked together by a character (uniquely designed carnivorous teeth) that diagnoses them as being exclusive of the third animal (the monkey).

How the characters, and even the animals, are arranged on the cladogram, is controlled by the *choice* of characters. Let's try something different:

- shortened snout
- large eyes

Based upon these characters, the cladogram in Figure 3.7 contradicts the cladogram in Figure 3.6.

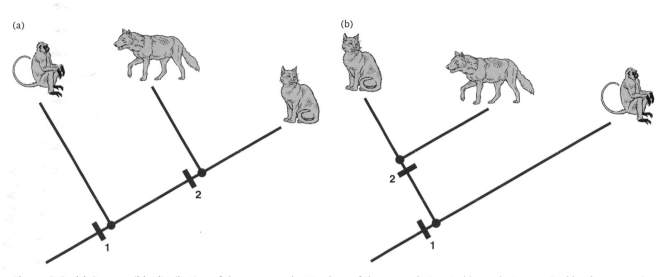

Figure 3.6. (a) One possible distribution of three mammals. Members of the group designated by node **1** are united by the possession of fur and warm-bloodedness; that group could be called Mammalia. Within the group Mammalia is a subset united by possession of carnivorous teeth (for example, "mammalian carnivores"). That subset is designated at node **2**. (b) The same cladogram, with the taxa reversed at nodes **1** and **2**. In terms of information conveyed, there is no difference between (a) and (b).

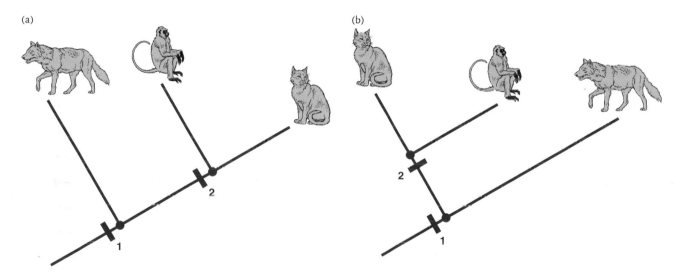

Figure 3.7. (a) An alternative distribution of three mammals. The characters selected at node **2** (see Figure 3.6) suggest that the cat and monkey have more in common with each other than either of them does with the dog (see the text). (b) The same cladogram, with the taxa reversed at nodes **1** and **2**. In terms of information conveyed, there is no difference between (a) and (b). There is no difference between the cladograms shown in Figure 3.6, nor is there any difference between the cladograms shown here. But, there is a significant difference between the cladograms shown in Figure 3.6 (both of which say that the cat and the dog are more closely related to each other than either is to the monkey), and those shown here (both of which say that the monkey and the cat are more closely related to each other than either is to the dog).

How do we choose? *The cladogram that is most likely correct is the one that doesn't change when new characters are added.* All characters that apply to these mammals support the cladogram in Figure 3.6. And we can infer from the cladogram in Figure 3.6 that a dog shares many more characters with a cat than it does with a monkey.

Cladograms as tools in understanding the evolution of organisms

So how does this apply to evolution? Using character hierarchies portrayed on cladograms, we establish **clades** or **monophyletic groups**[2]: *groups that have evolutionary significance because the members of each group are more closely related – by genealogy – to each other than they are to any other creature.* If a group is monophyletic, it also implies that all members of that group share a more recent common ancestor with each other than with any other organism. The cladogram in Figure 3.6 suggests that the cat and the dog are more closely related to each other than either is to the monkey.

Now that we're speaking in terms of organic evolution, the specific characters that we said characterize groups can now be treated as homologous among the groups that they link. Mammalian fur once again (!) provides a convenient example. We conclude that mammals are monophyletic based upon the fact that mammals all share a unique type of fur (among many other characters). If all mammals are fur-bearing (and mammals are monophyletic), the implication is that the fur found in bears and that found in horses can in fact be traced back to fur that must have been present in the most recent common ancestor of bears and horses.

Because now we're working with the evolution of organisms (and not just hierarchies of characters), the terms **derived**, or **advanced** can be used along with "diagnostic," and the term "general" ("non-diagnostic") is replaced by the words **primitive** or **ancestral**. "Primitive" certainly does *not* mean worse or inferior, just as "advanced" certainly does *not* mean better or superior; these refer only to how much the character has been changed by evolution. Primitive specifies the condition of a particular feature in the ancestor; advanced specifies an evolved condition of that character in its descendant.

Mapping the course of evolution with the cladogram

Because the form of the cladogram (a branching stick diagram) is hierarchical, it is an excellent way to map the hierarchical distributions of characters in nature. *Derived characters are evidence of monophyletic groups because, as newly evolved features, they are potentially transferable from the first organism that acquired them to all its descendants*: they therefore characterize the bifurcations at each node on the cladogram. Primitive characters – those with a much more ancient history – provide no such evidence of monophyly.

To illustrate this, we resort for the last time (!) to mammals and their fur. Mammalian fur, we said, is among the shared, derived characters that unite mammals as a monophyletic group. On a cladogram, therefore, we look for characters that mark a node in the diagram. All organisms characterized by shared, derived characters are linked by the cladogram into monophyletic groups. Reflecting the hierarchy of character distributions in nature, the cladogram documents monophyletic groups within larger monophyletic groups. In Figure 3.8, a small part of the hierarchy is shown: humans (a monophyletic group, possessing shared, derived characters) are contained in mammals (another monophyletic group possessing other shared, derived characters). Notice that the character of warm-bloodedness is primitive for *Homo sapiens*, but derived for mammals. As we have seen, features can be derived or primitive, all depending upon what part of the hierarchy one is investigating.

The cladogram need not depict every organism within a monophyletic group. If we are talking about humans and carnivores, we can put them on a cladogram and show the derived characters

---

[2] To pile on the nomenclature, these are sometimes termed "natural groups."

that diagnose them, but we might (or might not) include other mammals (for example, a gorilla). So we said, with regard to Figures 3.6 and 3.7, if the hierarchical relationships that we have established are valid, the addition of other organisms into the cladogram should not alter the basic hierarchical arrangements established by the cladogram. Figure 3.9 shows the addition of one other group into the cladogram from Figure 3.8. The basic relationships established in Figure 3.8 still hold, even with the new organism added. The cladogram is likely correct.

How to read evolution in the cladogram

We identified monophyletic groups using derived characters, and that the hierarchies of characters designate hierarchies of groups. So, looking at Figure 3.9, the distribution of shared, derived characters suggests that humans and gorillas are more closely related to each other than either is to a bear. It also suggests that all three are more closely related to each other than they are to something that does not posess the derived character of bearing fur or hair.

And how does that apply to evolution? The evolution of the derived character of fur is associated with the evolution of the group Mammalia. As we currently understand their relationships, what we call mammals were invented when that character – among others – first evolved. And, the cladogram tells you that sometime thereafter – the cladogram does not specify when – a character that unites both humans and gorillas evolved, a character that we now recognize diagnoses a new group within Mammalia.[3]

Here is a fundamental difference between a cladogram and the more familiar tree of life that we discussed above. The cladogram does not incorporate time, nor does it tell you who the ancestors were. Instead, it can indicate the sequence of the evolutionary events and, more importantly, it specifies the characteristics that the ancestor possessed. So, in the case of the cladogram in Figure 3.9, we aren't told who was the ancestor of bears, gorillas, and humans, but we infer that the earliest mammal was fur-bearing (among the many other characters that diagnose Mammalia).

Used as a tool to reconstruct evolution, then, a cladogram is actually a **hypothesis of relationship**; that is, a hypothesis about how closely (or distantly) organisms are related, and

[3] That new group is called "Hominoidea," as it happens, and is diagnosed by lots of characters, among which are a series of specializations in the arms and trunk associated with walking bipedally and swinging through trees.

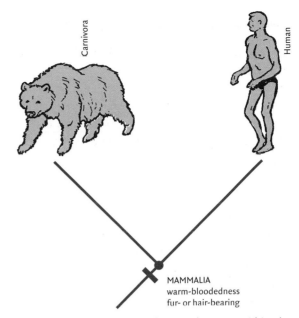

Figure 3.8. A cladogram showing humans within the larger group Mammalia. Mammalia is diagnosed by warm-bloodedness and possession of fur (or hair); many other characters unite the group as well. Carnivora, a group of mammals that includes bears and dogs (among others) is shown to complete the cladogram. Carnivores all uniquely share a special tooth (the carnassial) and humans all uniquely share, among many other features of the skull and skeleton, a large cranium. Note that all mammals (including humans and carnivores) are warm-blooded and have fur (or hair), but only humans have a large cranium, and only members of Carnivora have the carnassial tooth.

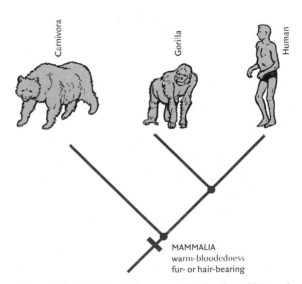

Figure 3.9. Addition of the genus *Gorilla*. The addition of gorillas to the cladogram does not alter the basic relationships outlined on the cladogram shown in Figure 3.8.

about what the sequence of the appearance of different diagnostic characters must have been. But how to test this hypothesis?

Parsimony

As we have seen, it is possible to construct several possible cladograms which would represent different evolutionary sequences. Which to choose? We choose using the principle of **parsimony**. Parsimony, a sophisticated philosophical concept first articulated by the fourteenth-century English theologian William of Ockham, states that *the explanation with the least necessary steps is probably the best one.* Why resort to complexity when simplicity is equally informative? Why suppose more steps took place when fewer can provide the same information?

Figure 3.10 shows two cladograms that are possible with birds, a human, and a bat and the characters of wings, fur, feathers, and mammary glands. In (b), the bird has to lose ancestral mammary glands and it has to replace fur with feathers. In (a), wings must be invented by evolution twice. Cladogram (a) is the simpler of the two because it requires fewer evolutionary events or steps. It is uncomplicated by the addition of more characters. In contrast to cladogram (a), the addition of virtually any other characters that are shared by humans and bats to cladogram (b) (for example, the arrangement, shape, and number of bones, particularly those in the skull and forelimbs, the structure of the teeth, the biochemstry of each organism) requires that each of these shared characters evolved independently: once in bats and once in humans. That considerably complicates the number of evolutionary steps, leaving us with the conclusion that cladogram (b), the hypothesis that birds and bats are more closely related to each other than either is to a human, is a less parsimonious alternative.

Indeed, as a hypothesis about the *evolution* of these vertebrates, it is much more likely that bats and humans share a more recent common ancestor than that either does with a bird (which, obviously, is why bats and humans are classified together here as mammals). In this case, the use of shared, derived characters has led us to the most parsimonious conclusion with regard to the evolution of these three creatures.

Figure 3.10. Two possible arrangements for the relationships of birds, bats, and humans. (a) The left-hand cladogram requires wings to have evolved two times; (b) the right-hand cladogram requires birds to have lost fur and mammary glands. These as well as many other characters suggest that (a) is the more parsimonious of the two cladograms.

Cladograms are science

Cladograms are in effect scientific hypotheses concerning phylogenetic relationships. Like all good science, they make testable predictions about the distributions of characters in organisms, from which evolutionary relationships can be inferred. Any organism – living or extinct – can test an existing cladogram-based phylogenetic hypothesis. With living organisms, not only do we use their anatomy on the cladogram; we can also use their genetic material. Parsimony is then used to determine which cladogram most likely approximates to the course of evolution. As will become evident, cladograms are among the most powerful tools available for learning about what occurred in ages long past.

Leaving Linnaeus: introducing an evolutionary classification – by definition

We regularly encounter the groups that Linnaeus established in his classification (terms like *Canis* [dog], Reptilia, Aves [birds], etc.), but recall that Linnaeus diagnosed his groups on the basis of the overall similarity of their characters; overall similarity, as we have seen, is not a suitable basis for reconstructing evolutionary events. Moreover, properly considered, characters in and of themselves do not really encompass the idea of ancestors giving rise to descendants (e.g., evolution). They are just particular qualities either possessed, or not possessed, by an organism. So how to bring an evolutionary perspective into the names of the groups that we discuss?

In the early 1990s, cladists Jacques Gauthier (see Figure 15.12) and Kevin de Queiroz established a subtle but important distinction between *diagnosis* (the familiar characters that we introduced earlier in this chapter that are shared – or not! – by groups) and phylogenetic **definition**, which explicitly involves the process of organic evolution in the membership of a group. We can define a group by considering two of its members, and then defining it as *all organisms stemming from the most recent common ancestor of those two organisms*. Naturally, this is most easily visualized in conjunction with a cladogram (Figure 3.11).

Consider, for example, humans. A phylogenetic (evolutionary) definition for the group might be "*Homo sapiens sapiens* and *Homo sapiens neandertalensis* (neanderthals), and all members of the group stemming from their most recent common ancestor." This is an evolutionary definition, and while it does not tell us what characters to look for in humans, it effectively delimits on the cladogram what can and what cannot be considered a human. Identifying the group by its *definition*, common ancestry – the fingerprint of organic evolution – becomes fundamental to the evolutionary meaning of the name "human." Defining taxa this way distinguishes our use of *any* taxon from the non-evolutionary origin of many still-useful Linnaean terms presently in use. We will use the important concept of definition to identify all of the key groups proposed in this book; for many, there will be clear-cut, diagnostic characters, but, for a few, there will not. Our first close encounter with the brave new world of definition occurs in Chapter 4.

Figure 3.11. Definition. Here, two subgroups ("A" and "B") belong to a group "C." The *definition* of Group C is "the clade that includes A and B and all the descendants of their most recent common ancestor."

SUMMARY

Relationship is essential to understanding the identities of organisms. To reconstruct relationships, branching diagrams called cladograms are used. Organisms are grouped on these using the presence of shared, derived (or diagnostic) characters; the groups of organisms that result are all inferred to be more closely related to each other than to anything else; that is, they are monophyletic. The evolution of new types of organisms is represented on the cladogram by the appearance of new features, which distinguish descendants from their ancestors. These novel features are the diagnostic (or shared, derived) characters.

Cladograms differ from "trees of life" in several fundamental aspects. Cladograms are based upon shared derived characters; they do not show time; they do not show ancestors (although they specify what the ancestral condition of an organism must have been like); and, most importantly, they are testable. Falsification consists of the prediction on the cladogram that a character will be present that is not (or vice versa). It also occurs when the addition of a new organism upsets the character distributions on a cladogram.

With several hypotheses of relationship (cladograms) to choose among, the cladogram requiring the least number of steps (that is, the most parsimonious) is the preferred one.

SELECTED READINGS

Cracraft, J. and Eldredge, N. (eds.) 1981. *Phylogenetic Analysis and Paleontology.* Columbia University Press, New York, 233 pp.

Darwin, C. 2008 [1859]. *On the Origin of Species: A Facsimile of the First Edition.* Harvard University Press, Cambridge, MA, 540 pp.

Eldredge, N. and Cracraft, J. 1980. *Phylogenetic Patterns and the Evolutionary Process: Method and Theory in Comparative Biology.* Columbia University Press, New York, 349 pp.

Nelson, G. and Platnick, N. 1981. *Systematics and Biogeography, Cladistics and Vicariance.* Columbia University Press, New York, 567 pp.

Futuyma, D. J. 2005. *Evolution.* Sinauer Associates, Sunderland, MA, 603 pp.

Stanley, S. M. 1979. *Macroevolution.* W. C. Freeman and Company, San Francisco, 332 pp.

Wiley, E. O., Siegel-Causey, D., Brooks, D., and Funk, V. A. 1991. *The Compleat Cladist: A Primer of Phylogenetic Procedures.* University of Kansas Museum of Natural History Special Publication no. 19, Lawrence, KS, 158 pp.

TOPIC QUESTIONS

1. Define: phylogeny, morphology, homologous, analogous, a tree of life, hierarchy, characters, diagnostic, cladogram, node, derived, advanced, monophyletic groups, primitive, ancestral, parsimony, test, falsify, science.
2. Why is it that the direct ancestor of any organism isn't easy to identify?
3. If you have several possible cladograms, how do you determine which is the preferred one?
4. Construct a cladogram of something that particularly interests you (examples: guitars, music, sports, shoes, etc.). Be sure to show the diagnostic characters in their correct hierarchical locations.
5. If you don't know the ancestor, how can you hope to understand how something evolved?
6. We have said that we could "reconstruct" the course of evolution with a cladogram. How do cladograms allow us to do this?
7. How is a cladogram a "hypothesis of relationship"?
8. Explain the difference between the definition of a group, and its diagnosis.

APPENDIX 3.1 WHAT IS "EVOLUTION"?

Charles Darwin, via his book *On the Origin of Species* (1859), is generally regarded as the starting point of modern evolutionary theory. Yet, the most common understanding of the word "evolution," that organisms on Earth have changed through time, was *not* the point of Darwin's work. That organisms have changed through time had been well established by savvy natural historians (or natural philosophers, as they were sometimes called) for well over 200 years before Darwin. Darwin's contribution was to propose the *means* by which such changes occurred. His hypothesis was constructed in the following way:

(1) Domesticated animals and plants show a wide range of variation.

(2) A similarly wide range of variation exists among wild animals and plants as well.

(3) All living creatures are engaged in a "struggle" to survive and ultimately reproduce, and that struggle is most severe among those individuals that are most closely related.

(4) The struggle to survive in combination with the variation that exists among individuals leads to the survival and, most importantly, production of viable progeny in some variants as opposed to some others, a process that Darwin called **natural selection**.

(5) The viable progeny of some variants as opposed to others ensures that the characteristics of the successfully reproducing variants make it into the next generation.

(6) This process, repeated over hundreds or even thousands of generations, is evolution by natural selection, sometimes called "Darwinian evolution."

The variants that survived to produce viable offspring are said to be more fit than those that did not. And successive generations of "fit" offspring would, in a manner analogous to breeding, eventually produce a descendant very different from its ancestor (for example, a new species). So, if higher fitness in some hypothetical lineage of organisms was conferred by having longer legs, then that lineage might show an evolutionary trend toward increasing leg length until the long-legged descendant was sufficiently different from its ancestor to be considered a different species. So Darwin's contribution to ideas about evolution, therefore, was actually the hypothesis of *evolution by natural selection*.

Because the science of genetics was not known to Darwin (having been invented, so to speak, during his lifetime), Darwin had no mechanism to explain what exactly was meant by "closely related," although he knew that in some physical way parents, for example, are closely related to their children. The explicit meaning of relationship came with the understanding of chromosomes, genes, alleles, and, some 70 or so years later, DNA.

Nonetheless, all of those ideas were, from the 1920s onward, integrated into Darwin's original hypothesis (except, of course the molecular basis of inheritance, which came somewhat later) in an intellectual movement called the "Evolutionary" or "Modern Synthesis." The Evolutionary Synthesis applied the then-burgeoning fields of population ecology, genetics, paleontology, and statistics to Darwinian evolution, to understanding the precise mechanisms by which particular combinations of genes (**genotypes**) are selected and passed on to succeeding generations. This led to the hypothesis of an unbroken genetic chain of successive changes in the physical appearance (**phenotype**) and behavior of organisms, to the origin of new species, as life perpetuated itself on Earth.

In the intervening years since the Evolutionary Synthesis, the theory of evolution by natural selection has been refined and now incorporates very significant advances in our understanding of genetics, embryology, and molecular biology. The origin of new features is recognized as gradual in some cases, as postulated by Darwin, but also as abrupt in other cases. Sometimes changes in genotype produce new species, but sometimes they may produce phenotypic changes that are either smaller- or larger-scale than the development of merely new species. Natural selection then acts upon phenotypes in the current generation favoring, as Darwin hypothesized, the reproductive viability of some phenotypes and not others.

The fitness – or lack of it – of an organism is determined by an immense number of variables, including, but not limited to, the environment in which it evolves, the other organisms (plant and animal) with which it must interact, and ultimately the viability of its progeny. Obviously these variables are utterly unpredictable, and so evolution by natural selection, as currently understood by evolutionary biologists, does not involve the possibility of predicting future evolutionary events.

As we cannot use it to predict the future, should we consider the hypothesis of evolution by natural selection to be unscientific? Recall that science requires a hypothesis with explicit predictions that can be tested. Although the hypothesis of evolution does not predict the future, it certainly makes testable predictions. For a few examples:

- It predicts that we will find organisms that contain mixes of characters of older organisms and younger organisms occurring intermediate in time between them. (We do.)
- It predicts that life's diversity takes the form of many variations on a basic design, with modifications upon modifications that take us to the present. (It does.)
- It predicts that the biochemical building blocks of life will be present – if slightly modified – in all organisms. (They are.)
- Of particular relevance in this book, it predicted that a creature that mixed bird and "reptile" features would exist. (It does; see Chapter 8.)

Geneticist Theodosius Dobzhansky, one of the pioneers of the Evolutionary Synthesis, once wrote in *The American Biology Teacher*, "Nothing in biology makes sense except in the light of evolution." We concur.

Yet for all that there is a pesky irony. The word "evolution" strictly speaking can refer to an unfolding to a predetermined and inevitable end, such as the *evolution* of a tragedy. Since the evolution of life – organic evolution – does *not* unfold along predetermined or inevitable pathways, it is not surprising that the word "evolve" was avoided by Darwin until the very last word of the very last sentence of the very last paragraph of the very last page of *On the Origin of Species* (1859).

Who are the dinosaurs?

4

CHAPTER OBJECTIVES

- Learn basic relationships among tetrapods – particularly amniotes
- Understand something about the course of tetrapod evolution
- Learn who dinosaurs are (and are not)
- Provide a phylogenetic definition of Dinosauria
- Introduce some of the characters that diagnose Dinosauria

Figure 4.1. *Pikaia gracilens*, a presumed chordate from the Middle Cambrian of Canada.

1 cm

Finding the history of life

In the preceding chapter, we explored the methods that scientists use to learn the identity and origin of all organisms. Now we will apply those techniques – diagnostic characters hierarchically distributed on cladograms – to properly position dinosaurs within the biota. The history of life will unfold as we systematically encounter each bifurcation in the cladistic road, reconstructing the path of evolution until we reach Dinosauria. We'll go with Glinda the Good's suggestion that "it's always best to start at the beginning."

In the beginning

Modern life is generally understood to be monophyletic. It's united by the possession of RNA, DNA, cell membranes with distinctive chemical structure, a variety of amino acids (proteins), the metabolic pathways (that is, chemical reactions) for their processing, and the ability to replicate itself (not simply grow).[1] Notice we said "modern" life – for who knows how many forms of molecular life arose, proliferated, and died out very early in Earth's history – before the thing that we now call "life" finally prevailed?

While it is theoretically possible to construct a cladogram to develop the full history of modern life, we'd have to summarize about 3.8 *billion* years of biotic evolution. Instead, let's cut to the chase, to Animalia, and to a mere 510 Ma, where we first meet *Pikaia gracilens*, a 5 cm, flattened, miniature anchovy fillet of a creature that represents an early data point in the ancestry of vertebrates (Figure 4.1).[2]

Chordata

Pikaia reveals characters that are diagnostic of the clade to which we (and the dinosaurs) belong: **Chordata** ("nerve cord-bearing"). Although *Pikaia* provides an inkling about our distant relatives, what we know about the early evolution of vertebrates and their forebears among Chordata comes principally from living organisms, with some input from a few fossils. The living chordates consist of living urochordates (*uro* – tail; popularly called "sea squirts"), cephalochordates (sometimes called "lancets"), and, most important for our story, vertebrates (Figure 4.2). All of these groups are united within Chordata on the basis of a distinctive suite of characters – pharyngeal gills, notochord, and

---

[1] This statement is not strictly true, because viruses don't have intrinsic membranes, amino acids, and metabolic pathways (they hijack the molecules and mechanisms of the cells they invade), nor do they reproduce themselves; retroviruses don't even have DNA. The origin of viruses remains shrouded in mystery; but one hypothesis is that they lost these early in their evolutionary history.

[2] But not the first. That honor is shared by two primitive chordates from Chengjiang, China, which are believed to date to about 515–520 Ma.

Figure 4.2. Two primitive chordates. (a) A cephalochordate (the lancet *Amphioxus*); (b) urochordate (the sea squirt *Ciona*); and (c) larval sea squirt. Cephalochordates (a) and urochordates (b) share with the vertebrates a host of derived features, including segmentation of the muscles of the body wall, separation of upper and lower nerve and blood vessel branches, and many newly evolved hormone and enzyme systems. The juvenile sea squirt (c) is free-swimming and has a notochord running down its tail. When it metamorphoses into the stationary adult, it parks on its nose and rearranges its internal and external structures.

nerve cord[3] – that appears to have evolved only once, thus uniting these animals as a monophyletic group. We see Chordata and its diagnostic characters at the base of the cladogram in Figure 4.3. We – and the dinosaurs – appear to have chordate relatives as far back as the Cambrian.

Vertebrata

Most interesting for us, within Chordata is also found the familiar **Vertebrata** (*vertere* – to turn, referring to joints in the spine that bend). The diagnostic characters of vertebrates, with the exception of Chondrichyes (sharks, skates, and rays[4]), include a mineralized (and thus, hardened) internal skeleton (that is, bone) divided into discrete pieces called **elements**,[5] and a variety of other characters (Figure 4.3).

Among Vertebrata, we're naturally interested in the subset **Gnathostomata** (*gnathos* – jaw; *stome* – mouth), vertebrates with true jaws (among other diagnostic features), which includes us and most of the vertebrates that might come readily to mind. But, we'll focus directly on the subset of gnathostomes that we call the bony fishes.

---

[3] Pharyngeal gills are the familiar gills of the throat region; a notochord is a stiffening rod that runs down the back; the nerve cord is, of course, a precursor to the familiar spinal cord.

[4] Chrondrichthyans are thought to have lost through evolution their original mineralized skeletons.

[5] In anatomy, the word "element" confusingly has a different meaning from the way it is used in chemistry (see Chapter 2 and Glossary).

Figure 4.3. Cladogram of Chordata. Because this is a book about dinosaurs (and not all chordates), we provide diagnoses for only some of the groups on the cladogram. Bars denote the locations of shared, derived characters of the groups. The characters are the following: at **1**, pharyngeal gill slits, a notochord, and a nerve cord running above the notochord along its length; at **2**, segmentation of the muscles of the body wall, separation of upper and lower nerve and blood vessel branches, and new hormone and enzyme systems; at **3**, bone organized into elements, neural crest cells, the differentiation of the cranial nerves, the development of eyes, the presence of kidneys, new hormonal systems, and mouthparts; at **4**, true jaws; at **5**, bone in the endochondral skeleton; at **6**, ray fins; at **7**, appearance of distinctive bones in fleshy pectoral and pelvic fins (see Figure 4.4); at **8**, bones acquiring a more tetrapod appearance and function; elongate rays still present in fin; at **9**, skeletal features relating to mobility on land – in particular, four limbs with stable (unchanging) element patterns. Consistent with a cladistic approach, only monophyletic groups are presented on the cladogram. Some of the groups may not be familiar, but are included to fill out the cladogram. *Cephalaspis* was a primitive, jawless, bottom-dwelling, swimming vertebrate, and *Eusthenopteron* was a predaceous lobe-finned fish, bearing many characters present in the earliest tetrapods. *Tikaalik* is an extremely important fossil fish, which shows a striking intermediate morphology between lobe-finned fin morphology and the legs of modern tetrapods.

Here, now, is a major split in the cladogram, representing a major evolutionary branching point. On the one hand is the lineage of bony fish (**Osteichthyes**; *osteo* – bone; *ichthys* – fish) leading to old friends such as goldfish, tuna, and salmon. But on the other branch is **Sarcopterygii** (*sarco* – flesh; see Figure 4.3), a not-so-familiar group that is diagnosed by, among other things, the presence of distinctive *lobed* fins, fleshy places where the fin attaches to the body.

Here too is demonstrated the power of homology: those lobes contain bones that are recognizable as – and thus homologous with – bones in the limbs of **Tetrapoda** (*tetra* – four; *pod* – foot): vertebrates with four limbs, including dinosaurs and ourselves (Figure 4.4). What's more, bones of the pelvis, vertebral column, and even bones in the skulls of lobe-finned fishes can also be recognized within tetrapods. All these diagnostic characters strongly indicate that it is here, among the lobe-fins, that the ancestry of Dinosauria – as well as our own ancestry – is to be found (Box 4.1).

(a)   (b)   (c)

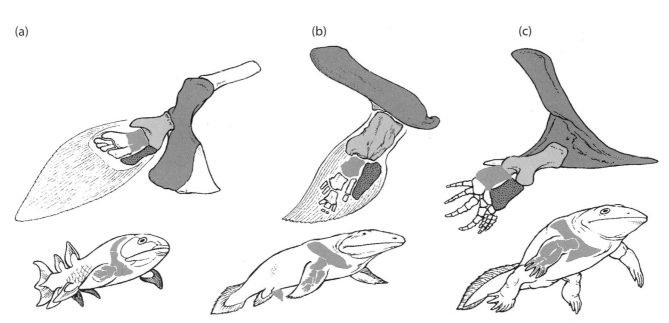

Figure 4.4. Some homologous features linking lobe-finned fish and tetrapods. (a) The fin and shoulder bones (shoulder girdle) of *Eusthenopteron*, an extinct lobe-fin fish; (b) the fin and shoulder girdle of *Tiktaalik*, an animal snugly positioned, morphologically speaking, between lobe-finned fish and tetrapods; and (c) the front limb and shoulder girdle in early tetrapods. Because aspects of the forelimb in different early tetrapods are incomplete, the forelimb shown here is a composite prepared from two early tetrapods (*Acanthostega* and *Ichthyostega*). Key homologous bones are highlighted in both sets of drawings.

Tetrapoda

Remembering what is meant by definition (Chapter 3), modern Tetrapoda is *defined* as the clade that includes salamanders, mammals, and all the descendants of their most recent common ancestor. It is *diagnosed* by the appearance of limbs with the distinctive arrangement of bones shown in Figure 4.4. So now let's take a closer look at Tetrapoda and, because we're interested in dinosaurs, we'll try to understand the part that's generally best preserved: the skeleton.

The tetrapod skeleton made easy

Figure 4.5 shows a typical tetrapod skeleton – in this case a prosauropod dinosaur – exploded. Not surprisingly (because Tetrapoda is monophyletic), tetrapods are all built on the same basic plan: a vertebral column is sandwiched by paired forelimbs and paired hindlimbs. The limbs are attached to the vertebral column by groups of bones called **girdles**. At the front end is the head, composed of a **skull** and **mandible**, or lower jaw. At the back end is the tail. It's that simple!

Vertebral column

The vertebral column is composed of distinct, repeated structures (the **vertebrae**), which consist of a lower spool (the **centrum**), above which, in a groove, lies the spinal cord (Figure 4.5, inset on p. 66). Planted on the centrum and straddling the spinal cord is the **neural arch**. Various **processes**, that is, bumps on bone that are commonly ridge-, knob-, or blade-shaped, may stick out from each neural arch. These can be for muscle and/or ligament attachment, or they can be sites against which the ends of ribs can abut. The repetition of vertebral structures, a relic of the segmented condition that is primitive for chordates, allows flexibility along the length of the animal.

Box 4.1   Fish and chips

As 1978 turned to 1979, a provocative and entertaining letter and reply were published in the scientific journal *Nature*, discussing the relationships of three gnathostomes: the salmon, the cow, and the lungfish. English paleontologist L. B. Halstead argued that, obviously, the two fish must be more closely related to each other than either is to a cow. After all, he argued, they're both *fish*! A coalition of European cladists disagreed, pointing out that, in an evolutionary sense, a lungfish is more closely related to a cow than to a salmon. In their view, if the lungfish and the salmon are both to be called "fish," then the cow must also be a fish. Can a cow be a fish?

The vast majority of vertebrates are what we call "fishes." They all make a living in either salt or fresh water and, consequently, have many features in common that relate to the business of getting around, feeding, and reproducing in a fluid environment more viscous than air. But, as it turns out, even if "fishes" describes creatures with gills and scales that swim, "fishes" is not an evolutionarily meaningful term because there are no shared, derived characters that unite all fishes that cannot also be applied to all non-fish gnathostomes. The characters that pertain to fishes are either characters present in all gnathostomes (that is, primitive in gnathostomes) or characters that evolved independently.

The cladogram in Figure B4.1.1 is universally regarded as correct for the salmon, the cow, and the lungfish. In light of what we have discussed, this cladogram might look more familiar using groups to which these creatures belong: salmon are ray-finned fish, that is, fish with long rays made of a distinct protein supporting their pectoral and pelvic fins; cows are tetrapods; and lungfishes are lobe-finned fishes. Clearly, lobe-finned fishes share more derived characters in common with tetrapods than they do with ray-finned fishes. Thus there are two clades on the cladogram:

Clade (1) lobe-finned fishes + tetrapods; and
Clade (2) lobe-finned fishes + tetrapods + ray-finned fishes.

Clade 1 is familiar as Sarcopterygia. Clade 2 occurs at the level of all fish (and the descendants of fish) and looks like part of the cladogram presented in Figure 4.3 for gnathostome relationships. If only the organisms in

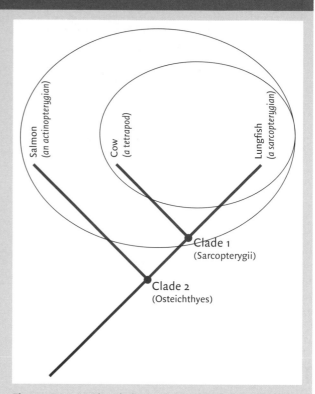

Figure B4.1.1. The cladistic relationships of a salmon, a cow, and a lungfish. The lungfish and the cow are more closely related to each other than either is to the salmon.

question are considered, the only two monophyletic groups on the cladogram must be (1) lungfish + cow; and (2) lungfish + cow + salmon (that is, representatives of the sarcopterygians and Actinopterygii, respectively).

Which are the "fishes?" Clearly the lungfish and the salmon. But the lungfish and the salmon do not in themselves form a monophyletic group unless the cow is also included. The cladogram is telling us that the term "fishes" has *phylogenetic* significance only at the level of Osteichthyes (or even below). But we can and do use the term "fishes" informally. Fish and chips will never be a burger and fries.

Halstead, L. B. 1978. The cladistic revolution – can it make the grade? *Nature*, **276**, 759–760.

Gardiner, B. G., Janvier, P., Patterson, C., *et al.* 1979. The salmon, the cow, and the lungfish: a reply. *Nature*, **277**, 175–176.

Girdles

Sandwiching the backbone are the **pelvic** and **pectoral girdles** (Figure 4.5). These are sheets of bone (or bones) against which the limbs attach for the support of the body. The pelvic girdle is the attachment site of the hindlimbs; the pectoral girdle is the attachment site of the forelimbs.

Each side of the pelvic girdle is made up of three bones: (1) a flat sheet of bone, called the **ilium**, which is fused to the **sacrum**, a block of vertebrae between the iliac blades; (2) a piece that points forward and down, called the **pubis**; and (3) a piece that points backward and down, called the **ischium**. Primitively, the three bones come together in a depressed area of the pelvis called the **acetabulum**: the hip socket.

By contrast, the pectoral girdle consists of a flat sheet of bone, the **scapula** (shoulder blade), on each side of the body, attached to the outside of the ribs by ligaments and muscles. The scapula is connected with the **coracoid**, a shield-like bone, and together these two bones support the attachment of the front limb.

Chest

Some chest elements deserve mention. The breastbone (**sternum**) is generally a flat or nearly flat sheet of bone that is locked into its position on the chest by the tips of the thoracic (or chest) ribs. Paired collar bones – the **clavicles** – sit just above and between the coracoids.

Legs and arms

Limbs in tetrapods show the arrangement pioneered in their sarcopterygian ancestors (see Figure 4.4). All limbs, whether fore or hind, have a single upper bone connecting to a pair of lower bones. In a forelimb, the upper arm bone is the **humerus**, and the paired lower bones (forearms) are the **radius** and **ulna**. The joint in between is the elbow. In a hindlimb, the upper bone (thigh bone) is the **femur**, the joint is the knee, and the paired lower bones (shins) are the **tibia** and **fibula**.

Beyond the paired lower bones of the limbs are the wrist and ankle bones, termed **carpals** and **tarsals**, respectively. The bones in the palm of the hand are called **metacarpals**, the corresponding bones in the foot are called **metatarsals** and collectively they are termed **metapodials**. Finally, the small bones that allow flexibility in the digits of both the hands (fingers) and the feet (toes) are called **phalanges** (singular phalanx). At the tip of each digit, beyond the last joint, are the **ungual phalanges**.

Tetrapods primitively had as many as eight digits on each limb. Early in the evolutionary history of tetrapods, this number rapidly reduced to, and stabilized at, five digits on each limb, although many groups of tetrapods subsequently reduced that number even further (Figure 4.5).

Head

At the front end of the vertebral column of chordates are the bones of the head, composed, as we have seen, of the skull and mandible (Figure 4.6). Primitively, the skull has a distinctive arrangement: the **braincase**, a bone-covered box containing the brain, is located centrally and toward the back of the skull. At the back of the braincase is the **occipital condyle**, the knob of bone that connects the braincase (and hence the skull) to the vertebral column. A rear-facing opening in the braincase, the **foramen magnum**, allows the spinal cord to attach to the brain. Located on each side of the braincase are openings for the **stapes**, the bone that transmits sound from the **tympanic membrane** (eardrum) to the brain. Finally, covering the braincase and forming much of the upper rear part of the skull is a curved sheet of interlocking bones, the **skull roof** (inset to Figure 4.6).

The skull has two familiar pairs of openings. Located midway along each side of the skull is a large, round opening – the eye socket, or **orbit**. At the anterior tip of the skull is another pair of

Dorsal (top)

Ventral (bottom)

Cervical vertebrae

Gastralia

Scapula

Coracoid

Humerus

Clavicles

Scapula

Coracoid

Carpals
(wrist
bones)

Humerus

Carpals

Radius

Metacarpals
(hand bones)

Ulna

V   IV

III   II

I

Phalanges
(finger bones)

Figure 4.5. Exploded view of a tetrapod skeleton exemplified by the saurischian dinosaur *Plateosaurus*.

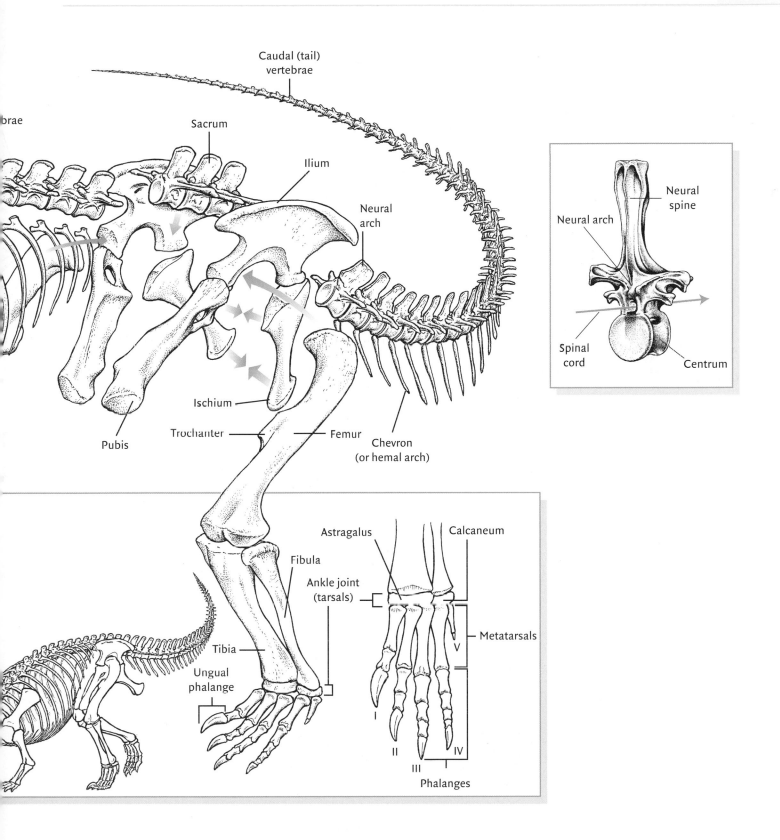

Caudal (tail) vertebrae

Sacrum

brae

Ilium

Neural arch

Ischium

Trochanter

Femur

Chevron (or hemal arch)

Pubis

Fibula

Ankle joint (tarsals)

Astragalus

Calcaneum

Metatarsals

Tibia

Ungual phalange

V

I

II

III

IV

Phalanges

Neural spine

Neural arch

Spinal cord

Centrum

Figure 4.6. Skull and mandible of the primitive saurischian dinosaur *Plateosaurus*, exemplifying the general arrangement of bones in the skull and mandible. (a) Skull elements exploded; (b) cross-section through braincase; (c) exploded elements of mandible (lower jaw); (d) rear view of skull.

openings – the **nares** (singular, naris), or nostril openings. Finally, flooring the skull, above the mandible, is a paired series of bones, organized in a flat sheet, which forms the **palate**.[6]

Within Tetrapoda

Tetrapods share a variety of derived features (Figure 4.7). We have seen many of these in the tetrapod skeleton: the distinctive morphologies of the girdles and limbs, as well as the fixed patterns

---

[6] In many amniotes, including all mammals, a passage forms between the floor of the nasal cavity and the roof of the oral cavity (mouth), so that air breathed in through the nostrils is guided to the back of the throat, bypassing the mouth. As a result, it is possible for chewing and breathing to occur at the same time. Similar kinds of palates (called secondary palates) are known in other tetrapods besides mammals, but primitively the nostrils lead directly to the mouth. So, if food were to be extensively chewed in the mouth, it would quickly get mixed up with the air that is breathed in. For this reason, chewing is not a behavior of primitive tetrapods.

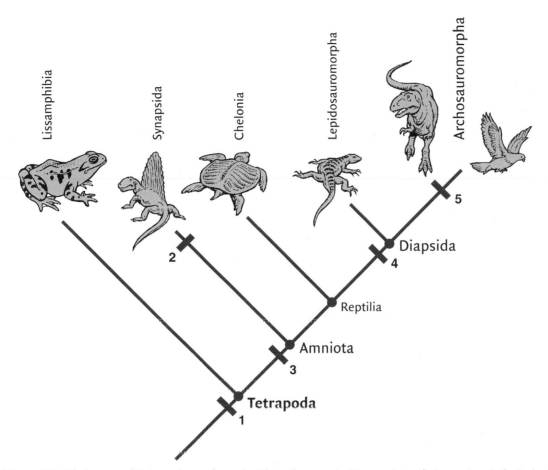

Figure 4.7. Cladogram of Tetrapoda, configured with turtles as primitive amniotes (but see Box 4.2). Derived characters include: at **1**, the tetrapod skeleton (see Figure 4.5); at **2**, lower temporal fenestra (see Figure 4.9); at **3**, the presence of an amnion (see Figure 4.8); at **4**, lower and upper temporal fenestrae (see Figure 4.9); and at **5**, antorbital fenestra (see Figure 4.12). Lepidosauromorpha is a monophyletic group, the living members of which are snakes, lizards, and the tuatara. Chelonia – turtles – are reptiles whose primitive, completely roofed skulls historically suggested a position near the base of Amniota; however, considerable question remains where they properly belong.

of skull roofing bones. The hypothesis that all of these shared similarities evolved separately in distantly related organisms is not parsimonious; for this reason, these characters reaffirm Tetrapoda as a monophyletic group. Now we continue our journey to find out exactly where to situate dinosaurs.

Amniota

A subset of the tetrapods, Amniota is defined as the monophyletic group that includes birds, mammals, and all the descendants of their most recent common ancestor. Amniota is characterized by the invention of a special membrane for the egg-bound, developing embryo called an **amnion** (Figure 4.8, and see below). The tetrapods without an amnion – **anamniotes** – are today represented only by frogs, salamanders, and a very rare, limbless, tropical amphibian known as a caecilian (almost not worth mentioning, unless you're

Figure 4.8. Amniote egg.

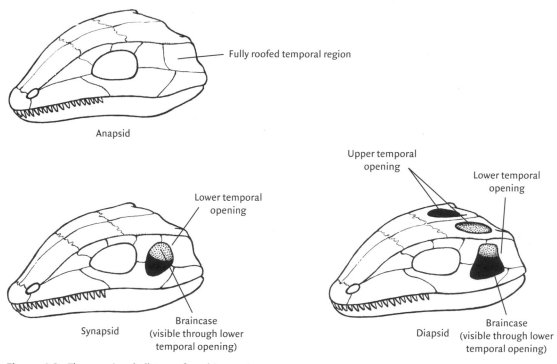

Figure 4.9. Three major skull types found in amniotes.

a caecilian, of course). If the living amphibians are any guide, the life cycles of anamniotes are intimately associated with water, as the eggs require, and likely required, an external source of moisture.

Amniotes, by contrast, are fully terrestrial, and need a means of retaining moisture within the egg. The semi-permeable amnion allows gas exchange but retains water, which permits the embryo to be continuously bathed in liquid. The evolutionary appearance of the amnion occurred in conjunction with several other features including a calcified shell, a large yolk for the nutrition of the developing embryo, and a special bladder for the management of embryonic waste. Amniotic eggs can thus be laid on land without drying out, which allowed amniotes to sever all ties with water (other than for drinking). This was a key step in the evolution of a completely terrestrial lifestyle, and is commonly associated with the advent of reptiles.

There are three great groups of amniotes – primitive amniotes, sometimes termed **anapsids** (*a* – without; *apsid* – arch), **Synapsida** (*syn* – with), and **Diapsida** (*di* – two). They're most easily distinguished by the number and position of the openings in the skull roof behind the eyes, called **temporal fenestrae** (*fenestra* – window; Figure 4.9). Our main interest is in diapsids, but we'll detour briefly to look at some basal amniotes and Synapsida.

Anapsids and Synapsida

The anapsid condition represents what is thought to have been the original morphology of the skull roof in amniotes. In these amniotes, the skull behind the eyes is completely roofed; they therefore have no temporal fenestrae. The anapsid condition is seen in many long-extinct, bulky

30 cm

Figure 4.10. *Dimetrodon grandis*, a fin-backed synapsid from the late Paleozoic of eastern Texas, USA.

quadrupeds that do not concern us here, and *may* persist today only in turtles. These had always been assumed to be the only living example of an anapsid amniote. Recently, considerable work – both morphological and molecular – has suggested that turtles are actually a monophyletic group of highly derived diapsids. This, if true, has major consequences for the meaning of the word "Reptilia" (see Box 4.2).

Synapsida is one of two great lineages of amniotic tetrapods (the other is Diapsida). All mammals (including ourselves) are synapsids, as are a host of extinct forms, traditionally (and ironically; see Box 4.2) called "mammal-like reptiles" (Figure 4.10). The split between the earliest synapsids and the earliest representatives of the other great lineage, Diapsida (including dinosaurs), likely occurred between 310 and 320 Ma. Since then, therefore, the synapsid lineage has been evolving independently, genetically unconnected to any other group.

Synapsids are united by having a skull roof that departs from basal amniotes: the skull roof has developed a low opening behind the eye – the **lower temporal fenestra** (see Figure 4.9). Jaw muscles pass through this opening and attach to the upper part of the skull roof. Synapsids are a remarkable and diverse group of amniotes and could easily fill a book just like this; indeed, this sentence is currently being read by a synapsid. But because we're interested in dinosaurs, we'll regretfully move right on past them.

Diapsida

The other great clade of amniotes is Diapsida (see Figures 4.7 and 4.9). The living diapsids include about 15 000 total species including snakes, lizards, crocodiles, the tuatara (a reptile found only in New Zealand), and birds; extinct diapsids include dinosaurs as well as many other forms. Nobody really knows how many members of this clade have come and gone.

Box 4.2   What, if anything, is a "reptile?"

As we consider the cladistic climb through amniotes to dinosaurs, the meaning of an old, familiar group, Reptilia, becomes very murky.

Recall from Chapter 3 that Linnaeus developed his classification by grouping similar-looking things. His Reptilia (*reptere* – to crawl; Linnaeus coined the name) denoted a group of scaly, four-legged creatures that "crawl on their bellies like a reptile," as the song[a] goes.

But if the classification is all about relationship, are the living reptiles – snakes, lizards, crocodilians, and the tuatara – monophyletic? Are they really more closely related to each other than they are to anything else? The cladogram in Figure 4.11 and those in Chapter 8 demonstrate that, on the basis of shared derived characters, birds are more closely related to crocodiles than crocodiles are to lizards. And that's just not possible unless a bird is a reptile. But how can a bird be a reptile?

The simplest answer is that clearly we have a decidedly different Reptilia from Linnaeus' traditional motley crew of crawling, scaly, non-mammalian, non-bird, non-amphibian creatures that were once tossed together as reptiles. Because crocodiles and birds are more closely related to each other than either is to snakes and lizards, a monophyletic group that includes snakes, lizards, and crocodiles must also include birds.

So where does that leave reptiles? Reptilia has no diagnostic characters that are not those of Amniota, and so an obvious conclusion that one might reach is that Reptilia = Amniota. It turns out that, in a strange way, the whole thing hinges on turtles.

The turtle problem

Legendary stalwarts of the world, turtles are unique: these venerable creatures with their portable houses, in existence

[a] Ray Stevens' "Little Egypt."

since the Late Triassic (~210 Ma), are a phylogenetic conundrum, and their position in the cladogram dramatically changes the meaning of the word "Reptilia."

If we look exclusively at living organisms, turtles are the only extant anapsids, and Reptilia can be properly defined as containing turtles (Chelonia), birds (Aves), and all the organisms that stem from their most recent common ancestor. This definition of Reptilia is shown in Figure B4.2.1 and, with the exception of the inclusion of birds, is only mildly mind-stretching.

- BUT -

Still looking only among the *living* amniotes, if, on the other hand, turtles are diapsids (as is indicated by equally impressive molecular and morphological evidence), then in effect Reptilia = Amniota. This was a common viewpoint in pre-cladistic days, when "reptiles" were identified as a group that could lay eggs on land, through the gentle offices of (no surprise!) an amnion. Thought about in those terms, birds, crocodilians, lizards, and dinosaurs are of course amniotes, but then so are mammals, when last we checked. This perspective left systematists with no logical alternative but to do something they were utterly unwilling to to do: call mammals reptiles (a moniker applied exclusively to one's ex in those days).

But here the fossil record saves the day: If we also include *extinct* organisms in our cladogram (an example of one such group, the small, superficially lizard-like Captorhinomorpha, then it can be seen that there exists a clade which we call Reptilia that includes some extinct anapsid amniotes, but which does not include synapids (and thus, mammals). In a phylogenetic telling, the sister clade of Reptilia is Synapsida (including Mammalia), which together form the clade Amniota. Figure B4.2.2 shows the cladogram with turtles as diapsids.

Diapsida is defined as birds, lizards, and all the descendants of their most recent common ancestor. It is united by a suite of shared, derived features, including having two temporal openings in the skull roof: the **upper and lower temporal fenestrae** (Figure 4.9). The upper and lower temporal fenestrae are thought to have provided accommodation for the bulging of contracted jaw muscles, as well as increased surface area for the attachment of these muscles.

There are two major clades of diapsids. The first, **Lepidosauromorpha** (*lepido* – scaly; *morpho* – shape), is composed of snakes and lizards and the tuatara (among the living), as well as a number of

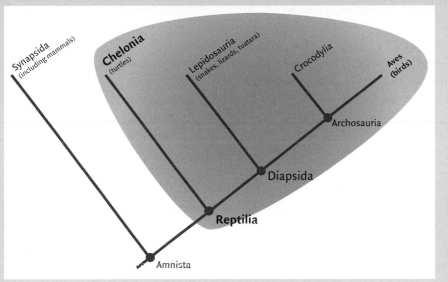

Figure B4.2.1. Turtles (Chelonia) as anapsids. Here, Reptilia is conveniently defined as Chelonia (turtles) and Aves (birds) and all the organisms that evolved from their most recent common ancestor. With the exception of Aves – who are there because, as we see in Figure 4.11, crocodiles are more closely related to birds than either is to lizards – this is a familiar group of animals to call reptiles.

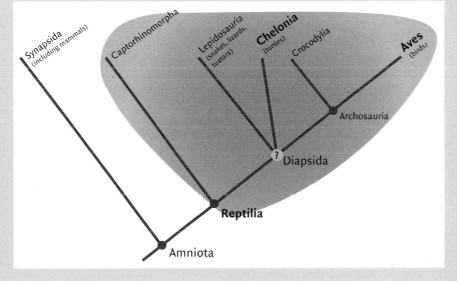

Figure B4.2.2. Turtles (Chelonia) as diapsids. Their position within Diapsida is uncertain; for this reason, a question mark denotes their relationships to Archosauria and Lepidosauria. The inclusion in the cladogram of extinct anapsid amniotes, such as Captorhinomorpha, gives us a more reasonable Reptilia, in which Amniota ≠ Reptilia. Here, all reptiles are amniotes, but not all amniotes are reptiles.

extinct lizard-like diapsids;[7] the second, **Archosauromorpha** (*archo* – ruling), brings us within striking distance of dinosaurs.

---

[7] Two marine groups, ichthyosaurs and plesiosaurs (see Figure 16.11), have also been placed within Diapsida; mosasaurs, the magnificent marine-adapted predatory lizards of the Cretaceous, belong within Lepidosauria. Turtles, long thought to represent the anapsid condition, may be diapsids (see text).

Archosauromorpha

Archosauromorpha is supported by many important, shared, derived characters (Figure 4.11). Within archosauromorphs are a series of basal members that are known mostly from the Triassic. Some bear a superficial resemblance to large lizards; others look like beefed-up crocodiles; a few even have snouts that look like reptilian pigs (see Figure 14.4).

A subset of archosauromorphs possesses a number of significant evolutionary innovations (Figure 4.11), most notably an opening on the side of the snout, just ahead of the eye, called the **antorbital fenestra** (Figure 4.12). This is the key character that unites **Archosauria**, the group that contains crocodilians, birds, and dinosaurs. It is ironic that, for all its phylogenetic importance, the function of the antorbital fenestra is still unknown.

Crocodilians and their close relatives belong to a clade called **Crurotarsi** (*crus* – the joint between the ankle and fibula; *tarsi* – ankles; ironically named because while some things that only *look* like crocodiles belong to this group, true crocodiles belong as well!) about which we won't be too concerned here; dinosaurs and their close relatives constitute a clade called **Ornithodira** (*ornith* – bird; *dira* – neck; see Figure 4.11).

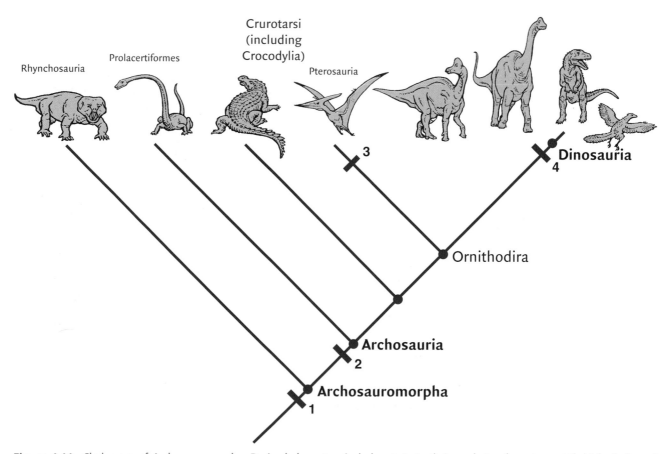

Figure 4.11. Cladogram of Archosauromorpha. Derived characters include: at **1**, teeth in sockets, elongate nostril, high skull, and vertebrae not showing evidence of embryonic notochord; at **2**, antorbital fenestra (see Figure 4.12), loss of teeth on palate, and new shape of articulating surface of ankle (calcaneum); at **3**, a variety of extraordinary specializations for flight, including an elongate digit IV; at **4**, perforate acetabulum (see complete list of shared derived characters of Dinosauria in Figure 4.14 and 4.15).

Ornithodira brings us quite close to the ancestry of dinosaurs. This group is composed of two monophyletic groups, Dinosauria (*deinos* – terrible) and Pterosauria (*ptero* – winged; Figure 4.13). That pterosaurs are unapologetically Mesozoic archosaurs has led to their being called "dinosaurs"; that they had wings and flew has led some to mistake them for birds; but in fact they were something utterly different from either dinosaurs or birds. They were unique, magnificent, and now, sadly, very extinct.

Dinosaurs

This climb up the cladogram leaves us wheezing and gasping for air, but at long last situated at the subject of our book: Dinosauria. Dinosauria is formally defined as *Triceratops horridus*, *Passer domesticus* (sparrow; a representative derived bird), and all members of the clade descended from their most recent common ancestor.

Dinosaurs can be diagnosed by a host of shared, derived characters. These are anatomically quite detailed, and a few are listed in Figure 4.14 and illustrated in Figure 4.15. Among the most accessible are the perforate acetabulum (an opening in the hip socket), large muscle attachment scars revealing that the jaw closing muscles attached to the top of the skull (unusually large musculature suggesting a distinctively powerful bite), and a well-developed, elongate deltopectoral crest on the humerus (upper arm), indicating unusual forelimb power.

Standing tall

While these characters appear to fall within the realm of dinosaur minutiae, there is one quality dinosaurs possess that does not: an **erect**, or **parasagittal** stance; that is, a stance in which the plane of the legs is parallel to the vertical (mirror image) plane of the torso (see Figure 4.11). This stance is also present in a few archosaurs very close to Dinosauria (see Chapter 5) but because it is truly emblematic of Dinosauria, we include it here.

In dinosaurs, an erect stance consists of a suite of anatomical features with important behavioral implications (Box 4.3). The head of the femur (thigh bone) is oriented at approximately 90° to the shaft. The head of the femur itself is barrel-shaped (unlike the familiar ball shape seen in a human femur), so that motion in the thigh is restricted largely to

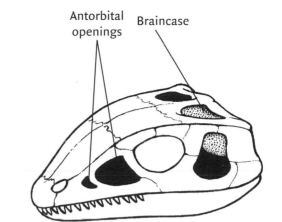

Figure 4.12. An archosaur skull with the diagnostic antorbital fenestra.

Figure 4.13. A close relative of Dinosauria: Pterosauria, as represented by a pterodactyloid pterosaur, from the Cretaceous of Asia.

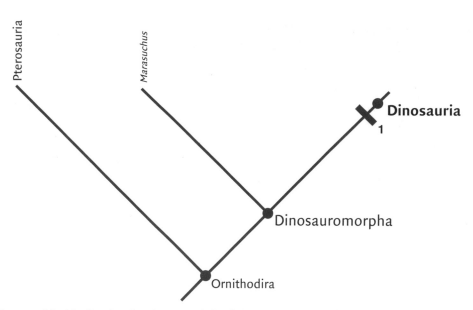

Figure 4.14. Cladogram of Ornithodira showing the monophyly of Dinosauria. Derived characters at **1**, elongate deltapectoral crest on humerus; perforate acetabulum; fibula contacts ≤ 30% of astragalus; temporal musculature that extends anteriorly onto skull roof; epipophyses on the cervical vertebrae; and asymmetrical fourth trochanter on femur (see also Figure 4.15). *Marasuchus* is a close relative of dinosaurs, both of which we shall meet in Chapter 5.

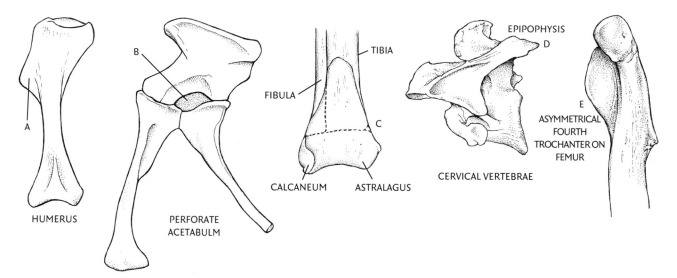

Figure 4.15. Some of the derived characters uniting Dinosauria. (A) Elongate deltopectoral crest on humerus; (B) perforate acetabulum; (C) fibula contacts ≤30 per cent of astragalus; (D) epipophyses on cervical vertebrae; and (E) asymmetrical fourth trochanter on femur.

forward and backward, within, as we've seen, a plane parallel to that of the body (see Figure 4.16). The ankle joint is modified to become a single, linear articulation. This type of joint, termed a modified **mesotarsal joint**, allows movement of the foot only in a plane parallel to that of the body: forward and backward (Figure 4.16). Note that again this situation differs from that in humans, in which the foot is capable of rotating.

With a design so powerfully directed towards running, their anatomy tells the story: Dinosaurs are terrestrial beasts through and through.

Box 4.3   Stance: it's both who you are and what you do

Tetrapods that are most highly adapted for land locomotion tend to have an erect, or upright, stance. This clearly maximizes the efficiency of the animal's movements on land, and it is not surprising that, for example, all mammals are characterized by an erect stance. Tetrapods such as salamanders (which are adapted for aquatic life) display a sprawling stance, in which the legs splay out from the body nearly horizontally. The sprawling stance seems to have been inherited from the original position of the limbs in early tetrapods, whose sinuous trunk movements (presumably inherited from swimming locomotion) aided the limbs in land locomotion.

Some tetrapods, such as crocodiles, have a semi-erect stance, in which the legs are directed at something like 45° downward from horizontal (Figure B4.3.1). Does this mean that the semi-erect stance is an adaptation for a combined aquatic and terrestrial existence? Clearly not, because a semi-erect stance is present in the large, fully terrestrial monitor lizards of Australia (goanna) and Indonesia (Komodo dragon). If adaptation is the only factor driving the evolution of features, why don't completely terrestrial lizards have a fully erect stance, and why don't aquatic crocodiles have a fully sprawling stance? The issue is more complex and is best understood through adaptation to a particular environment or behavior, *as well as* through inheritance. If we consider stance simply in terms of ancestral and derived characters, the ancestral

condition in tetrapods is sprawling. An erect stance represents the most highly derived state of this character, but are animals with sprawling stances not as well designed as those with erect stances? In 1987, D. R. Carrier of Brown University, Rhode Island, USA, hypothesized that the adoption of an erect stance represents the commitment to an entirely different mode of respiration (breathing) as well as locomotion (see Chapter 12 on "warm-bloodedness" in dinosaurs). Those organisms that possess a semi-erect stance may reflect the modification of a primitive character (sprawling) for greater efficiency on land, but they may also retain the less-derived type of respiration. Dinosaurs (see Figure 4.11) and mammals both have fully erect stances, which represent a full commitment to a terrestrial existence as well as to a more derived type of respiration. The designs of all these organisms are thus compromises among inheritance, habits, and mode of respiration. Who can say what other influences are controlling morphology?

Interestingly, the cladogram (see Figure 4.7) shows that the most recent common ancestor of dinosaurs and mammals – some ancestral amniote – was itself an organism with a sprawling stance. Because dinosaurs and mammals (or their precursors) have been evolving independently since their most recent common ancestor, an erect stance must have evolved at least twice in Amniota: once among the synapsids and once in dinosaurs.

Figure B4.3.1. Stance in four vertebrates. To the left, the primitive amphibian and crocodile (behind) have sprawling and semi-erect stances, respectively. To the right, the human and the dinosaur (behind) both have fully erect stances.

Figure 4.16. The fully erect posture in dinosaurs. Unlike in, for example, a human, the bones of the leg restrict movement to only forward and backward, effectively restricting movement to just one plane.

SUMMARY

In this chapter, cladograms are used to map the relationship of Dinosauria to the rest of the biota. Dinosaurs fall within Vertebrata, backbone-bearing animals within the larger group Chordata (bilateral creatures bearing a dorsal nerve cord and a notochord). Within vertebrates, dinosaurs are tetrapods, or animals with four limbs. Dinosaurs are amniotic tetrapods (or amniotes), which means that, like ourselves, they possess an amnion.

Amniotes are generally identified by the number and type of temporal fenestrae. Synapsida (the group that includes mammals) is united by the possession of lower temporal fenestrae, while

Diapsida (the group that includes snakes, lizards, crocodilians, birds, and possibly turtles) all possess lower and upper temporal fenestrae. The split between these two major groups of reptiles took place about 320 Ma.

Dinosaurs are diapsids and among living diapsids, dinosaurs are most closely related to crocodiles and birds, both of which are archosaurs, a group of reptiles united by the presence of an antorbital opening. Calling birds reptiles is, of course, contrary to conventional Linnaean classification, which in this case fails to accurately depict their evolutionary relationships as revealed by cladistic analysis.

Finally, this chapter contains a brief summary of basic diapsid bone morphology. Elements of the post-cranial skeleton are presented, including centra, neural arches, hemal arches, cervical vertebrae, caudal vertebrae, ribs, gastralia, pelvic and pectoral girdles. Limb skeletal structure is presented, including humerus, radius, and ulna; femur, tibia, and fibula; carpals, tarsals, metapodials, and phalanges. The elements of the skull and mandible are also presented.

SELECTED READINGS

Bakker, R. T. and Galton, P. M. 1974. Dinosaur monophyly and a new class of vertebrates. *Nature*, 248, 168–172.

Benton, M. J. 1997. Origin and early evolution of the dinosaurs. In Farlow, J. O. and Brett-Surman, M. K. (eds.) *The Complete Dinosaur*. Indiana University Press, Bloomington, IN, pp. 204–214.

Benton, M. J. 2004. Origin and relationships of Dinosauria. In Weishampel, D. B., Dodson, P., and Osmólska, H. (eds.) *The Dinosauria*, 2nd edn. University of California Press, Berkeley, pp. 7–20.

Brusatte, S. L. 2102. *Dinosaur Paleobiology*. Wiley-Blackwell, Oxford, 322 pp.

Carrier, D. R. 1987. The evolution of locomotor stamina in tetrapods: circumventing a mechanical constraint. *Paleobiology*, 13, 326–341.

Gauthier, J. A., Kluge, A. G., and Rowe, T. 1988. Amniote phylogeny and the importance of fossils. *Cladistics*, 4, 105–209.

Parrish, M. J. 1997. Evolution of the archosaurs. In Farlow, J. O. and Brett-Surman, M. K. (eds.) *The Complete Dinosaur*. Indiana University Press, Bloomington, IN, pp. 191–203.

Sereno, P. C. 1991. Basal archosaurs: phylogenetic relationships and functional implications. *Journal of Vertebrate Paleontology*, 11 (suppl.), 1–53.

TOPIC QUESTIONS

1. Define: Chordata, notochord, gnathostome, Sarcopterygii, skull, mandible, girdles, neural arch, centrum, process, ilium, ischium, pubis, acetabulum, sternum, humerus, femur, radius, ulna, tibia, fibula, phalanx, ungual, metacarpals, metatarsals, metapodials, occipital condyle, foramen magnum, stapes, skull roof, tympanic membrane, nares, orbit, palate, amniote, anamniote, amnion, anapsid, synapsid, diapsid, upper temporal fenestra, lower temporal fenestra, archosaur, Archosauromorpha, lepidosaur, Crurotarsi, Ornithodira, Dinosauria, pterosaur, mesotarsal, Ornithischia, Saurischia.

2. Explain the importance of the amnion in the evolution of terrestrial tetrapods.

3. Describe the basic structure of the vertebrate limb.

4. Draw a skull and lower jaws, indicating the skull roof, braincase, temporal region, orbit, nares, and snout.

5. For how long have the mammal and bird lineages been evolving separately?
6. Why is it that we said that a bird was more closely related to a crocodile than a crocodile is to a lizard?
7. Construct a cladogram with only Vertebrata, Diapsida, Dinosauria, Synapsida, and Lepidosauria marked on it.
8. Construct a cladogram with just Dinosauria, Ornithischia, Archosauria, and Crocodylia marked on it.
9. How can birds be reptiles?
10. Within Amniota, warm-bloodedness and flight occur in bats, in birds, and in pterosaurs. Use a cladogram to show how many separate evolutionary events this required.

Dinosaurs
In the beginning

5

CHAPTER OBJECTIVES

- Understand the origin and early evolution of dinosaurs
- Learn how we study the origin of a major group of animals

In the beginning...

How did it all begin? Was there something special, or different, or *better* about that first dinosaur, that led to all the rest? Did it show up, modest but fetching, on a half-shell? Do we even know who the first dinosaur was? Shouldn't the ancestor of all dinosaurs have been something *different* and *unique*?[1] Ultimately, it is possible that the importance of that first dinosaur was just the proliferation of all its descendants. Maybe the first true dinosaur was just a slightly modified Middle Triassic archosaur doing its Middle Triassic archosaur thing.

"Middle Triassic" may be right; as we'll see below, many undoubted dinosaurs are known from the Late Triassic from nearly everywhere around the world. So a good guess for the age of the very first dinosaur could be, perhaps, towards the end of the Middle Triassic. Still, even if such rocks and fossils were extremely abundant, the likelihood of finding the actual *ur*-dinosaur – the very beast that was the great granddaddy (or mommy) of a lineage – is vanishingly small; even if you did find the remains of that animal, how would you know it was *the one* (and not, for example, its first cousin or next-door neighbor)?

So we'll resort again to the cladogram: the hierarchy of characters that specifies what features ought to be present in an ancestor. It is then a question of finding a creature that most closely matches the expected combinations of characters and character states. We'll search among primitive Archosauria (sometimes termed **basal archosaurs**) for fossils that embody the unique features, specified by the cladogram, of the hypothetical dinosaur ancestor. In the case of dinosaurs, we have several small early archosaurs that are very good candidates.

Archosauria
Crurotarsi vs. Ornithodira

Recall (Chapter 4) archosaurs: that group of reptiles bearing an antorbital fenestra (see Figures 4.11 and 4.12), whose living representatives are crocodiles and birds. Crocodiles and birds represent a fundamental split early in the evolution of archosaurs, with two groups emerging: Crurotarsi, the crocodile line, and Ornithodira (Figure 5.1), the bird line. These two can are distinguished by their ankle bones (see the inset foot and ankle, shown at the bottom of Figure 4.5): in crurotarsans, the two bones of the ankle, the calcaneum and the astragalus, are more or less equally sized; this is called, unsurprisingly, the **crurotarsal** ankle. In ornithodirans, however, the astragalus is somewhat larger, and the ankle allows movement in only one plane: this is the familiar (again from Chapter 4) mesotarsal ankle. These differences reflect different modes of walking; crurotarsans tend to have a sprawling, or semi-erect stance, while ornithodirans generally – but not always – have an erect stance (Figure 5.2; see also Figure 4.16).

Crurotarsi is composed of a variety of ground-hugging, quadrupedal, armored behemoths, several of which are showcased in Figure 5.3. Most crurotarsans didn't make it into the Jurassic. The exceptions, of course are crocodilians and their relatives (Crocodylomorpha), which still reside among the living.

Ornithodira consists of two groups: Dinosauromorpha (*morph* – shape), which includes true dinosaurs and their near-relatives, and Pterosauria, the magnificent flying reptiles that aren't birds (Figures 4.13 and 14.7). Birds of course are theropod dinosaurs, and are sadly the only living ornithodiran representatives (Figure 5.1; see Chapters 7 and 8).

---

[1] Answers : Notsure; Notsure; No; No; Notnecessarily!

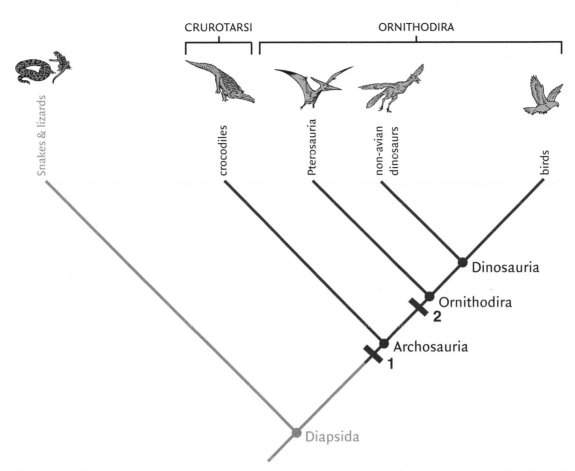

Figure 5.1. Cladogram of Archosauria, showing the fundamental split between Crurotarsi and Ornithodira. Derived characters include: at **1**, antorbital opening; at **2**, distal ends of neural spines not expanded; second phalanx of hand digit II longer than first phalanx; tibia longer than femur; compact metatarsus.

Dinosauromorpha

Paleontology is notoriously terminology-rich, and as we get successively closer to Dinosauria, the names come thick and fast. Some paleontologists (including us) use Dinosauromorpha as a formal term encompassing dinosaurs *and* their close relatives; some use it to refer to just those close relatives.[2] Here we keep it simple with just the clades Ornithodira and, within it, Dinosauromorpha, and of course Dinosauria (Figure 5.4).

So who are dinosauromorphs? The earliest, most primitive forms, non-dinosaur (basal) dinosauromorphs, were small (about 1 m or less, including tail), lightly built, likely insectivorous or carnivorous creatures, many of whom walked on all fours but ran bipedally, and could have starred in the opening scene of *Jurassic Park II*. From Argentina come *Marasuchus* (once known as "*Lagosuchus*") and *Lagerpeton*. First from Poland, and then Africa, Argentina, and North America, are the diminutive silesaurids, including *Silesaurus*, *Asilisaurus*, and *Pseudolagosuchus*, respectively. Also among the silesaurids is the poorly understood

---

[2] To add to the confusion, dinosauromorphs exclusive of dinosaurs were once called "pseudosuchians." Some paleontologists recognize a group – not quite true dinosaurs – snuggled even closer to Dinosauria than dinosauromorphs, called Dinosauriformes (*formes* – in form of). Dinosauromorphs exclusive of pterosaurs are grouped within the clade Avemetatarsalia ("bird feet"). We stick to the KISS Principle here.

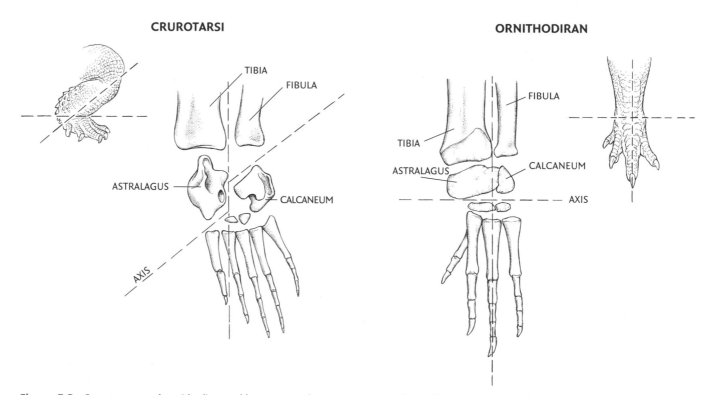

CRUROTARSI

TIBIA
FIBULA
ASTRALAGUS
CALCANEUM
AXIS

ORNITHODIRAN

FIBULA
TIBIA
ASTRALAGUS
CALCANEUM
AXIS

Figure 5.2. Crurotarsan and ornithodiran ankles compared. In crurotarsans, the ankle must accommodate a complicated range of motion between the moving leg – which is held at a constantly changing acute angle to the ground, and the foot, which is parallel with the ground. The mesotarsal ankle, on the other hand, transmits the single-plane movement of the legs to the same, single-plane movement of the foot.

(a)

(b)

(c)

(d)

Figure 5.3. Assorted primitive archosaurs. (a) A phytosaur (*Rutiodon*); (b) an aetosaur (*Stagonolepis*); (c) a rauisuchian (*Postosuchus*); and (d) a primitive crocodile (*Protosuchus*).

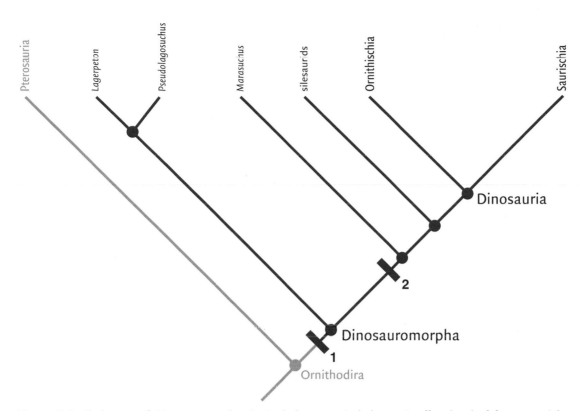

Figure 5.4. Cladogram of Dinosauromorpha. Derived characters include, at **1**, offset head of femur, straight cnemial crest, longest metatarsal > 50% of length of tibia, metatarsal V without phalanges and pointed; at **2**, elongate deltopectoral crest on humerus; tibia with transversely expanded subrectangular distal end (see also Figure 4.15).

Lewisuchus[3] from Argentina. Only recently recognized, silesaurids apparently had a world-wide distribution. Silesaurids were evidently herbivorous, with teeth that superficially resemble those of ornithischians. Several non-dinosaur basal dinosauromorphs are shown in Figure 5.5.

Dinosauria

The line between dinosaur and non-dinosaur turns out to be a finely graded one. Yet our definition of Dinosauria from Chapter 4 leaves no ambiguity: dinosaurs are *the monophyletic group (e.g., the clade) that contains both* Passer *and* Triceratops *and all organisms descended from their most recent common ancestor* (Figure 5.6). This, of course, precisely limits who can and who can't be considered a dinosaur regardless of how closely they superficially resemble a dinosaur. And this also removes some animals – the non-dinosaur basal dinosauromorphs – from Dinosauria.

Early Dinosauria

The oldest record of dinosaurs *might* be some tantalizing footprints from Middle Triassic rocks in Europe and South America (in the range of 242–237 Ma). These prints appear to match what is known of feet of the earliest dinosaurs, and could be a reasonable age for the first appearance of dinosaurs on Earth.

---

[3] There is some evidence that *Lewisuchus* is the same animal as *Pseudolagosuchus.* If so, because *Pseudolagosuchus* is the younger (more recently named) name, all specimens of *Pseudolagosuchus* will come to be called *Lewisuchus.*

Figure 5.5. Basal dinosauromorphs – *Silesaurus*, *Marasuchus*, *Pseudolagosuchus*.

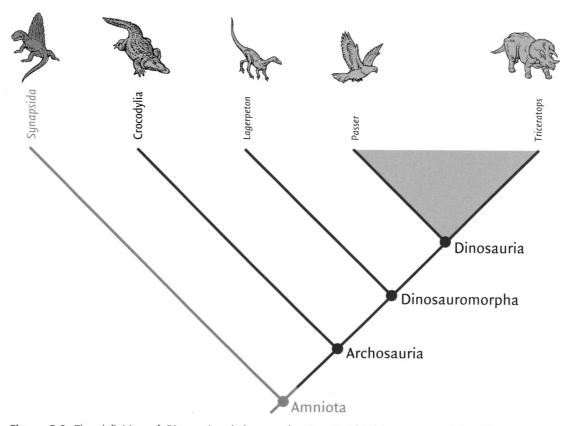

Figure 5.6. The definition of Dinosauria: cladogram showing Ornithischia, represented by *Triceratops*, and Saurischia by a sparrow (*Passer*), in Archosauria. The shaded triangle on the cladogram is, of course, a monophyletic Dinosauria.

(a)

(b)

Figure 5.7. (a) *Eoraptor;* (b) *Herrerasaurus*.

The oldest known confirmed dinosaur material, preserved as body fossils, comes from the Upper Triassic Ischigualasto Formation of Argentina. The Ischigualasto Formation has not been definitively dated: it could be as old as (or even slightly older than) 229 Ma; equally, it might be as young as 218 Ma. The evidence suggests that it likely does not span that full time interval, however.

The Ischigualasto produced two primitive bipedal dinosaurs, *Eoraptor* and *Herrerasaurus* (Figure 5.7). Not only are these the oldest known dinosaur body fossils, but they are by their characters among the most primitive.

Their very primitiveness has made them difficult to classify definitively. They were once believed to be generic dinosaurs that belonged to neither of the two great groups of Dinosauria (Saurischia and Ornithischia; see below); then they came to be identified as primitive theropods (a type of saurischian); recently, it was suggested that *Eoraptor* is a very primitive sauropodomorph (yet another type of saurischian); while *Herrerasaurus* and a newly discovered saurischian, *Eodromaeus*, are both theropods! From the same rocks also came a frustratingly fragmental possible ornithischian: *Pisanosaurus*, as well as another sauropodomorph: *Panphagia*, and the theropod *Sanjuansaurus*.

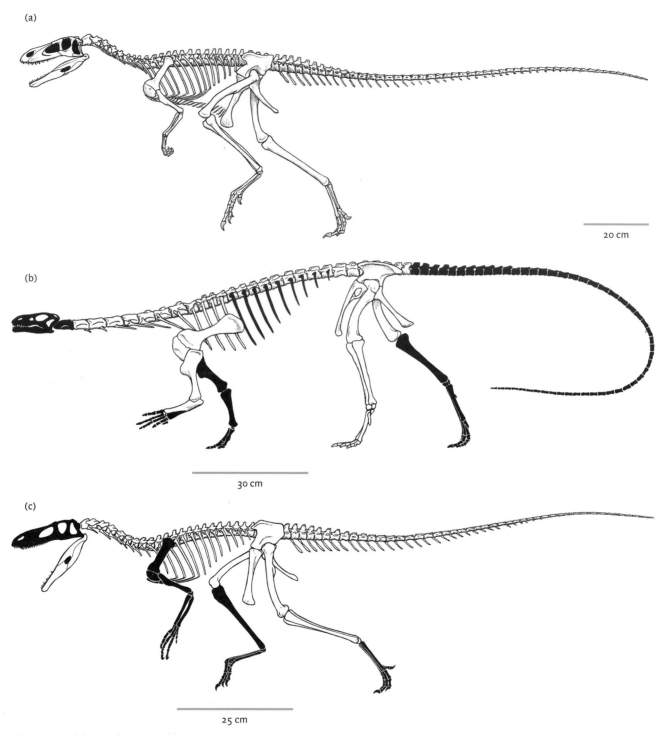

Figure 5.8. (a) *Staurikosaurus*; (b) *Saturnalia*; (c) *Guaibasaurus*.

From nearby rocks (in Brazil) of presumed equivalent age, come the basal saurischian *Staurikosaurus* and sauropodomorph *Saturnalia*. In rocks younger than those that produced *Staurikosaurus*, also in Brazil, the incompletely known *Guaibasaurus* was discovered (Figure 5.8). Originally described as a basal saurischian, it was reinterpreted as a theropod, and then again interpreted as a saurischian too primitive to be properly considered either a theropod or sauropodomorph. We're hopeful that some of this confusion will abate with more discoveries from these important rocks.

Ornithischia and Saurischia

With the Late Triassic appearance of these dinosaurs, it's high time we formally met the two great clades of Dinosauria: Saurischia and Ornithsichia. In 1887, English paleontologist Harry Seeley first recognized a fundamental division among dinosaurs. **Saurischia** (*saur* – lizard; *ischia* – hip) were those that had a pelvis more like a lizard, in which the pubis is directed anteriorly, and slightly downward (Figures 5.8 and 5.10). **Ornithischia** (*orni* – bird) were all those dinosaurs that had a bird-like pelvis, in which at least a part of the pubis runs posteriorly, along the lower rim of the ischium (Figure 5.9). This pelvic distinction has held sway ever since 1887.

Poor Seeley! He described the shape of the pelvises exactly as he (correctly) saw them: the saurischian pelvis looks rather similar to that found in a crocodile or lizard; while the ornithischian pelvis, with part of the pubis pointing backwards and running along the base of the the ischium, is superficially reminiscent of a bird pelvis. Ironically, birds are *saurischian* dinosaurs[4]; from our (living) vantage point, Ornithischia is, evolutionarily speaking, a dead end.

That dinosaurs had one or the other kind of pelvis was of great importance to understanding the evolution of these animals and, for Seeley, it implied that the ancestry of Ornithischia and Saurischia was to be found separately and deep within a heterogeneous group of primitive archosaurs once called "**Thecodontia**."[5] Dinosauria was not monophyletic to Seeley.

All that changed in 1986, when cladistic analysis, in the skilled hands of J. A. Gauthier, provided powerful evidence for a monophyletic Dinosauria. And, since Gauthier's studies, numerous other analyses by other paleontologists, using both newly discovered and familiar taxa, have confirmed that dinosaurs are monophyletic.

Is Saurischia more primitive than Ornithischia?

The distinction between Ornithischia and Saurischia is valid and both groups are diagnosed by suites of well-established characters (Figure 5.11). Indeed, the split between ornithischians and saurischians is the fundamental division within Dinosauria. But is one more primitive? Saurischians, with their claws, and teeth, appear to be a lot like their basal dinosauromorph forebears – claws, teeth, etc. – and in many books, including this one, they are treated first, suggesting the intuitive notion that saurischians are more primitive and that somehow ornithischians must have evolved from a saurischian ancestor.

No evidence for this, however, has ever surfaced. Figure 5.11 shows that, cladistically speaking, Saurischia is no less (or more) derived than Ornithischia; thus, we really don't know which came first.

---

[4] The pelvis is thought to have secondarily evolved the rearward-facing pubic branch, independently, in birds.

[5] The group "Thecodontia" is based on the same characters that diagnose all archosaurs. If we're discussing archosaurs, we can't cherry-pick a few basal ones; we need to include *all* members of the group (including dinosaurs and pterosaurs) that bear the diagnostic characters. In short, the term "Thecodontia," though venerable, has not withstood cladistic scrutiny (see Chapters 8 and 15).

Figure 5.9. The pelvis of the ornithischian *Stegosaurus*, viewed from the left side. A piece of the pubis is pointing posterior, along the base of the ischium, exemplifying the ornithischian condition.

Figure 5.10. The pelvis of the saurischian *Apatosaurus*, viewed from the right side. The pubis is directed anterior only, exemplifying the saurischian condition.

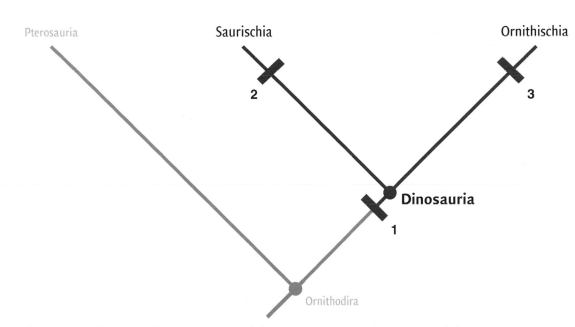

Figure 5.11. Cladogram of Dinosauria. Derived characters at **1**, see Figure 4.14; Derived characters at **2**, elongate cervical vertebrae, fossa expanded into the anterior corner of the external naris, lacrimal expanded over the rear part of antorbital fenestra, a concave facet on the axial intercentrum for the atlas, elongation of the centra of anterior cervical vertebrae, distinctly large hand, loss of distal carpal V, twisting of the first phalanx of manual digit I, metatarsals overlapping. Derived characters at **3**, opisthopubic pelvis, predentary bone, toothless and roughened tip of snout, reduced antorbital opening, palpebral bone, jaw joint set below level of the upper tooth row, cheek teeth with low subtriangular crowns, at least five sacral vertebrae, ossified tendons above the sacral region, small prepubic process along the pubis, long and thin preacetabular process on the ilium.

All that is clear is that the earliest dinosaur body fossils record show that the two great clades of dinosaurs, Ornithischia and Saurischia, go back to the Late Triassic.

The evolution of Dinosauria

The earliest days of dinosaur evolution remain shrouded in a bit of mystery. As we have seen, the oldest dinosaurs known are from South America; Argentina and, possibly, Brazil. In North America, dinosaurs are also found in the southwest part of the United States, in rocks that range in time from nearly as old as, to up to 10 million years younger than, the Late Triassic dinosaurs of South America.

The dinosaurs from the American southwest are not primitive; animals such as the small carnivorous bipeds *Coelophysis bauri* and *Tawa hallae* are unambiguous examples of theropods, a type of saurischian that we shall formally meet in Part II of this book.

The relationships among dinosauromorphs, dinosaurs, and their near relatives, pterosaurs (see Chapter 4; Figures 4.11 and 4.13) have been somewhat confusing – mainly because the characters of basal dinosauromorphs and early dinosaurs are too primitive, and those of pterosaurs are too advanced! This leaves us, on the one hand, with a cloud of primitive organisms lacking distinctive diagnostic characters upon which to build the phylogeny, but, on the other hand, with a group (pterosaurs) with so many derived characters, where they came from and how they evolved is equally difficult to figure out. Nonetheless, within the past few years, the outline of a consensus has been reached regarding the relationships of these two groups.

It had been thought, as recently as the early 1980s, that pterosaurs – otherwise highly modified for flight – might be the closest archosaurian relatives to dinosaurs, together sharing derived features as ornithodirans (see Figure 4.11). More recently, however, with the discovery of silesaurids, as well as a deepened understanding of creatures like *Marasuchus* and *Lagerpeton*, the non-dinosaur basal dinosauromorphs are likely closest to dinosaur ancestry, with pterosaurs distant within the clade Ornithodira. Among non-dinosaur basal dinosauromorphs, silesaurids are believed to be closest to dinosaur ancestry, with *Marasuchus* and *Lagerpeton* successively more distant (Figure 5.4).

The general outlines of this consensus didn't come easily; paleontologist M. Langer and colleagues have shown that, between the years 1986 and 2007, new hypotheses of the relationships of non-dinosaur basal dinosauromorphs to dinosaurs were published at the rate of about one a year! Part of the reason for this is that, as we have seen, all of these early dinosauromorphs have many primitive characters. It is also due to the abundance of new material that was discovered during those years. For example, once silesaurids were identified, paleontologists began to correctly recognize silesaurid fossils all over the world. Then, too, many of these fossils were, and are still, lacking skull material (note that the diagnostic characters listed in Figure 5.4 are not cranial). Skull material has been tremendously useful in sorting out the relationships of fossil taxa, and we can hope that when complete skulls for some of these taxa become known, their phylogenetic relationships will be clearer. For now, we offer Figure 5.12 as a current (but not likely final!) summary of the relationships among dinosauromorphs and many of the early dinosaurs mentioned in this chapter.

There is an interesting and perhaps surprising consequence of this phylogeny. With archosaurs like *Marasuchus* and the silesaurids close to dinosaurian ancestry, apparently dinosaurs were primitively **obligate bipeds**. This means that the earliest dinosaurs were creatures that were completely bipedal. Because the primitive stance for archosaurs is quadrupedal, and because Dinosauria

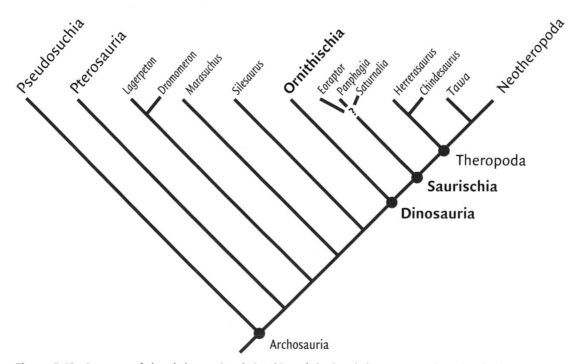

Figure 5.12. Summary of the phylogenetic relationships of the basal dinosauromorph and early dinosaur taxa discussed in this chapter.

Box 5.1   No dates; no rates

Samuel A. Bowring, a geochronologist (scientist who determines the numerical ages of rocks), reminds us, "No dates; no rates!" By this they mean that if we don't know when, numerically, events occurred, we can't really determine how quickly or slowly they took place. In this chapter, we tell a story of rapid early dinosaur evolution; yet, without dates, our story could be a house of cards. Numerical dates from unstable isotopes, such as those discussed in Chapter 2, are rare for this time interval (one geochronologist called the Late Triassic a "black hole"), and those few that are known are not particularly precise.

Historically, in the absence of numerical ages, relative dating was carried out using a combination of lithostratigraphy and biostratigraphy (see Chapter 2; Figure 2.3). These techniques were used to divide the Triassic into three formal Epochs: Early, Middle, and Late. Each of these, in turn, was subdivided using biostratigraphy into smaller blocks of time called Stages. The key Stages for our story are the Carnian, the Norian, and the Rhaetian, the three Stages of the Late Triassic (Figure B5.1.1).

Numerical dates for these three Stages are very rare indeed; so while we know the sequence of time, exactly when that sequence occurred can be uncertain. For example, stratigraphers can still get a rude shock: in 2007, following a careful reanalysis of marine dates and faunas, the Norian gained almost 7 million years (at the expense of the Carnian!).

The situation is more of a problem in the terrestrial realm. Dinosaurs, as we have seen, are exclusively terrestrial beasts, and geographically dispersed dinosaur faunas around the globe have been correlated with each other. Yet, it has not been easy to securely tie these dinosaur faunas to the Norian and Carnian Stages, because these Stages are based upon marine organisms.

Still, there is progress: as we have seen, some equivocal dates (229–218 Ma) exist for the Ischigualasto Formation of Argentina (Chapter 13), the unit that produced *Eoraptor* and *Herrerasaurus*. But we have no dates for the presumably equivalent rocks in nearby Brazil, which produced *Guaibasaurus* and *Staurikosaurus*. Recently a suite of extremely precise dates has been obtained for the Chinle Formation of the North American southwest, a rock unit that produced Late Triassic dinosaurs such as the theropod *Coelophysis* and the primitive saurischian *Chindesaurus*. For the time being, though, those dates stand alone, because their exact relationship to the Carnian and Norian Stages is poorly understood – since

Figure B5.1.1. The Late Triassic time scale. Numerical dates from below the base of the Carnian approximate the base of the Carnian to slightly younger than 236 Ma. With the exception of a date from the upper Carnian of Italy, there are generally no reliable isotopic ages for the Late Triassic at this time. Therefore, ages for the Carnian–Norian boundary (227–228 Ma) and the Norian–Rhaetian boundary (208–209 Ma) have been interpolated using magnetostratigraphy, a technique based upon global magnetic polarity reversals detected in key stratigraphic sections. The Triassic–Jurassic boundary, on the other hand, has been precisely dated by isotopic methods at 201.3 Ma – but in the marine realm, ecologically and geographically far from dinosaurs.

the numerical ages of those Stages are still not well understood. And this means that exactly how these North American faunas tie into other North American Late Triassic faunas, let alone global Late Triassic faunas, is not so clear, either. Stay tuned!

is monophyletic, it follows that creatures like *Triceratops, Ankylosaurus,* and *Stegosaurus,* all *quadrupedal* dinosaurs, must have **secondarily evolved** (or re-evolved) their quadrupedal stance. That is, they must have (phylogenetically) gotten back down on four legs, as it were, after having been up on two. In fact, you can see the remnant of bipedal ancestry when you look at a stegosaur or a ceratopsian, in which the back legs are quite a bit longer than those at the front. Even silesaurids may have flirted with partial quadrupedality.

Let the games begin!

Our information is limited about both the place and pace of the spread of dinosaurs. The oldest dinosaur body fossils known (above) are all from a chunk of time of the Late Triassic called the Carnian Stage (Box 5.1). Because the Carnian-aged dinosaurs described above are all known from Brazil and Argentina, paleontologists have suggested that the cradle of Dinosauria was South America. Yet, terrestrial Carnian rocks that preserve dinosaur remains are not so common globally. So we end up asking, are the South American dinosaurs truly the oldest dinosaurs, and thus South America was the place where dinosaurs originated? Or, alternatively, are the terrestrial Carnian rocks of South America simply the only game in town, and thus they are most of what we know of the terrestrial Carnian Stage globally? We have no satisfactory answer for those questions. At this point all we can really say is that the evidence is strong that dinosaurs first appeared either in the latter part of the mid-Triassic (remember the footprints?) or, without doubt, based upon their body fossils, during the Carnian Stage.

What a difference a Stage makes

The Carnian Stage is thought to have ended at 228 Ma, which marks the beginning of the Norian Stage, the block of time that followed it (Box 5.1). By Norian time, dinosaurs clearly had acquired a startlingly pervasive global presence, and apparently populated all of the supercontinent of Pangaea (Figure 5.13).

What is striking, even among these very early Norian dinosaurs, is their diversity: examples of ornithischians, sauropodomorphs, and theropods are recognizable, suggesting that, if we understand the timing correctly (Box 5.1), the evolution and global dispersal of these groups took place relatively quickly after the first appearance of dinosaurs (wherever that was). Ornithischians, later a pervasive group of dinosaurs, *may* be represented in the Carnian by the enigmatic *Pisanosaurus*, but are definitely present later in the Late Triassic, as represented by the meter-long *Eocursor parvus* (Figure 5.14).

This is not to say that the basal dinosauromorphs politely stepped aside the first time that a true dinosaur showed up at the party. Both likely did similar kinds of things ecologically and, indeed, the earliest dinosaurs were rather primitive representatives of their group/clade, looking superficially (and in some cases not-so superficially) similar to the basal dinosauromorphs from which they are thought to have derived. Yet a strange signal comes from the record. The rich Carnian deposits of Argentina contain both early dinosaurs and non-dinosaur basal dinosauromorphs, but rarely are the two found together; dinosaurs appear to supplant basal dinosauromorphs in South America. Yet, in the Norian of the North American southwest, by contrast, saurischian dinosaurs like *Coelophysis* and *Chindesaurus* lived for almost 12 million years with non-dinosaur basal dinosauromorphs such as *Drommomeron*, an animal closely related to *Lagerpeton*. Indeed, in North America at least, the Late Triassic was a time generally characterized by a relatively stable, long-lived global fauna consisting of dinosaurs and non-dinosaur dinosauromorphs co-existing. By the end of the Late Triassic,

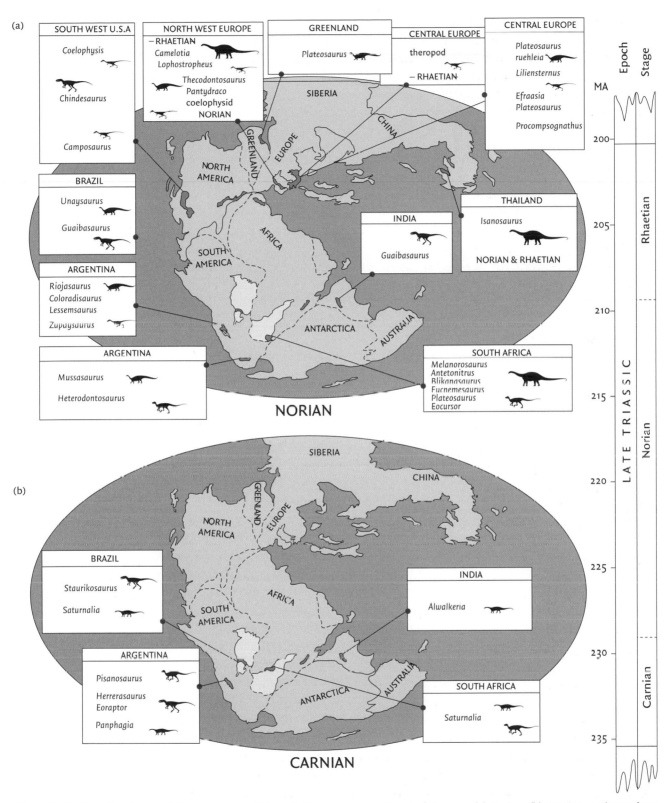

Figure 5.13. The distribution of dinosaurs around the Late Triassic supercontinent of Pangaea (a) Norian; (b) Carnian. Redrawn from Langer *et al.*, 2010.

Figure 5.14. The primitive, early undoubted ornithischian, *Eocursor parvus*, from South Africa.

however, non-dinosaur dinosauromorphs phased out of the picture globally, and true dinosaurs were off and cursorial!

Feathers

Before we begin our stroll through the various groups of dinosaurs, one last remarkable and unexpected subject intrudes here: feathers. Feathers? For previous generations of paleontologists and dinosaur lovers, life was simple: only birds have feathers, and dinosaurs, being reptiles, must have had scales. A few bumpy skin impressions (see Chapter 12) sufficed to make the point. Feathers were on *nobody's* radar.

Who invented feathers?

Thinking began to change when John Ostrom, in the early 1970s, showed that the oldest "bird" known – a winged wonder called *Archaeopteryx* (Chapter 7) was also a dinosaur[6] (see also Chapter 15). So, most (but not all) paleontologists then concluded, OK, a few highly derived primitive bird-dinosaurs must have had feathers; what's so amazing about that? But in the 1990s it became strikingly obvious that actually *nothing* was "amazing about that," since all kinds of non-flying, saurischian dinosaurs – even *T. rex* – evidently had feathers. Feathers appeared to be the invention of Theropoda... until feathers – or something like feathers – were found in Ornithischia! With feathers potentially on both saurischians and ornithischians, were feathers actually an invention of Dinosauria? We'll meet these various feathered dinosaurs as we explore each group, but regardless, it was clear that after 150 years of familiar smooth-skinned, shiny, green-brown dinosaurs, it was time to change the paradigm to scruffy, feathery, dander-spewing, brightly-colored dinosaurs (Figure 5.15)! Because feathers are present in both Saurischia and Ornithischia, and thus, having been invented by dinosaurs, may characterize the group, we'll introduce feathers here.

[6] Provoking great confusion in those pre-cladistic, Linnaeus-dominated days: if it is a bird, how could it *also* be a reptile (dinosaurs, of course, were known to be reptiles)? Linnaean taxonomy, as we have seen, does not really play very nicely with evolutionary relationships!

(a)   (b)

Figure 5.15. The carnivorous dinosaur *Deinonychus*. (a) A feathered model; (b) image without feathers. For dinosaurs like *Deinonychus*, the non-feathered appearance is, to this dinosaur, the way a plucked turkey is to the living, healthy bird!

Feathers without flight

Feathers in modern birds do lots of things: they insulate (think down sleeping bags and vests), they are used for display (think peacock), they are used as sensory organs, and, most famously, they permit flight (think airfoil). One anatomical structure; multiple uses.

Insulation

The insulatory properties of down feathers are well known. Could feathers have evolved first for insulation, and then only later gotten co-opted for flight? This idea is less far fetched when we remember that all living birds are warm-blooded; and most warm-blooded creatures, especially small ones, require insulation to keep from losing heat. It makes no sense for a cold-blooded animal, which needs an exterior source of heat to warm up, to be insulated.

If feathers were actually an adaptation for insulating a warm-blooded creature, then fossils of small, non-flying dinosaurs ought to be found with feathers. When evidence finally came that showed that feathers were used for insulation (and not for flight), it was spectacular (see Chapter 7). But did feathers *originate* for insulation?

Stylin'

Modern birds are brightly colored creatures, and it is no secret that, along with flight and insulation, birds use feathers for display. While shape is important (think about a lyrebird or a peacock), and excluding behavior and calls, color is probably the key criterion for reproductive success. So if feathers today have an important display function, did feathers also serve a display role in dinosaurs? Recently, paleontologists have begun to gain insights into the actual colors and patterns of ancient feathers, and blacks, browns, reds, iridescent colors, as well as color patterns, have been identified. We'll visit this in greater detail in Chapters 6 and 7 (see Chapter 6 "The wonderful world of color," and Figures 7.10 and 7.23). The murky crocodile-green of your parents' dinosaurs is a thing of the past: dinosaurs were colorful, brightly patterned animals. Did feathers therefore originate for display?

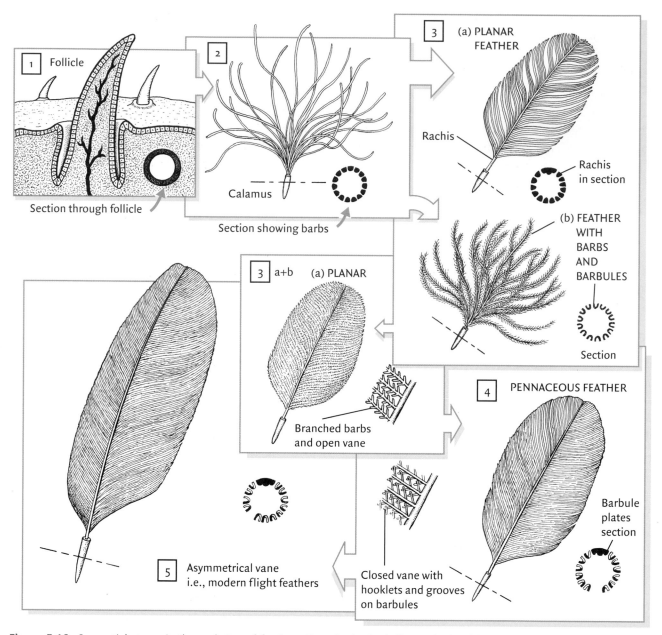

Figure 5.16. Sequential stages in the evolution of feathers. Type 1: simple, hollow, cylindrical filaments. Type 2: tufts of elongate, multiple filaments, attached at one end. Type 3: filament tufts align in a single plane (Type 3a) while also developing barbs and barbules (Type 3b). Eventually, a new planar (vaned) barbed form evolves (Type 3a+b). Type 4: vane becomes "closed"; that is, tiny hooks on the barbule attach to grooves on adjacent barbules, producing an integrated semi-rigid vane that does not allow much air to pass through. Type 5: vane becomes asymmetrical (for example, a flight feather).

Senses

Anatomist W. S. Persons and paleontologist P. J. Currie recently pointed out that feathers in birds have yet another key function: sensing. Like the whiskers on mammals or the little sensory projections (vibrissae) on insects, single-filament feathers on birds – particularly on the snout – occur in low concentrations and serve as sensing organs. The authors note that, for feathers to have insulated, they needed to be densely concentrated. Perhaps, these authors suggested, simple, widely scattered feathers originally began as tactile organs, eventually coming to serve display, insulation, and finally flight functions.

Embryology

Feathers were long thought to be an outgrowth of "reptilian" scales. Somehow the scales grew longer and divided into barbs and barbules. Work in the past ten years, however, suggests that the development of feathers may occur by the interaction of specialized follicles and a series of genes that control the onset and termination of growth. Four sequential stages of feather evolution have been identified, each stage a developmental modification of the previous stage, and each found in living birds (Figure 5.16). These stages are:

(1) Formation of a keratin-based, single filament, constructed as a hollow cylinder (**monofilamentous**, hair-like feathers).
(2) Loosely associated, unconnected, unhooked barbs (**downy** feathers).
(3) Hooked barbs on a symmetrical vane (**contour** feathers, such as wrap around the body).
(4) Hooked barbs on an asymmetrical vane (**flight** feathers).

That there are multiple stages of feather development explains how it comes to be that different feather types are found in living birds: the *embryological* sequence shown in Figure 5.16 may reflect the *evolutionary* sequence of feather development. Thus, for example, single filamentous feathers (Stage 1) may represent a very early stage of feather development, whereas the hooked barbs and asymmetrical vanes found in flight feathers likely evolved later. Feathers in some non-flying dinosaurs may have simply been a hair-like coating (monofilamenous) and in smaller, active non-flying forms, down may have been used as an insulator; perhaps covered with contour feathers. Dinosaurs that flew, of course, developed specialized flight feathers for that purpose.

So feathers are associated with sensory, display, insulation, and flight in all these animals, and may very well be the quintessential diagnostic character for dinosaurs. Alternatively, it may be a bit more complicated. In 1971, a small pterosaur called *Sordes pilosus* demonstrated unequivocally that pterosaurs were covered with a soft (?) fur-like covering. Were feathers actually the invention of Ornithodira, thus going even farther back in the phylogeny than Dinosauria?

The very earliest dinosaurs were evidently carnivorous, bipedal, and small- to medium-sized, and may have been covered with some kind of monofilamentous, or downy plumage. Unfortunately, to date, the fossil record is moot (and mute) on this point. With all that feathery integument, they might even have taken a tentative bipedal step or two down the road towards endothermy (warm-bloodedness; see Chapter 13).

SUMMARY

The earliest and morphologically most primitive dinosaurs were small, bipedal, carnivorous, and/or insectivorous creatures, known from deposits in Argentina and Brazil, dated at about 231–218 Ma. These animals lived with, as well as were preceded by, several other groups of archosaurs – non-dinosaur basal dinosauromorphs – of similar morphology and inferred behavior. The similarities of morphology suggest that dinosaur ancestry is to be found among the basal dinosauromorphs, likely, among the recently recognized silesaurid clade. These earliest known dinosaurs, although primitive, have been affiliated with somewhat derived dinosaurian groups, such as Ornithischia, Theropoda, and Sauropodomorpha. This suggests that some amount of evolution must have occurred preceding 229 Ma and the appearance of early dinosaurs such as *Herrerasaurus* and *Eoraptor*.

There are two major groups of dinosaurs: Ornithischia and Saurischia. These are identified by a variety of features among which only the orientation of the anterior position of the pubis is highlighted here. As the phylogeny is currently understood, ornithischians and saurischians are equally derived; thus it is not possible to say whether one is more primitive than the other.

Among the striking innovations of dinosaurs, among the most surprising may be the invention of feathers. Once thought only to pertain to birds, feathers are now known to have graced both saurischian and ornithsichian dinosaurs, with the possibility that feathers go as far back as basal Ornithodira on the cladogram. Early feather coats may have been monofilamentous and superficially fur-like; down likely evolved early as a means of insulation which, along with display, was the probable original function of feathers.

Once they made their Carnian (early Late Triassic) or even pre-Carnian appearance, dinosaurs radiated and diversified efficiently. By Norian (middle Late Triassic) time, they had achieved a global distribution, and were starting to dominate terrestrial vertebrate faunas.

SELECTED READINGS

Brusatte, S. L., Nesbitt, S. J., Irmis, R. B., *et al.* 2010. The origin and early radiation of dinosaurs. *Earth-Science Reviews*, **101**, 68–100.

Langer, M. C., Ezcurra, M. D., Bittencourt, J. S., and Novas, F. E. 2010. The origin and early evolution of dinosaurs. *Biological Reviews*, **85**, 55–110.

Nesbitt, S. J. 2011. The early evolution of archosaurs: relationships and the origin of major clades. *Bulletin of the American Museum of Natural History*, **352**, 292 pp.

Person, W. S. IV and Currie, P. J. 2015. Bristles before down: A new perspective on the functional origin of feathers. *Evolution*, **69**, 857–862.

Prum, R. O. and Brush, A. H. 2003. Which came first, the feather or the bird? *Scientific American*, **288**, 84–93.

Xu, X. and Guo, Y. 2009. The origin and early evolution of feathers: insights from recent and neontological data. *Vertebrata PalAsiatica*, **47**, 311–329.

TOPIC QUESTIONS

1. What is Crurotarsi? Ornithodira? How would one tell the difference?
2. Give an example of a crurotarsan archosaur. When did the last crurotarsan live?
3. What do those names (Crurotarsi; Ornithodira) have to do with Dinosauria?
4. What is a dinosauromorph?
5. Please name two non-dinosaurian dinosauromorphs.
6. What do those non-dinosauromorphs tell us about the morphology of the first dinosaurs?
7. Describe the earliest known dinosaurs. Are these also the most primitive dinosaurs? What is the difference between "earliest known" and "most primitive?"

8. People have discussed whether or not the evolution of Dinosauria was "fast" or "slow." Which do you think it was? Please explain your answer: (a) what do you mean by "fast" (or "slow"), and (b) how do you measure fast or slow evolution?

9. Should "presence of feathers" be a diagnostic character for birds? For Dinosauria? For Ornithodira? Please explain your answer to each.

10. Construct and justify an evolutionary sequence for the various uses of feathers.

PART II

SAURISCHIA: MEAT, MIGHT, AND MAGNITUDE

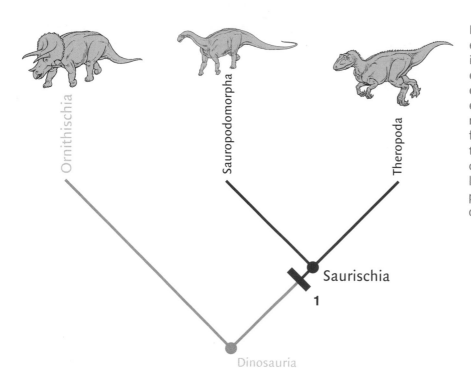

Figure II.1. Cladogram of Dinosauria, emphasizing the monophyly of Saurischia. Derived characters include: at **1**, elongate cervical vertebrae, fossa expanded into the anterior corner of the external naris, lacrimal expanded over the rear part of antorbital fenestra, a concave facet for the atlas on the axial intercentrum, elongation of the centra of anterior cervical vertebrae, distinctly large hand, loss of distal carpal V, twisting of the first phalanx of manual digit I, metatarsals overlapping. See Figure II.2.

Saurischians include the smallest of dinosaurs and the largest animals that ever lived on land; the most agile and ferocious of predatory dinosaurs and the most ponderous of plant-eaters; the brightest and, quite possibly, the dumbest of dinosaurs; the most Earth-bound and the most aerial. And those most familiar of saurischian dinosaurs, birds, remain with us today, very much alive and well.

Saurischians include the famous toothy carnivorous dinosaurs (Theropoda) and their equally famous long-necked herbivorous cousins (Sauropoda). There isn't an obvious family resemblance to look at these groups, and you could be forgiven if you concluded that Saurischia isn't monophyletic. But as we've seen it's not about superficial resemblance, and our best evidence suggests that Saurischa is a well-supported clade. Figure II.1 highlights the monophyly of Saurischia.

What makes a saurischian a saurischian?

Saurischia is diagnosed by many derived features (Figure II.1), two of which are shown in Figure II.2. In general, these characters suggest a general tendency in the group for well-developed, grasping hands and powerful feet. In the case of the hands, the large size (almost half of the arm length), long fingers, and a distinctive thumb that, a bit like a human thumb, appears to fall across the palm[1] instead of simply pointing outwards from the hand; this suggests an organ designed for grasping, presumably prey. In the case of the feet, the tightly articulating, close-fitted metatarsals may have supported more efficient walking, but in carnivorous forms (including the earliest saurischians), they may have also aided in prey manipulation. While not all saurischians were carnivorous, the group from the very outset appears to have had some tendencies toward carnivory.

---

[1] The distinctive position of the thumb is regarded as "semi-opposable" by some paleontologists.

(a)   (b)

10 cm

Figure II.2. (a) Overlapping metatarsals of the primitive saurischian foot (exemplified by the "prosauropod" *Ammosaurus*); (b) thumb (digit I of the hand) tends to angle across the palm.

Recall from Chapter 5 that, as early as 1887, H. G. Seeley recognized two great clades within Dinosauria: Ornithischia and Saurischia. Seeley's Saurischia originally consisted of Sauropodomorpha (see Chapter 9) and its sister-taxon Theropoda (*theros* – wild beast; Chapters 6, 7, and 8). To date, all known saurischian fossils sustain his concept. Figure II.2 illustrates two saurischian characters.

Here, we will follow the apparent morphological evolution of Saurischia and meet the carnivores – the Theropoda – first (Chapters 6–8), moving to the herbivores (Sauropodomorpha) afterwards in Chapter 9.

SELECTED READINGS

Gauthier, J. A. 1986. Saurischian monophyly and the origin of birds. *Memoirs of the California Academy of Sciences*, **8**, 1–55.

Langer, M. 2004. Basal Saurischia. In, Weishampel, D. B., Dodson, P., and Osmólska, H. (eds.) *The Dinosauria*, 2nd edn. University of California Press, Berkeley, pp. 25–46.

Langer, M. and Benton, M. J. 2006. Early dinosaurs: a phylogenetic study. *Journal of Systematic Palæontology*, **4**, 309–358.

Sereno, P. C. 1999. The evolution of dinosaurs. *Science*, **284**, 2137–2147.

Sereno, P. C., Forster, C. A., Rogers, R. R., and Monetta, A. M. 1993. Primitive dinosaur skeleton from Argentina and the early evolution of Dinosauria. *Nature*, **361**, 64–66.

Sereno, P. C. and Novas, F. E. 1992. The complete skull and skeleton of an early dinosaur. *Science*, **258**, 1137–1140.

Theropoda I
Nature red in tooth and claw

6

CHAPTER OBJECTIVES

- Introduce Theropoda
- Develop familiarity with current thinking about lifestyles and behaviors of theropods
- Develop an understanding of theropod evolution using cladograms, and an understanding of the place of Theropoda within Dinosauria

Figure 6.1. The saurian tyrant king, *Tyrannosaurus rex*.

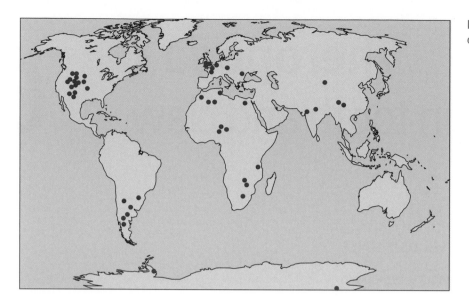

Figure 6.2. Global distribution of non-coelurosaurian Theropoda.

Theropoda
Eating meat the theropod way

When dinosaurs did carnivory, they did it the theropod way: with steak-knife teeth, sinewy haunches, and grasping hands and feet tipped with scimitar claws (Figure 6.1). The combination was at once formidable and successful, and produced a rainbow palette of carnivores as well as, it turns out, a surprisingly large pile of other-ivores. Carnivorous dinosaurs find their home among the theropods, but why stick to meat when there are such other good eats out there? So Theropoda (*thero* – beast; *pod* – foot) is *really* diverse, and counts among its many members coelophysoids, neoceratosaurs, carnosaurs, therizinosauroids, ornithomimosaurs, oviraptorosaurs, troodontids, spinosaurids, dromaeosaurids, tyrannosauroids. . . and birds.

These dinosaurs have had a long evolutionary history extending back from the Late Triassic right up until the end, 66 Ma. *Past* that "end," really, since birds are still very much with us. But in this chapter, we'll concentrate on **non-avian** (that is, non-bird) theropods, holding off on the avian side of the story until Chapter 7.

Non-avian theropods (for simplicity, "theropods") have been found on every continent including Antarctica (Figures 6.2 and 6.3). They have been collected from virtually every kind of depositional environment. And, they have been found as isolated finds, but also in rich, monospecific (single-species), theropod bonebeds (see "Social behavior" below).

Theropods ranged in size from less than a meter (*Microraptor*) to animals growing to upward of 15 m in length (*Tyrannosaurus*, *Carcharodontosaurus*, *Giganotosaurus*). For all the variety, though, non-avian theropod evolution seems to have been largely about tracking, attacking, and feeding.

Who are theropods?

Theropoda is monophyletic group within Saurischia; you might say that it includes all saurischians that are not sauropodomorphs. Using a formal definition – that in effect says exactly the same thing – we would say that it incudes *Passer* (our old friend the sparrow - again!) and all taxa sharing a more recent common ancestor with *Passer* than with any sauropodomorph (Figure 6.4). That, of course, also says that birds are theropods.

Why not use the bracketing definition we've seen previously? That is, why not define theropods by a derived theropod on the one hand (*Passer*), a primitive theropod on the other, and call "theropods" all the descendants of their most recent common ancestor? Because in this case, as we've seen, it is

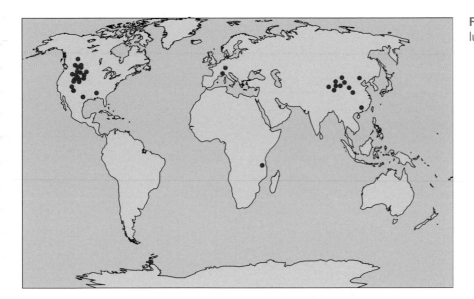

Figure 6.3. Global distribution of Coelurosauria (excluding birds).

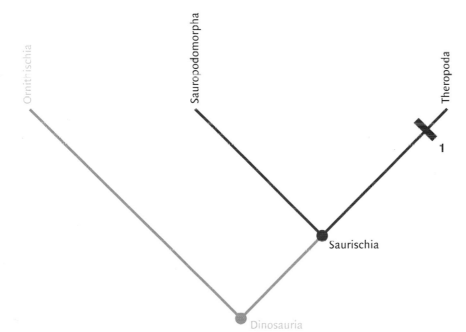

Figure 6.4. Cladogram of Dinosauria emphasizing the monophyly of Theropoda. With uncertainty surrounding the phylogenetic position(s) of the earliest dinosaurs such as *Eoraptor*, *Eodromaeus*, *Guaibasaurus*, and *Herrerasaurus*, appropriate diagnostic characters (e.g., those characterizing theropods as defined) can be tricky to find. Here are a few proposed in 2012 by theropod specialist Stephen Brusatte (University of Edinburgh): at **1**, knife-like teeth (thin, recurved, serrated; termed "**ziphodont**"), promaxillary fenestra on maxilla (an extra opening in front of the antorbital opening), large hands with advanced grasping ability.

rather difficult to be sure which, among the very earliest dinosaurs we've met, is a theropod and which is not. So this definition leaves the door open to new fossils and new interpretations, by including everything within Saurischia except Sauropodomorpha.

Viewed superficially, theropods are all clawed bipeds with sharp, flattened serrated teeth (Figure 6.5; but only when they haven't lost them, as has occurred in several groups), and distinctively hollowed vertebrae and limb bones.[1]

---

[1] The serrated teeth, claws, hollow bones, and bipedal stance of theropods all appear to have been primitive carry-overs from earlier in ornithodiran history. These characters give the *appearance* that theropods are somehow more primitive organisms than, say, ornithischians; the cladograms, however, show that they're not (see Figure 6.4).

Figure 6.5. Typical large theropod tooth. Isolated tooth of *Allosaurus*, a Late Jurassic carnivorous theropod; note the distinctive serration, most visible on the right side (posteriorly) of the specimen.

Figure 6.6. Skeleton of the medium-sized early Cretaceous carnivore *Deinonychus*, mounted in attack mode.

Theropod lives and lifestyles

Theropod imagery is haunted by solitary, drooling brutes like *T. rex* mauling herds of herbivores. But is that really how it was? Did biggest = baddest? Evidence suggests that packs of highly social, sharp-eyed, large-brained, agile small theropods, armed with grasping hands and slicing claws were likely the real nightmare terrors of Mesozoic landscapes (Figure 6.6). And there were other theropods as well – fleet-footed ornithomimids, hulking herbivorous therizinosaurs, and some predatory fish-eaters that likely would find a modern crocodile like an *hors d'oeuvre*. Theropods tried everything – even flight.

Running for life

We have seen how all dinosaurs had a fully erect stance; built for running (cursoriality). In theropods, this is taken to an extreme. Theropods were (and are) **obligate bipeds**, like us, unable to walk or run on anything but their hind legs. The body was balanced directly over the solidly constructed pelvis, with the vertebral column held nearly horizontal (see Figure 6.7), balanced at the pelvis. Evidence from the skeleton and trackways reflects the fact that the hind legs were held close to the narrow body, feet so close to the midline that it appears that one foot was placed *ahead* of the other, rather than along its side (Figure 6.8).

   In most theropods, the length of the leg was increased. The foot was supported by its toes, in what is termed a **digitigrade** stance (Figure 6.9). By contrast, our own foot rests not on its toes (except during ballet, of course!) but rather on the foot bones and heel. Our stance is termed **plantigrade**. With the digitigrade foot in theropods, it effectively extended the leg close to the length of the foot. In small- to medium-sized theropods, visible in the proportionately long, slender legs, is an interesting relationship: the thigh (femur) is short compared with the length of the dominant bone of the hindlimb (the tibia); this is a condition typical of fast-running bipeds (Figure 6.10).

50 cm

Figure 6.7. Reconstruction of *Allosaurus*. The legs attach at the pivot point between the long tail, and the deep, but narrow body and skull.

Figure 6.8. A theropod trackway from the Middle Jurassic Entrada Formation, Utah, USA. Note how closely the right and left prints are placed, with the steps very close to the animal's midline. When spectacular trackways like this are found, it is not hard to imagine the ghostly image of the trackmaker leaving a row of footprints in the soft mud.

Figure 6.9. Digitigrade stance in the large, Late Jurassic, North American theropod *Allosaurus*, spotted in the Museum of Natural History, Vienna.

Theropods were surely cursorial animals!

But just how cursorial is cursorial? Calculations of running speeds on the basis of hindlimb proportions indicate that the fastest theropods probably clocked 40–60 kmh. Some footprint evidence bears these numbers out; for example, a Cretaceous trackway in Texas was made by a theropod that thundered away at between 30 and 40 kmh (18–25 mph; see Box 13.3). Of course most of the time, most theropods weren't running full tilt, and leisurely lakeside strolls of about 4 kmh (2.5 mph) have been recorded from Early Jurassic trackways in Connecticut.

In the case of large theropods, outrunning jeeps was probably not an option (except in movies). No trackways are known that demonstrate large, speeding theropods, and computer modeling suggests that, to achieve high speeds (in the 50 kmh [31 mph] range), so much leg musculature would have been required that the animal would have been grotesquely overmuscled. The balance of dispassionate evidence, therefore, suggests that the largest theropods likely could not exceed about 40 kmh (25 mph).

We've seen that theropods were spectacularly cursorial; we'll see in Chapter 7 how they mastered flight; we saw in Chapter 1 that they could burrow – even if it was not that common, and we also know, from a trackway in Spain showing the imprints of a medium-sized theropod, that some, at least, could swim. Theropods got around.

Figure 6.10. Many small- to medium-sized theropods, such as the ostrich-mimic (ornithomimosaur) *Struthiomimus* shown here, have a thigh (femur) that is significantly shorter than the shin (tibia), suggesting that these dinosaurs were capable of running fast.

Paws and claws

As in modern birds, the grasping, powerful, clawed feet were not only for locomotion, but were an important part of the theropod predatory arsenal (Figure 6.11). This character reached unparalleled sophistication in dromaeosaurids and troodontids, in which the claw on the second digit of the foot was enlarged, curved, and sharp, and capable of a very large arc of motion. During normal walking and running, it would have been held back or up, to protect it from abrasion or breakage. But, when needed, it could be brought forward and, with the powerful kicking motion of the rest of the leg, used to eviscerate the bellies of hapless prey, lethally disemboweling in one rapid stroke (Figure 6.11).

Strong arms and dextrous, three-fingered hands characterized most theropods, although primitive theropods can show four digits (digit V is lost), and a few highly derived forms – most notoriously *Tyrannosaurus* – had only two (I and II). The digits, I (thumb), II, and III, were long and capable of extreme extension, and tipped with powerful claws. Digit I (the thumb) could fold across the palm in a semi-opposable fashion; that is, somewhat like a human thumb. There is no mistaking the function of this hand: it's all about grasping (Figure 6.12).

Even highly specialized large theropods such as *Tyrannosaurus*, *Tarbosaurus*, and *Carnotaurus*, with their notoriously short arms (the hands could not reach the mouth) had stout, yet powerful bones and fingers tipped with large claws, suggesting active use (note the hands and arms of *Tyrannosaurus* and *Carnotaurus* in Figure 6.13). But used for what? It has been suggested that perhaps the arms were short in order to balance an overly large head, or perhaps to help the animal get up when lying down.

It's hard to take these ideas terribly seriously. The size and design of the arm bones suggests some more compelling function. Indeed, it has been calculated that the arm of *Tyrannosaurus* could have lifted 300 kg. It is clear from the robust bones and large, stout claws at the tips of the strong fingers that the forelimbs aspired to greater purposes than getting up in the morning. Tearing flesh? Gripping prey? Lifting? Nobody quite knows.

Figure 6.11. Theropod feet. (a) Skeleton of *Allosaurus* foot; (b) reconstruction of (a) doing what non-avian theropod feet did best; (c) left foot of *Deinonychus* with its disemboweling second-toe claw. (d) Reconstruction of (c) in action.

Teeth and jaws

As with many other meat-eaters, carnivorous theropod heads tended to be proportionately large. In the case of the biggest, the heads could be upward of 1.75 m in length. In general, theropod skulls are superficially reminiscent of those of many non-dinosaurian ornithodirans. Yet there are differences among theropods: tyrannosauroids had robust, deep-jawed skulls, suggesting a powerful bite. Other theropods – even large ones like *Carcharodontosaurus* – had much more lightly built skulls (Figures 6.14 and 6.15). Non-carnivorous theropods, such as the toothless ornithomimosaurs and toothed, but non-carnivorous therizinosaurs, had proportionally smaller heads.

Figure 6.12. Three-fingered right hand of *Allosaurus*. Digit I was semi-opposable, conferring a grasping cap ability. In the highly evolved dromaeosaurs, this ability was even further developed.

(a) (b) (c)

Figure 6.13. Left forelimb of (a) *Struthiomimus*, (b) *Tyrannosaurus*, and (c) *Carnotaurus*. Human hand for scale.

Theropod teeth tend to be flattened from side to side, **recurved** (curved backward), pointed, and serrated (see Figure 6.5). With the jaw joint at the level of the tooth row, the effect was like that of a pair of scissors, slicing from back to front (Figure 6.16). While these teeth are clearly about puncturing and slicing, they are equally clearly *not* about chewing. Chewing, as we shall see in Part III, is a behavior independently invented by mammals and ornithischian dinosaurs, and the fact

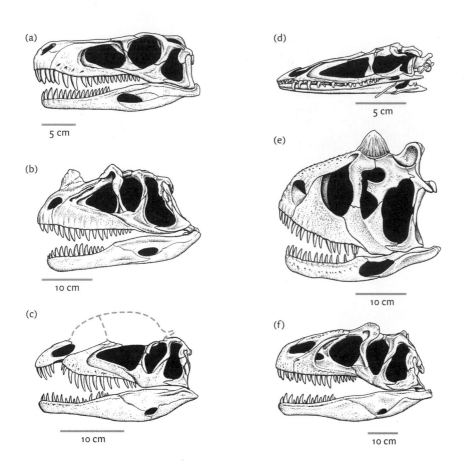

(a)

(b)

(c)

(d)

(e)

(f)

5 cm

10 cm

10 cm

5 cm

10 cm

10 cm

Figure 6.14. Left lateral view of the skull of (a) *Herrerasaurus*, (b) *Cerato-saurus*, (c) *Dilophosaurus*, (d) *Coelophysis*, (e) *Carnotaurus*, and (f) *Allosaurus*.

is that most tetrapods (and most vertebrates, generally) don't really chew their food to any great degree. So in this sense, it is no surprise that theropods didn't bother with chewing and with all of the tooth and mouth modifications that this behavior entails.

It was the sharply pointed, recurved, and serrated teeth in the upper and lower jaws that handled the prey. The recurved shape kept prey from escaping from the mouth, but also accommodated the geometry of the theropod jaw: the teeth tended to be more curved toward the back, and straighter towards the front, which seems to have been about properly positioning them for the most efficient puncturing as the jaw closed. The teeth of smaller theropods such as *Velociraptor*, with their prominently pointed serrations and narrow cross-sections, sliced like hacksaw blades. The mineral crystals that make up the enamel of these teeth are arranged in a simple pattern, suggesting greatest strength in only one dimension. This implies that the teeth were subjected to uniform stresses, perhaps associated with slicing, and were not subjected to complex stresses such as holding a struggling prey.

Tyrannosauroids, at the other end of the spectrum, with bulbous teeth and rounded serrations, had a weaker cutting ability, but greater bite power, suggesting that they could withstand complex, strong, and violent forces, such as might occur with a powerful, actively struggling prey (Figure 6.17). Interestingly enough, the enamel of tyrannosaur teeth has a complex pattern of crystal orientations, suggesting increased strength in multiple directions. They may have even been able to crush bone (see below). More enigmatic, perhaps, the seemingly hypercarnivorous troodontids – to judge from their powerful, well-clawed hands and feet – had distinctively large denticles on their teeth instead of the usual serrations, a morphology that has been linked by some paleontologists with a partially herbivorous diet (Figure 6.17c).

Figure 6.15. Left lateral view of the skull of (a) *Ornitholestes*, (b) *Oviraptor*, (c) *Albertosaurus*, (d) *Tyrannosaurus*, (e) *Saurornithoides*, (f) *Gallimimus*, (g) *Dromiceiomimus*, (h) *Deinonychus*, and (i) *Velociraptor*.

Figure 6.16. In its manner of slicing and in the placement of its hinge, the carnivorous theropod mouth was very similar to the design of a pair of scissors.

With the differences in skull and teeth, theropods evidently bit in different ways. Recent studies have paired CT scans and computer-modeled stress analyses to the architecture of theropod skulls (Figure 6.18). We now know, for example, that *Allosaurus*, with its relatively lightly built skull, used a "slash-and-tear" attack on its prey, in which powerful neck muscles drove the skull downward rather than delivering a crushing bite with the jaw muscles alone. When the head was retracted, the teeth sliced and tore flesh. Such a wound might not kill prey immediately – but blood loss and possible bacterial infection would work their relentless damage. Tracking and waiting may have been part of the killing technique of dinosaurs with lightly built skulls and thin, blade-like teeth.

This contrasts with tyrannosauroids (see Figures 6.15d and 6.19) and perhaps abelisaurids such as *Carnotaurus* (see Figure 6.14e), whose more bulbous teeth and larger, more heavily built skulls likely delivered a bone-crushing, heart-stopping bite. They may also have suffocated their victim by seizing their snout or neck between their jaws and clamping down. This kind of attack is consistent with the large gape, powerful jaws, and stout teeth of these predators. Powerful neck muscles supported the large skull. In all cases, however, the skull had considerable mobility on the neck because of a well-rounded occipital condyle and its articulation with the first part of the cervical (neck) vertebrae (see Box 6.1).

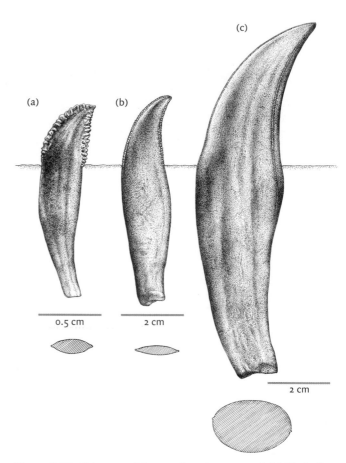

Figure 6.17. Extremes of theropod teeth compared. (a) Blade-like meat-slicing tooth of *Dromaeosaurus*; (b) bulbous bone-crunching (?) tooth of *Tyrannosaurus*; (c) *Troodon* tooth, with its large denticles in place of the serrations more commonly seen in theropods.

Toothless

At least two times in their history,[2] theropods drastically reduced or lost all of their teeth. With the exception of one primitive genus, all ornithomimosaurs, a group of small-skulled, long-legged theropods that look very much like ostriches with long tails (see Figure 6.10), lost all their teeth (see Figure 6.15f,g). Ornithomimosaurs had a beak, and in the case of *Gallimimus*, at least, the beak had a ridged feature that appeared almost sieve-like along its margin, provoking a controversial suggestion that it fed aquatically, much like a modern duck. Later interpretations of the beak edge suggest that it is less a sieve and more of a shearing feature, consistent with grinding fibrous plant matter.

Ornithomimosaurs are also known to have had **gastroliths**, stomach stones that are stand-ins for teeth. Instead of chewing (e.g., processing food in the mouth), gastroliths were swallowed and kept in a muscular gizzard; then, swallowed but unchewed plant matter was ground down in the gizzard by the stones to make digestion more efficient (Figure 6.20). These, in combination with powerfully clawed hands and superb running capability, suggest a terrestrial existence rather than a duck-like

---

[2] Three times, when we include birds (see Chapter 8 and 9).

Figure 6.18. Reconstructing musculature. (a) The skull of *Allosaurus* marked with its finite element analysis model, indicating the regions of stress that pass through it as a function of biting. (b) Reconstructed jaw musculature in the skull of the carnivorous theropod *Majungasaurus* showing (A) interior musculature blocked by the exterior muscles and (B) exterior musculature. Recalling that muscles can only retract, the jaw muscles are antagonistic: they work against each other, and one contracting will stretch the relaxed one. The adductor muscles close the jaw when they contract; the depressor muscles open the jaw when they contract. Not surprisingly, jaw-*closing* muscles are far larger than the jaw-*opening* muscles. Finite element analysis, in combination with Extant Phylogenetic Bracketing and CT scanning, allows the accurate positioning and reconstruction of muscle groups (see Box 6.1). From Holliday (2009) and Brusatte (2012).

lifestyle. Consistent with the beak, these animals were likely fast-running creatures that used their gastric mills to grind up fibrous plant matter, as do modern plant-eating birds.

Oviraptorosaurs were also toothless (see Figures 6.15b and 6.21; see also Chapter 7 and Figure 7.17). The skull was very short with apparent pneumaticity (see Chapter 8), and the jaw musculature was very well developed. Located between their shortened upper jaws is a pair of peg-like projections dead center in the middle of the palate. One analysis of the mechanics of the

Figure 6.19. The teeth of *Tyrannosaurus* as seen from the left side of the skull.

oviraptorosaur skull suggested that the jaws were designed to feed on hard objects that required crushing, such as clams, oysters, and mussels. Oviraptorosaurs presumably cracked them open by the brute force of their jaw muscles acting on the thick horny bill covering the margins of the mouth and the palate, and especially the stout pegs in the center. As we discuss in Chapter 7, however, what oviraptorsaurids did with their distinctive skulls is still an all-too-open question.

Senses

To locate and track their prey, theropods of all kinds needed a keen awareness of their environment. Thanks to brain endocasts and, more recently, CT-scan reconstructions of the interiors of braincases (Figure 6.22; see also Box 11.1), we've begun to learn a lot more about how these animals operated. Because the brain represents the center for the senses, we look for similarities to the brains of related living creatures,[3] such as birds and crocodiles, whose behavior is known, as well as to regions of enlargement, which suggested increased or heightened function of that part of the brain.

---

[3] Sound familiar? This is extant phylogenetic bracketing again, but applied to the brain instead of the musculature. The cranial nerves, which are consistent, conserved features among all vertebrates, serve as important landmarks for identifying key parts of the brain.

Box 6.1   Putting some meat on them bones

In this chapter and the ones that follow, there is a lot of glib talk about gnashing (teeth), slashing (claws), bashing (usually skulls); crunching (bone) and munching (plants). All of this takes muscles, of course; but how do we know where the muscles go – and how much muscle goes there?

We've all heard that skeletons provide support for the body; without them we would look like quivering gelatinous puddles on the ground. While this is in a very broad way more or less true, it doesn't really explain why, for example, your back muscles get tired if you take a very long walk; isn't your skeleton supporting you?

But of course it isn't actually your skeleton that does the majority of the supporting – it is your musculature. The skeleton provides well-positioned support and attachment for the muscles, and the muscles do the heavy lifting. And that turns out to be part of the key to understanding the musculature of extinct animals, whose actual muscles are long gone.

We start with homology – just as the bones show their phylogenetic history through homology, so do the muscles. That is, different muscles and muscle groups – sometimes unmodified and sometimes somewhat modified – are traceable via homology, just as the bones were.

That being the case, if you want to know about the muscles in non-avian dinosaurs, you begin with understanding the positions and functions of the muscles of the two closest living groups: birds and crocodiles, and groups successively distant from archosaurs (such as lizards and even mammals). This process, which is actually all about determining muscle homologies, is termed "extant phylogenetic bracketing."

The skeleton, as we now know, is for muscle support, and therefore its shape is commonly (but not always) responsive to stresses placed upon it by those muscles: where the stresses are highest, the bones are largest and thickest; commonly prominent ridges develop. The shape of the bone is, in part, a reflection of the stresses it undergoes. Moreover, where the muscles attach to the skeleton, the bone shows marks called "muscle scars." The skeleton, therefore, in combination with phylogenetic bracketing, provides a kind of guide about where muscle groups ought to lie, and thus how they most probably were in life. The keys to this process, therefore, are well preserved, undistorted skeletal elements. Here, novel techniques such as CT-scanning (see Figure 6.22) are very important, giving high-resolution, undistorted images of dinosaur skeletons.

Figure 6.20. Gastroliths.

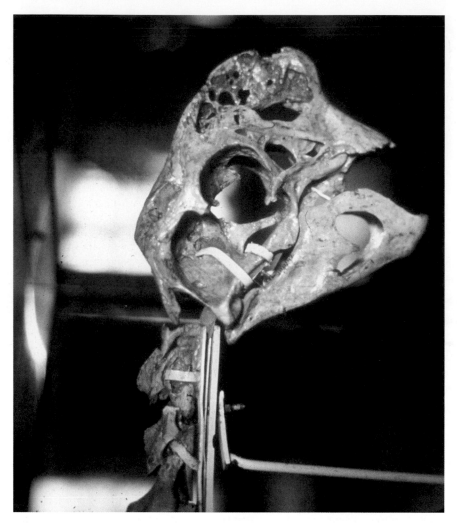

Figure 6.21. The box-like, toothless skull of the Late Cretaceous Mongolian theropod *Oviraptor*.

Clearly, however, sharp vision was key for theropods, so it is not surprising that their eyes were large. In deinonychosaurs generally, but especially in troodontids, the eyes have migrated to a more forward-looking position, indicating overlapping fields of vision. Overlapping fields of vision almost certainly means that these animals saw stereoscopically – that is, they merged the two separate independent images from each eye into a single image, much as humans and many modern carnivorous birds do today. Recent work suggests that the narrow snouts of even tyrannosauroids allowed a 55° range of binocular vision, not nearly as much as a human or an owl, but far exceeding what one might find in a herbivorous dinosaur, such as hadrosaurid or ceratopsian (Figure 6.23).

Hearing, too, is important to predatory animals and so it is not surprising that many theropods likely had good sound perception. Indeed, the inner ear cavity of troodontids and ornithomimosaurs was greatly enlarged, suggesting that these theropods were especially able to hear low-frequency sounds. In troodontids, detailed anatomical study of the outer and middle ears suggests that the group was capable of identifying the direction from which sounds came; knowledge that would have been of extreme use to a predator.

Making sense of tyrannosaurs
With nothing close to a living analog, the behavior and lifestyle(s) of tyrannosaurs have always been enigmatic, with the mighty *Tyrannosaurus* being perhaps the most perplexing – yet compelling – of all.

Figure 6.22. The detail that modern CT reconstructions ("visualizations") can achieve is astounding. (a) CT reconstruction of the brain of *T. rex* from left side of animal. Scale bar = 4 cm. (b) CT reconstruction of the skull of *T. rex*, showing elaborated sinuses suggesting a relatively well-developed sense of smell, top view; (c) bottom view; (d) left side. Abbreviations (for the cranial anatomists among you) above to right of figure. After Witmer and Ridgely, 2009.

Recent CT-scan-based work by L. M. Witmer, R. C. Ridgely, and associates at Ohio University has provided some exciting insights about tyrannosaurs, notably *Tyrannosaurus*, *Gorgosaurus*, and *Nanotyrannus*.[4] With the shape of the brain fully reconstructed (Figure 6.22), it is clear that despite enlarged cerebral hemispheres, tyrannosaur brains were not the functional – thinking – equivalent of that of a modern bird. No surprise there! Somewhat more surprising is that tyrannosaurs had well-developed senses of smell. By comparison with other theropods, tyrannosaurs had relatively large olfactory bulbs, as well as a particularly large nasal cavity. The likelihood is that these features reflect behaviors that involve the ability to smell.

Like other theropods, tyrannosaurs were particularly adept at hearing low-frequency sounds. In the case of tyrannosaurs, however, the design and construction of the ear bones were such that hearing was clearly an unusually important sense. Witmer and colleagues have reconstructed tyrannosaurs as predators, with the neural evidence suggesting an ability to track prey rapidly by sight, smell, and sound, using eyes, head, and neck.

Figure 6.23. Three-quarter view of the skull of *T. rex* showing a 55° range of binocular vision (see text).

Balance

The formidable armament of theropods, particularly the small- to medium-sized ones, must have been lethally coupled with exceptional balance. The nearly horizontal position of the vertebral column took advantage of the center of gravity being positioned near the hips.

Deinonychosaur balance was aided by a remarkable weapon in their arsenal: a tail stiffened by immensely elongate processes (zygopophyses) along the neural arches (Figure 6.24). The tail was flexible only at its base, just behind the pelvis; the rest was completely rigid. This stiffening allowed the rigid tail to move as a unit in any direction. It thus functioned as a dynamic counterbalance against the motions of the long arms, grasping hands, and large skull and jaws.

On the basis of their light, yet powerfully built skeletons, dromaeosaurids and troodontids must have had an extraordinary degree of agility. We imagine them flinging themselves at fleeing prey, kicking with great accuracy with one, or even both, of their dangerous feet (see Figure 6.6).

Thoughts of a theropod

All theropods for which there is brain-volume information have surprising (to humans!) cerebral powers: their brains are every bit as large as one would expect of a crocodilian or lizard "blown up" to the proper body size. Indeed, troodontids had the largest brains for their body size of any non-avian theropods, and were well within the bird range of inferred intelligence (see Box 13.5). This suggests that

---

[4] *Nanotyrannus* has been a controversial creature since its discovery in the middle of the twentieth century. Originally referred to as *Gorgosaurus*, it has since been called *Tyrannosaurus*, then a juvenile of *Tyrannosaurus*. Exactly what it is, taxonomically speaking, remains unclear.

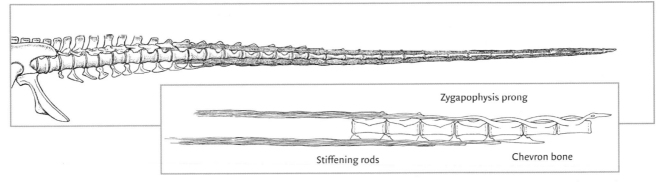

Figure 6.24. The rigid tail of *Deinonychus*. The elongate, intertwined zygapophyses on each neural arch give the tail its rigidity.

these animals probably had more complex perceptual ability and more precise motor–sensory control than some of their smaller-brained brethren. It also likely implies sophisticated inter- and intraspecific behavior, attributes that belie the image of irremediably stupid dinosaurs!

The skinny on skin

As we saw in Chapter 5, a modern conception of dinosaurs comes with feathers attached. Which dinosaur and which feathers is still something of an open question, but anyone who has been paying attention would agree that feather are becoming more widespread in Dinosauria rather than less. Ultimately they may prove to be ubiquitous.

When birds and other exceptionally well-preserved fossils began showing up in China in the mid 1990s, paleontologists first recognized feathers and feather-like structures on non-avian theropods. Among the most famous of these are *Caudipteryx*, *Protarchaeopteryx*, *Sinornithosaurus*, and *Microraptor* (see Chapter 7). Two other theropods from the same region – *Sinosauropteryx* and *Dilong* – are extensively covered with filaments that have been interpreted as feather precursors. But feathered dinosaurs do not come exclusively from Asia; ornithomimisaurids from North America have also been reported with feathers. And, unsurprisingly, the feathers are all colored and patterned, suggesting that these animals interacted visually (see "color" below).

We do not have skin impressions with feathers for all theropods. Still, some kind of insulatory covering (feathers) has been suggested for highly active theropods, such as dromaeosaurids and troodontids, and some paleontologists have speculated that, as juveniles, even large adult theropods may have been covered with a downy feather-like insulation.

Yet, the image of very large, powerful, heavy theropods, like *T. rex* or *Carcharodontosaurus*, covered with feathers is just a little outré. But cheerfully embrace it we must, because in 2011 Chinese paleontologist Xu Xing and colleagues announced the discovery of *Yutyrannus huali*, an 8 m tyrannosaur from China. *Yutyrannus* was evidently covered, at least partially, with 15 cm long filamentous feathers (see Figure 7.8). With feathers clearly present in its near relative *Yutyrannus*, the "crazy" probability that *T. rex* was also feather-covered becomes very high indeed. We'll revisit feathers and theropods in greater detail in Chapter 7.

Eaters and eatees

With all the different theropods (see Chapter 7), prey surely varied. The most dynamic and irrefutable evidence about the preferred prey of *Velociraptor* is the so-called "fighting dinosaurs" specimen: *Velociraptor* with its hindfeet half into the belly of a subadult *Protoceratops* and its hands grasping, or being held in, the jaws of the soon-to-be victim (Figure 6.25).

Figure 6.25. The famous fighting dinosaurs specimen; *Velociraptor* wrapped around *Protoceratops* from the Late Cretaceous of Mongolia.

A specimen of the Late Jurassic coelurosaur *Compsognathus* is known that contains most of the skeleton of a fast-running lizard. Not only did *Compsognathus* swallow nearly whole this delectable meal, but it must have captured its victim through its own speed and maneuverability. Other evidence of theropod stomach contents and diet comes from *Sinosauropteryx* (lizards and mammals), *Baryonyx* (fish remains), and *Daspletosaurus* (hadrosaurid bones).

Occasionally, the prey can be more easily identified than the predator. A 44 cm long, 13 cm high, and 16 cm wide coprolite was found, containing between 30 per cent and 50 per cent of bone

Box 6.2   *Triceratops* spoils or spoiled *Triceratops*

In 1917, L. M. Lambe of the Geological Survey of Canada suggested that the large carnivorous dinosaur *Gorgosaurus* was not so much an aggressive predator, but instead scavenged. The basis for his remarks was the apparent absence of heavy wear on the teeth of this theropod – these animals, he proposed, must therefore have fed primarily on the softened flesh of putrefying carcasses. This interpretation has appeared on and off again in discussions of theropod diet and hunting behavior, frequently enough to be something like a cottage industry in anecdotal "knowledge" about these animals.

The notion of theropod scavenging rests on the assumption that tooth wear was usually absent and that carcasses were readily available. It is further bolstered by the lack of a convincing account of why the forelimbs of these animals are so small. We take each of these in turn. Firstly, it turns out that tooth wear is present on the teeth of nearly all large theropods. This doesn't "prove" that tyrannosauroids and other large theropods had to have been active predators; both modern scavengers and active predators alike can have a high degree of wear on their teeth.

The commonness of carcasses, putrefied or otherwise, was probably dependent on the season – dry, stressful seasons probably claimed their share of hadrosaurids, ceratopsians, sauropods, and other creatures. Interestingly, this potentially great contribution of carcasses would have been in the form of tough, dry flesh – dinosaur jerky, really – not the kind of soft, ripe, malodorous predigested carrion postulated by Lambe.

Finally, short forelimbs were likely used to slice and dismember prey, and are fully consistent with a head-first attack.

Recently, the suggestion of tyrannosaurids as carrion eaters has come from the observation that these theropods had surprisingly broad, bulbous teeth. Modern scavengers such as the hyena likewise have broad teeth, which are used to crush the bones of carcasses. This notion has been pooh-poohed by scientists who cannot imagine a dinosaur with the size and obvious carnivorous equipment of *T. rex* being a scavenger.

What *is* clear is that *Tyrannosaurus* teeth clearly exceed the size increase that might be predicted from its enlarged body. For this reason, *T. rex* leaves paleontologists with a mouthful of confusion.

In the end, we think it likely that animals like *Tyrannosaurus* would have been a more than adequate, indeed terrifying, active predator, yet one that wouldn't turn up its nose at a lunch of carrion. Indeed, it is probably better to view all theropods, from the exceptionally large tyrannosauroids down to much smaller troodontids, dromaeosaurids, and coelophysoids, as equal-opportunity consumers, taking large herbivores, smaller flesh eaters, and even the occasional carcass.

---

fragments, thought to be the remains of limb bones or parts of a ceratopsian frill. The source of this coprolite? Age, geography, and, uh, size, point to *Tyrannosaurus rex* as the perpetrator. The specimen, in combination with other information about theropod diets, provides physical evidence that tyrannosauroids, at least, crushed, consumed, and incompletely digested large quantities of bone.

Along those same lines, toothmarks attributed to *Tyrannosaurus* are known from a pelvis of *Triceratops*, and from a thoroughly crunched tail of the duck-billed dinosaur *Edmontosaurus* (Box 6.2). Evenly spaced grooved toothmarks on the bones of the sauropods *Apatosaurus* and *Rapetosaurus* have been attributed to the local large theropods: *Allosaurus* and *Majungatholus* (from the USA and Madagascar, respectively). And finally, the bite patterns of the large Late Jurassic theropod *Allosaurus* have been tied to marks on *Stegosaurus* neck plates and, conversely, *Stegosaurus* spikes have been implicated in a partially healed *Allosaurus* puncture shaped wound. These dinosaurs lived together; were they also predator and prey?

There had been hints for a long time that the large Cretaceous theropod *Spinosaurus* was fish-eating (**piscivorous**): equipped a crocodile-like snout, high nostrils, and tall conical teeth (hallmarks of fish-eaters; Figure 6.26; see also Chapter 7), it was not a completely improbable idea, although swimming is not a common theropod adaptation.

In 2014, newly discovered material[5] suggested the interpretation that *Spinosaurus* spent a lot of time in the water, and that it might have been amphibious. The thing was huge – up to 15 m – and showed a lot of unusual adaptations, including surprisingly long, powerful arms for a theropod that size (see Chapter 7). Moreover it had dense bone, a most unusual character in a theropod, animals that normally have hollow bones. It had flattened toe bones suggestive of a foot used for paddling (could it have been webbed?). Remarkably, it has a relatively weak, small pelvis (recall that the strong pelvis is the balance point for this very terrestrially adapted group), and nostrils towards the middle (rather than the front) of the skull. All evidence suggested that the animal was amphibious.

If *Spinosaurus* was an aquatic predator, as all of these adaptations appear to suggest, it would likely have been piscivorous – but which fish? *Spinosaurus* teeth have been found in the same deposits as those of the large sawfish *Onchopristis*; this and the fact that identifiable *Onchopristis* fragments were found embedded in the tooth sockets of *Spinosaurus* have led to media[6] speculation that *Spinosaurus* dined on *Onchopristis*. Could the sawfish fragments simply have lodged like pebbles in the empty sockets when the bones were washed together during burial? Of course; but the idea of *Spinosaurus* as a gigantic fish-eating amphibious predator is becoming more and more viable.

It took us about 100 years to begin to interpret the unusual morphology of *Spinosaurus* in a way that generates a coherent behavioral picture. But what about its most striking feature: the row of 2 m

---

[5] The original material was partially destroyed during the bombing of Munich by the Royal Air Force during World War II. The specimens were destroyed, and thus lost for about 70 years, until new specimens were excavated (see Box 15.7).

[6] The British television show *Planet Dinosaur*.

Figure 6.26. *Spinosaurus* rediscovered! The newly reconstructed *Spinosaurus*, mixing elements from the original specimen and from newly discovered specimens (see Box 15.7).
Photograph courtesy of Paul Sereno.

high neural spines? Although it has been suggested that these were for display, you may have to check back in another 100 years for anything definitive about those!

Cannibals

Some dinosaurs didn't shy away from a bit of cannibalism: for example, *Majungatholus* apparently didn't avoid the odd conspecific snack: grooved toothmarks matching the spacing of its own teeth have been found on *Majungatholus* specimens (Figure 6.27). With only one carnivore known from

Figure 6.27. Cannibalism in theropods. Above: the teeth of the large Malagasy theropod *Majungatholus*. Below: toothmarks on *Majungatholus* bone. Note that the spacing of the grooves in the bone matches that of the teeth; note also the scratches next to each groove, reflecting the serrations of the teeth.

Madagascar from that time with that tooth spacing, the evidence is circumstantial, but damning. Of course, we still don't know if such cannibalism occurred on the hoof, or whether it was the scavenging of sick or dead individuals (or both).

Some dinosaurs got undeserved reputations for cannibalism. It was once thought that the Late Triassic *Coelophysis*, based upon the supposed presence of its own juveniles within the rib cage, was cannibalistic; not a great recipe for species survival. However, recent work indicates that the juveniles in the rib cage were actually non-dinosaurian archosaurs; a better strategy.

Who ate whom?

Beyond these few direct observations of dietary preferences, we are left to speculate on who ate whom. Our best guesses are informed by the known theropods and potential prey in a particular place and time – we might pair *Tyrannosaurus* with *Triceratops* (the classic latest Cretaceous confrontation-of-the-titans), *Tarbosaurus* with *Saurolophus*, or *Troodon* with small ornithopods or juveniles of much larger co-existing dinosaurs (such as hadrosaurids).

The invention of pack-hunting could have allowed relatively small animals to bring down much larger co-existing prey. Bones of different dinosaurs are often found together, leading to media-fueled speculation about diners and dinners. Teeth of *Deinonychus* (a ~3 m dinosaur) have been found in association with the large ornithopod *Tenontosaurus* (~7 m), suggesting the former ate the latter. But the size difference between the two made this proposal a tall order, unless *Deinonychus* hunted in packs. Thus, the genesis of the idea of dromaeosaur pack-hunting, distorted almost beyond recognition in the Jurassic Parks! Whether true or not, these are the often the best available data that can be used to address the question of theropod diets (see also Chapter 13).

The most successful attacks come from where they are least expected, and recent work by J. R. Horner and colleagues suggests that the mighty *T. rex* was under regular, and potentially crippling, attack by a lowly protozoan. These researchers noted that the jaws of many specimens of *Tyrannosaurus* show numerous lesions (Figure 6.28). Once thought to be the work of other tyrannosaurs, these lesions are very similar in morphology to those found in the jaws of modern chickens, turkeys, and falcons. In the modern birds they are the handiwork of a parasitic protozoan, *Trichomonas gallinae*, an organism that infects jaws, eventually eating out chunks of the jaw bone, deforming and/or damaging the jaw. Similarities in the morphology of the lesions strongly suggest that this same organism, or a very close relative, was at work on Cretaceous non-avian theropods as well. The handiwork of other bacteria is also evident in theropods. The redoubtable *Tyrannosaurus* "Sue" (Box 1.1) shows the results of a chronic bacterial bone infection, as does "Big Al," an *Allosaurus* at the Field Museum in Chicago.

Social behavior: sex and the single *rex*

Single-species bonebeds remain strong suggestions that theropods functioned in packs. Many of these mass graveyards – which include both juveniles and adults – pertain to the coelophysoid group of theropods (see Chapter 7) and include such luminaries as *Syntarsus* (Zimbabwe, South Africa, and Arizona) and *Coelophysis* (New Mexico).

Other theropods, though, are also known in bonebeds: *Allosaurus* (Utah), as well as *Giganotosaurus* and *Mapusaurus* (Patagonia). Could it have been that some theropods were gregarious, living in large family groups that perished together? Or perhaps each accumulation represents a communal feeding site? Do the bonebeds represent pack-hunting? As new finds with two or more individuals get discovered, even those most supposedly solitary of killers – the *über*-tyrannosaurs

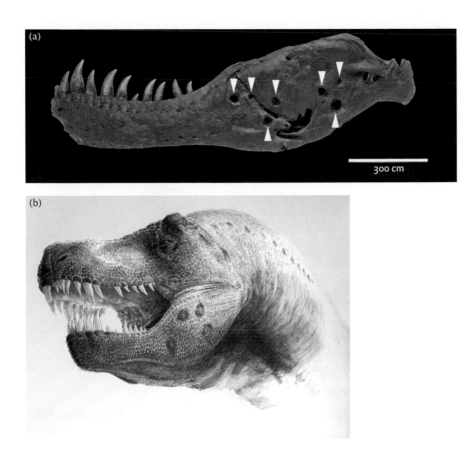

Figure 6.28. (a) Left lower jaw exhibiting multiple lesions (indicated by arrows). (b) Hypothesized reconstruction of the infection in a living tyrannosaur. Lesions within the lower jaw perforate the bone. Reconstruction based on photographs of living birds with similar pathologies.

Daspletosaurus and *Tyrannosaurus* – are potential candidates for the comforts of at least family. Theropods as gregarious beasts – even the largest theropods – is becoming more and more the expected, and less and less the anomalous.

Sexual dimorphism and its role in sexual selection

Many vertebrates are strongly **sexually dimorphic**, that is, males and females are not identical, but instead have different attributes. Commonly, one gender is larger than the other, or they differ by other features. In diapsids generally, and modern birds in particular, coloration between males and females can be strikingly different, with the males commonly assuming the brightest hues. These traits are generally related to **sexual selection**, which can be described as selection not occurring equally on all members of a particular population or species (such as one sees in natural selection – see Appendix 3.1), but rather a kind of selection based upon gender. Male deer clashing for reproductive rights with an insouciant female would be a familiar example of sexual selection. In birds, sexual dimorphism is deeply tied into sexual selection, and commonly goes hand in hand with fancy, bright plumage, virtuosic vocalizations, and complicated dances, all types of **display**. To tell the truth, how different is this from humans?!

What can the skeletons of theropods tell us about how these animals related to each other socially? Like frills and horns in neoceratopsians (Chapter 11), or crests in hadrosaurids (Chapter 12), quite a number of predatory dinosaurs – including *Syntarsus*, *Dilophosaurus*, *Proceratosaurus*, and possibly *Ornitholestes*, to *Ceratosaurus*, *Cryolophosaurus*, *Alioramus*, and *Oviraptor* – flaunted highly visible cranial crests (see Figures 6.14c and 6.29). Some are made of thin sheets of bone, while others are hollow, presumably part of the cranial air-sinus system. Beyond these, theropods such as *Yangchuanosaurus*, *Allosaurus*, *Acrocanthosaurus*, and the tyrannosauroids bore slightly elevated upper margins on the snout and raised and roughened bumps over the eyes. These structures are believed to have been cores for **hornlets** (small horns) made of keratin, which must have given the face a spiky punk-rock look (Figures 6.14 b, e, f, and 6.15 a, c).

The crests and hornlets assuredly functioned in display and – at least for the latter – may have been used occasionally in head-butting squabbles over territories and mates. If crests and hornlets functioned in visual display, particularly in those theropods that lived in large groups (see above), we might expect them to be species-specific and probably sexually dimorphic so as to signal a given animal's identity and sex. And likewise we might expect crests to show their greatest development in reproductively mature individuals; youngsters should have small, poorly developed crests and hornlets.

Are these expectations met in any theropods? To a degree; sexual dimorphism is found in *Syntarsus* and *Coelophysis*. In these two theropods, one of the two morphs had a relatively long skull and neck, thick limbs, and powerfully developed muscles around the elbow and hip, while the other form had a shorter skull and neck,

Figure 6.29. The skull of the Late Jurassic southern hemisphere polar theropod *Cryolophosaurus,* with its distinctive crest.

and slender limbs. In *Tyrannosaurus*, the case has been made for sexually dimorphic ornamentation. The larger, more robust morph was hypothesized to be the female. In the cases of other theropods, we just don't yet know; what we would give to see one (or two, or a pack) alive (Box 6.3)!

The wonderful world of color

With the evidence (above) for sexual selection so strongly imprinted in the remains of extinct theropods (as well as in the behavior of living theropods), could color have played a role in non-avian theropod sexual selection, as it does today in *avian* sexual selection?

Incredibly, some patterns have actually survived the process of fossilization. For example, a striped pattern in the tail of *Sinosauropteryx* is clearly visible with the naked eye (Figure 6.30). Beyond rudimentary patterns, however, color in any dinosaur – including those covered with feathers – has been, until recently, very elusive.

In modern birds, feather color is based in part upon microscopic capsule-shaped organelles called **melanosomes**. These organelles have different shapes, sizes, and distributions, which correlate in modern birds with colors such as blacks, reds, browns, and yellows. In 2007, using a high-powered electron microscope, melanosomes identical to those found in modern birds were identified in the vaned feathers of a primitive bird, *Confuciusornis* (see Figure 8.12), as well as in the primitive feathers of non-flying theropods *Sinosauropteryx* (Figure 6.30) and *Sinornithosaurus* (see Chapter 7; Figure 7.20). Using the shapes and distributions in modern birds as a guide, it was possible, for the first time, possible to learn something about the actual colors present in these

Box 6.3   Dinosaur zombies

In western African lore, a *bokor*, or wizard, can revive the dead (the word "zombie" is thought to have came from *nzumbe*, a word in the North Mbundu language of western Africa). Could scientists ever revive a (very) dead, (very) fossilized dinosaur?

The idea was vetted by Michael Crichton in *Jurassic Park* (1990). In that yarn, dinosaur blood, sucked by mosquitoes, was extracted from the insects when they themselves were trapped in amber. The dinosaur blood contained dinosaur blood cells; the blood cells were cloned, and Sam Neill faced off against *Velociraptor*. At the time, cloning dinosaurs from preserved cells received a bit of play in the serious scientific community, but in the end, DNA was found to degrade within a million or so years, and amber turned out to be semi-permeable, and things could get in and grow and contaminate the samples. And the idea was buried, until the late 1990s when, zombie-like, it was re-exhumed in an entirely different form.

Mary H. Schweitzer, at the time just beginning a PhD at Montana State University, observed red structures, each with a black dot in the center, in a blood vessel channel in a slice of bone from a particularly well-preserved *T. rex*. Her thought was – counter to everything that everybody "knew" about the process of soft-tissue (not bone) preservation – these looked like red blood cells; they were located in the right place; could they actually *be* red blood cells from the dinosaur? As time progressed, she ended up recovering a lot of soft tissue: blood vessels, bone cells, and keratin from the claws of *Rahonavis* (Chapter 11) and from the feathers of *Shuvuuia* (Chapter 10).

Could this stuff be *real*? One way to test it was by checking immune responses. Living bodies, of course, react to foreign objects by creating antibodies. So Schweitzer and her colleagues used mice to generate antibodies to the supposed dinosaur proteins she had extracted. Sure enough, when the antibodies that had been generated were exposed to hemoglobin from rats and turkeys, the antibodies bound to the hemoglobin, suggesting to Schweitzer and her colleagues that the proteins they had extracted from the dinosaur were something very much like hemoglobin. The keratin, too, was tested in the same way: keratin-specific antibodies were generated using the modern claw and feather proteins; these then bound to the ancient keratin. It appeared Schweitzer had actually isolated soft tissue from very extinct, long-gone dinosaurs. It got even crazier when another *T. rex*, astoundingly well

preserved over the 65.5+ million years that had elapsed since the animal's death, was found. Schweitzer, who had by then moved to North Carolina State University, found medullary tissue in the fossil bones – tissue that is diagnostic of an animal that is producing eggs. Today medullary tissue is produced only by ovulating female birds; a clear indication that this *T. rex*, at least, must have been female – and ovulating (see Chapter 10). Schweitzer's reaction to her discovery, quoted in the pages of *Scientific American*, was priceless: "'Oh, my gosh, it's a girl – and it's pregnant!' I exclaimed to my assistant, Jennifer Wittmeyer. She looked at me like I had lost my mind."[a] From this *T. rex*, she extracted hollow, transparent, flexible tubes, some filled with some red material, which she interpreted as blood vessels with dinosaur blood.

Schweitzer's work was not accepted uncritically, and there is still some concern that she is actually looking at the remains of modern organisms, rather than dinosaur soft tissue. Underlining the difficulty of this type of work, some authors have suggested that she is looking at modern invading microbes or bacterial biofilms growing on the fossil bone, rather than actual dinosaur tissue. That possibility became less likely very recently, however, when collagen polypeptide sequences were obtained from a hadrosaur (*Brachylophosaurus*) and subjected to the same rigorous suite of tests undergone by the *T. rex* polypeptides. Moreover the work has been duplicated in separate laboratories. Indeed, recent work on fossil collagen has begun to elucidate aspects of the structure of this molecule that led to its unusual resistance to long-term degradation.

It won't end up happening quite as Michael Crichton described it, but the possibility of building backwards from the small fragments of DNA preserved in these fossils has become the kind of thing that some paleontologists dream about. Jack Horner, for one (see Box 15.9), teaming up with molecular biologist colleagues, attempted to grow the tail of a non-avian dinosaur from long-dormant genes in modern birds (a chicken). His idea was that the ancestral dinosaur-tail genes, left over from when theropods regularly grew tails, are still there, waiting to be activated. They never quite got their tail, but their tale, and the dream that motivated it, are described in Horner's book *How to Build a Dinosaur* (2009).

The possibility of seeing a living non-avian dinosaur becomes less and less fantastic, as our knowledge of molecular biology and genetics becomes deeper.

[a] Schweitzer, M. H. 2010. Blood from a stone. Scientific American, December, 62–69.

long-extinct dinosaurs (see Figure 6.31). The banding in the tail of *Sinosauropteryx* appears to have been black; a crest along its back may have had tan-red-dish-brown coloration. Even iridescence has been recognized in fossil non-avian theropods! The colors and patterns present in these non-avian theropods reinforce the idea that non-avian theropods probably used color and pattern for a variety of complex social behaviors, including display and sexual selection, just as we see in living (avian) theropods. We'll see other examples of non-avian dinosaur color in Chapter 7.

Mama's (?) little theropod

Regardless of sexual dimorphism in non-avian theropods, what we know about their reproductive biology has been greatly enhanced by the discovery of brooding oviraptorosaurs. Several recently discovered articulated oviraptorosaur skeletons are preserved overlying nests of eggs, laid in a circular pattern of as many as 22 eggs. The embryos are the same species as the adult skeleton overlying them.[7] The oviraptorosaur skeleton (Mom? Dad?) is positioned directly above the center of the nest, with its limbs arranged symmetrically on either side and its arms spread out around the perimeter as if protecting the nest (Figure 6.32). These specimens indicate that incubating eggs on open nests evolved well before the origin of modern birds.

Recently, however, scientists, took a step farther. Knowing certain compounds responsible for the coloration in eggs, a team of scientists from the University of Bonn were able to isolate pigments in the Late Cretaceous eggshell genus *Macroolithus* – attributed to the Chinese oviraptorosaur *Heyunannia*. They reconstructed the color of these eggs as ranging from blue to green, presumably matching the background of the substrate on which they were laid. This work reinforced conclusions reached from the fossil shown in Figure 6.32: the eggs were likely laid in an open nest and incubated by a concerned parent.

Figure 6.30. Patterns of color of the feathered, non-avian theropod *Sinosauropteryx*. Filamentous features – interpreted to be early, non-flight feathers – are preserved down the back of the animal, extending the length of the tail. Arrows indicate color banding, particularly well preserved on the tail.

What we know of the post-hatching growth of non-avian theropods mostly comes from bonebeds (for example, Ghost Ranch (New Mexico) and Cleveland-Lloyd (Utah)), as well as from some of the Upper Cretaceous localities of the Gobi Desert in Mongolia. For *Coelophysis* and *Syntarsus*,

---

[7] The eggs were first attributed to the ceratopsian *Protoceratops*. Because they were found with theropod skeletons, it was assumed that the theropods were stealing the "*Protoceratops* eggs" and were given the name *Oviraptor* (= egg stealer). *Oviraptor* languished, falsely accused, for 70 years, until the mid 1990s discovery of the unquestionably nesting specimens.

(a)

(b)

Figure 6.31. Melanosomes. This specimen (a), attributed to *Anchiornis huxleyi*, yielded melanosomes, shown in an electron micrograph (b). Melanosomes are cells that produce color in feathers, whose shape gives an indication of some of the colors that graced these non-avian theropods.

(a)

Figure 6.32. *Oviraptor* eggs and babies. (a) *Oviraptor* eggs. (b) Specimen of an oviraptorosaur adult on its haunches, nesting its eggs. Most of the body of the adult is gone, but the legs and arms indicate where it once was, arms cradling the eggs.

(b)

Figure 6.32. (cont.)

apparently there was a 10- to 15-fold increase in body size from hatchling to adulthood, and this growth is thought to have been quite rapid. We'll revisit this question about the rate of dinosaur growth in Chapter 13.

For now, we leave the details of theropod lives, and in the next chapter, we explore the extraordinary diversity of theropods.

SUMMARY

Theropods are among the most iconic of dinosaurs, including beasts such as *Tyrannosaurus rex*. Clawed bipeds all, with distinctive hollow bones, the earliest known dinosaurs were theropods, and it is theropods that are still living (as birds). Most of the Mesozoic forms had claws with a semi-opposable thumb on a grasping three-fingered hand; possessed recurved, serrated, laterally compressed teeth; and were carnivorous. As a group, they were cursorial through and through.

Theropods were the most intelligent of dinosaurs, with highly developed sensory and motor skills for tracking and killing their prey. All had very good vision – in many cases, stereoscopic – and *T. rex*, at least, enjoyed a highly developed sense of smell. A strong impression of high agility in small- to medium-sized theropods is conveyed by the skeletons.

Many – all? – theropods were feather covered; and likely rather colorful animals. With modern birds as living (if highly derived) examples, social behavior can be inferred in theropods. A variety of facial features such as hornlets adorned theropods, and some evidence suggests that many of them hunted in packs. Particularly bird-like is theropod maternal behavior: in those forms in which it is known, non-avian theropod mothers (?) incubated clutches of eggs, very much like avian theropods.

SELECTED READINGS

Abler, W. L. 1992. The serrated teeth of tyrannosaurid dinosaurs, and biting structures in other animals. *Paleobiology*, 18, 161–183.

Brusatte, S. L. 2012. *Dinosaur Paleobiology*. Wiley-Blackwell, New Jersey, 322 pp.

Carpenter, K., Sanders, F., McWhinney, L. A., and Wood, L. 2005. Evidence for predator-prey relationships: examples for *Allosaurus* and *Stegosaurus*. In Carpenter, K. (ed.) *Carnivorous Dinosaurs*, Indiana University Press, Blooomington, IN, pp. 325–350.

Clark, J. M., Maryańska, T., and Barsbold, R. 2004. Therizinosauroidea. In Weishampel, D. B., Dodson, P., and Osmólska, H. (eds.) *The Dinosauria*, 2nd edn. University of California Press, Berkeley, pp. 151–164.

Currie, P. J. 1990. Elmisauridae. In Weishampel, D. B., Dodson, P., and Osmólska, H. (eds.) *The Dinosauria*, 1st edn. University of California Press, Berkeley, pp. 245–248.

Currie, P. J., Trexler, D., Koppelhus, E. B., Wicks, K., and Murphy, N. 2005. An unusual multi-individual tyrannosaurid bonebed in the Two Medicine Formation (Late Cretaceous, Campanian) of Montana (USA). In Carpenter, K. (ed.) *Carnivorous Dinosaurs*. Indiana University Press, Blooomington, IN, pp. 313–324.

D'Amore, D., 2009. A functional explanation for denticulation in theropod dinosaur teeth. *Anatomical Record*, 292, 1297–1314.

Holliday, C. M. 2009. New insights into dinosaur jaw muscle anatomy. *Anatomical Record*, 292, 1246–1265.

Holtz, T. R. 2004. Tyrannosauroidea. In Weishampel, D. B., Dodson, P., and Osmólska, H. (eds.) *The Dinosauria*, 2nd edn. University of California Press, Berkeley, pp. 111–136.

Holtz, T. R. Jr. 2012. Theropods. In Brett-Surman, M. K., Holtz, T. R. Jr., Farlow, J. O., and Walters, R. (eds.) *The Complete Dinosaur*, 2nd edn. Indiana University Press, Bloomington, pp. 346 – 378.

Holtz, T. R., Molnar, R. E., and Currie, P. J. 2004. Basal Tetanurae. In Weishampel, D. B., Dodson, P., and Osmólska, H. (eds.) *The Dinosauria*, 2nd edn. University of California Press, Berkeley, pp. 71–110.

Horner, J. and Gorman, J. 2009. *How to Build a Dinosaur*. Dutton, NY, 246 pp.

Makovicky, P. J. and Norell, M. A. 2004. Troodontidae. In Weishampel, D. B., Dodson, P., and Osmólska, H. (eds.) *The Dinosauria*, 2nd edn. University of California Press, Berkeley, pp. 184–195.

Makovicky, P. J., Kobayashi, Y., and Currie, P. J. 2004. Ornithomimosauria. In Weishampel, D. B., Dodson, P., and Osmólska, H. (eds.) *The Dinosauria*, 2nd edn. University of California Press, Berkeley, pp. 137–150.

Norell, M. A. and Makovicky, P. J. 2004. Dromaeosauridae. In Weishampel, D. B., Dodson, P., and Osmólska, H. (eds.) *The Dinosauria*, 2nd edn. University of California Press, Berkeley, pp. 196–209.

Osmólska, H. and Barsbold, R. 1990. Troodontidae. In Weishampel, D. B., Dodson, P., and Osmólska, H. (eds.) *The Dinosauria*, 2nd edn. University of California Press, Berkeley, pp. 259–268.

Osmólska, H., Currie, P. J., and Barsbold, R. 2004. Oviraptorosauria. In Weishampel, D. B., Dodson, P., and Osmólska, H. (eds.) *The Dinosauria*, 2nd edn. University of California Press, Berkeley, pp. 165–183.

Rega, E. 2012. Disease in dinosaurs. In Brett-Surman, M. K., Holtz, T. R. Jr., Farlow, J. O., and Walters, R. (eds.) *The Complete Dinosaur*, 2nd edn. Indiana University Press, Bloomington, pp. 666 – 712.

Sereno, P. C., Martinez, R. N., Wilson, J. A., *et al.* 2008. Evidence for avian intrathoracic air sacs in a new predatory dinosaur from Argentina. *PLoS ONE*, 3, e3303. doi:10.1371/journal.pone.0003303.

Stokosa, K. 2005. Enamel microstructure variation within the Theropoda. In Carpenter, K. (ed.) *Carnivorous Dinosaurs*. Indiana University Press, Blooomington, IN, pp. 163–178.

Sues, H. D. 1990. *Staurikosaurus* and Herrerasauridae. In Weishampel, D. B., Dodson, P., and Osmólska, H. (eds.) *The Dinosauria*, 2nd edn. University of California Press, Berkeley, pp. 143–147.

Therrien, F., Henderson, D. M., and Ruff, C. B. 2005. Bite me: biomechanical models of theropod mandibles and implications for feeding behaviour. In Carpenter, K. (ed.) *Carnivorous Dinosaurs*. Indiana University Press, Blooomington, IN, pp. 179–237.

Tykoski, R. S. and Rowe, T. 2004. Ceratosauria. In Weishampel, D. B., Dodson, P., and Osmólska, H. (eds.) *The Dinosauria*, 2nd edn. University of California Press, Berkeley, pp. 47–70.

Witmer, L. M. and Ridgely, R. C. 2009. New insights into the brain, braincase, and ear region of tyrannosaurs (Dinosauria, Theropoda), with implications for sensory organization and behavior. *The Anatomical Record*, **292**, 1266–1296.

TOPIC QUESTIONS

1. Describe the general features of theropods.
2. Describe how we know that running – cursoriality – was a key feature of theropod behavior.
3. What features of the theropod hand strongly suggest dexterity and grasping ability?
4. What kinds of evidence exist for what theropods ate?

5. Can a toothless animal be carnivorous?

6. Describe the range of skull and tooth design in theropods. How do those relate to our understanding of how theropods bit and killed?

7. What are the key clues that suggest that, as a group, theropods were carnivorous?

8. Highlight the evidence for pack-hunting by theropods.

9. Consider the design of ornithomimosaurs. Why is it difficult to explain the lifestyle of these animals?

10. We said that the most unambiguous features of a carnivorous lifestyle are seen in small- to medium-sized theropods. What features caused us to say this? Speculate on why this might be the case.

11. Theropod evolution was complex, so here is an exercise to help you better understand their relationships. Construct a single cladogram with the groups Theropoda, Coelophysoidea, Neoceratosauria, Tetanurae, Avetheropoda, Coelurosauria, Carnosauria, Maniraptora, Eumaniraptora, and Aves on it.

Theropoda II
Meet the theropods

7

CHAPTER OBJECTIVES

- Understand theropod diversity
- Refine our use of cladograms as tools for investigating evolution

The outline of theropod relationships

By now you know that Theropoda encompasses a *lot* of dinosaurs, some familiar and some not-so-familiar. In this chapter we'll try to sort out all of these different animals, by highlighting important groups and their relationships to each other. We begin with a simple framework onto which we can "hang" a bit of theropod diversity. As the complexity mounts, the cladogram is our friend; it helps us distill the welter of dinosaur names into a few large, simple, and connected groups. The basic relationships of these large groups are key, because these allow us to identify the dominant themes in theropod evolution.

Major events in theropod evolution

Figure 7.1 shows the core relationships of the major groups of theropods. We start with **Neotheropoda** (*neo* – new) – all theropods excepting some basal ones we met in Chapters 5 and 6.

Within neotheropods, there is a fundamental split between the Late Triassic *Coelophysis* and some near relatives on the one hand, and **Tetanurae** (*tetanus* – stiff; *uro* – tail). And within Tetanurae, we'll see yet another division between the megalosaur group (of whom we met *Spinosaurus* in Chapter 6), and **Avetheropoda** (*aves* – birds, denoting the fact that this group is on the evolutionary road to birds). Finally within Avetheropoda, we meet **Coelurosauria**, a group that includes all kinds of dinosaurs, and is as diverse *Tyrannosaurus* and birds. Here the cladogram maps out, simply, these basic relationships (Figure 7.1) Among other things, the phylogeny shows the non-intuitive fact that *Tyrannosaurus* and its relatives are not so distantly related to birds!

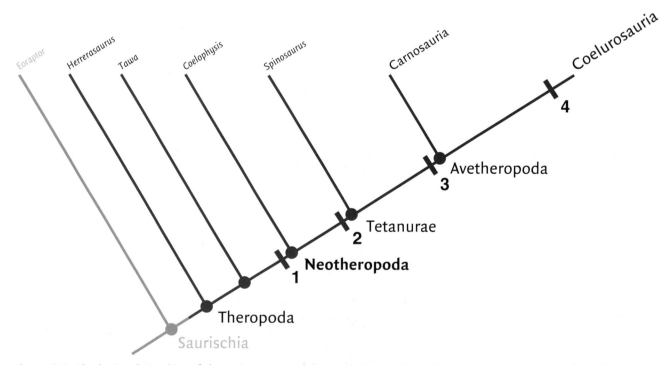

Figure 7.1. The basic relationships of the major groups of theropods. Again, these characters are taken largely from the work of theropod specialists Tom Holtz (2012) and Stephen Brusatte (2012; University of Maryland and University of Edinburgh, respectively): at **1**, furcula present; digit V on hand lost; $\geq$ 5 sacral (pelvic) vertebrae; foot functionally three-toed (tridactyl); at **2**, teeth restricted to the front of the jaws; large hands (proportionately); interlocking zygapophyses on tail vertebrae; maxillary fenestra (a second antorbital opening, anterior to the first); at **3**, three-fingered hand (loss of digit IV); complex chambers in vertebrae suggesting elaborate air sac system; at **4**, enlarged brain; long, narrow foot.

Neotheropoda

Neotheropoda is a well-diagnosed clade, among whose many characters are the development of the familiar "three-toed" theropod foot (four, actually, however digit I is very reduced and was evidently held above the ground, a bit like the "dew" claw on dogs). Another distinctive neotheropod character is the **furcula**, a single bone composed of the two clavicles (collar bones) fused together. The furcula is the familiar "wishbone" of chickens and turkeys. In *living* tetrapods, a furcula is known only in birds: and all living birds have a furcula. But the cladogram tells us that rather than being invented by birds, it is a carryover from the neotheropod ancestors of birds. Figure 7.1 also lists other neotheropod characters.

Famous neotheropods include the Late Triassic small biped *Coelophysis* (Figures 7.2a, 6.14d), the larger Early Jurassic *Dilophosaurus* (Figures 7.2b, 6.14c), and the large, stubby, carnivorous, south American abelisaurids such as *Carnotaurus* (Figures 7.2c, 6.14e), and their near relatives, the ceratosaurids (including most famously the Late Jurassic *Ceratosaurus*; Figure 7.2d), and of course tetanurans (see below).

Tetanurae

Linking the members of the great dinosaur clade Tetanurae – and the source of its name – is the possession of an inflexible tail: beyond its base, the tail simply doesn't bend as it does in other tetrapods. This is because the back (distal) half of the tail, towards the tip, is stiffened by interlocking **zygapophyses**, fore-and-aft projections from the neural arches (Figure 7.3). Recall that we saw an extreme example of this tendency in *Deinonychus* (Figure 6.24). The tail is an important counterbalance in theropod architecture (Chapter 6), and through the evolution of tetanuran functional design, there is a marked tendency to *decrease* its flexibility (except at its base).

It was in Tetanurae that some of the most remarkable and important theropod evolution took place. For starters, tetanurans evidently took breathing to new places. All saurischians, primitively, appear to have had an internal system of cavities in their skeletons, called **pleurocoels**, as well as small openings called **pneumatic foramina**. These same features are found in modern birds, and they augment the volume of air in their lungs by a complicated series of interconnected auxillary air sacs (see Box 7.1). The auxillary air sacs in modern birds reach (and expand into) the internal cavities via the pneumatic foramina. Such bone is called **pneumatic** in modern birds, and it would seem that tetanurans, too, had truly pneumatic bone. A basal tetanuran, *Aerosteon*, shows the openings (pneumatic foramina; see Glossary and Chapter 13) and hollowed bones indicative of a complexly branching suite of air sacs like those found in modern birds. These adaptations only increased in complexity as theropods evolved, so that highly derived theropods had (and have) some of the most efficient breathing ever invented. This doubtless enabled them to sustain high levels of activity, and led to a respiratory system capable of supporting the sustained flight of modern birds.

Tetanurans share a large number of other derived features (Figure 7.1), and the group includes a host of large theropods – including megalosaurids such as *Megalosaurus* and *Torvosaurus* (Figures 7.4a and b, respectively) and spinosaurids such as *Baryonyx* and *Spinosaurus* (see Figures 7.4c and d, respectively; also Chapter 6). Finally, Avetheropoda, our next major group, is also a member of Tetanurae.

Avetheropoda

Avetheropoda (*avis* – bird; a reference to bird-like features of many members of this group) is that group of theropods more closely related to birds than are the preceding genera and their near

Box 7.1　Every breath you take

In animals that pass air bidirectionally into and out of the lungs (that is, during inhalation and exhalation) like a bellows (for example, as in mammals), the trachea creates physiological dead space: some portion of the inhaled air never reaches the lungs. It is simply brought into the respiratory system and returned without being involved in oxygen–carbon dioxide exchange. In animals with long necks, like giraffes, this can be a lot of air!

By contrast, birds have unidirectional air flow, in which when inhaled air passes into the lungs, nearly all the oxygen is absorbed into the bloodstream, and the now oxygen-depleted air is run through a series of air sacs around the lungs and back into the trachea for exhalation. Obviously, the avian system wrings more oxygen out of the air than the bidirectional, bellows-style lungs found in mammals (Figure B7.1.1). So unidirectional breathing is to be expected in animals that have highly active lifestyles, in which a need to maximize the efficiency of oxygen consumption is very important. Along with birds, crocodiles are the only living unidirectional breathers, and birds have a complex system of air sacs and other adaptations associated with maximizing the efficiency with which they consume oxygen (see Chapter 8).

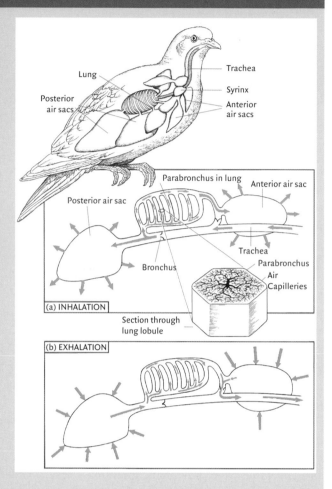

Figure B7.1.1. Unidirectional respiration, shown diagrammatically. As the animal inhales (a) air enters the trachea and posterior air sacs (here represented by a single sac), which expand. Air is then moved to the lungs, where it is deoxygenated, and then stored in the anterior air sacs (here represented by a single sac), which expand and fill with deoxygenated air. As the animal exhales (b), the deoxygenated air, in the lungs and anterior air sack, is expelled via contraction out of the trachea.

relatives. Avetheropods share many derived features, and consist of two clades, Carnosauria (*carn –* meat) and Coelurosauria (*coel –* hollow; Figure 7.1).

Carnosauria and Coelurosauria

There was a time when people thought of carnosaurs as sluggish, thuggish meat eaters, and coelurosaurs as delicate bird-like snipers. While cladograms show that the dichotomy between the two is real (see Figure 7.1), the terms now have rather different meanings.

Carnosaurs include some really large dinosaurs, such as the allosaur and carcharodontosaur groups. The well-known *Allosaurus* (Figure 7.5a) and *Sinraptor* (Figure 7.5b) are typical allosaurs; carcharodontosaurs are here represented by the truly gargantuan *Carcharodontosaurus* and *Giganotosaurus* (Figures 7.5c and 7.5d, respectively). At ~15 m, these last two are contenders for the title of World's Biggest Land Carnivore Ever.

Figure 7.2. Some representative (non-tetanuran) neotheropods. (a) *Coelophysis* (Late Triassic, Arizona, USA); (b) *Dilophosaurus* (Early Jurassic, Arizona, USA); (c) *Carnotaurus* (Early Cretaceous, Argentina); (d) *Ceratosaurus* (Late Jurassic, western USA). Human for scale.

Coelurosauria includes both small and large forms, most famously the tyrannosauroids. It also includes, as we shall see, some of the most interesting dinosaurs that ever lived. But coelurosaurs are sufficiently compelling that we'll give them their own separate section (with private cladogram) a few paragraphs down the road. For now, we take a moment to think about the Big Boys.[1]

---

[1] In a quirk of human psyche, even the most socially enlightened non-professionals commonly use the pronoun "he" when referring to large theropods. But of course this gender designation can only be right ~50% of the time.

Fins on posterior process

Figure **7.3.** Zygapophyses in tetanurans. Note how these processes extend across the adjacent vertebrae both anteriorly and posteriorly, hindering flexibility.

Convergent evolution in large theropods

Looking within Tetanurae, we see a striking quality of theropod evolution. Superficially, big theropods all resemble each other, and by their size and evident carnivorous habit, carnosaurs are the stuff of endless over-heated speculation, and in "Who's-the-biggest-baddest-dinosaur?"[2] comparisons, carcharodontosaurs such as *Giganotosaurus* are commonly contenders with the mighty *T. rex*.

Clearly, however, as theropods evolved to large sizes, lineages *independently* developed some of the same features. Such similar, although independent, evolution is called **convergent**. In the case of large theropods, features such as proportionally large heads and a tendency toward shorter arms occurred convergently. The same features occur *independently* in neoceratosaurs (for example, *Carnotaurus*), in primitive tetanurans (*Szechuanosaurus*, *Megalosaurus*, and *Afrovenator*), and in avetheropods (*Giganotosaurus*, *Allosaurus* and, again independently, in all the tyrannosauroids). Here, then, is the explanation of why the gigantic *Spinosaurus*, with its longish arms, appeared so aberrant to paleontologists!

Despite their superficially convergent morphology, all of these big-headed, short-armed dinosaurs likely behaved very differently from each other; as we have seen, some are lightly built – perhaps more cursorial – and others are ponderous and tank-like. To judge from the teeth and jaws, they likely attacked and killed their prey differently too. Why, then, did they independently develop similar gross morphologies?

The answer was once thought to reside in the logistics of growing really BIG. It was thought that to increase size yet maintain reasonable agility as bipeds, compromises had to be made in the size and power of various body parts, and it was assumed that shortened arms, reduced numbers of fingers, and large heads were part of such compromises. All that changed in 2009, however, with the discovery of the Early Cretaceous *Raptorex* from China: a 3 m tyrannosaurid about a fifth the length of the mighty *T. rex* (Figure 7.6)! Here was a small, Early Cretaceous tyrannosaur, indeed, a primitive member of the group, which had short arms and a large head. This suggested that these adaptations, once thought to be part and parcel of life as a generic large theropod, were, in the case of tyrannosaurs at least, really about life as a tyrannosaur. Yet here, even an early, relatively primitive member of the group still shows the typical *T. rex* morphology of an exceptionally large head and exceptionally short arms. The "why" of this particular morphology remains unanswered.

Coelurosauria

By almost any measure, coelurosaurs are among the most remarkable group of dinosaurs that ever lived. They are startlingly diverse: there were all kinds (hunters and herbivores); they came in all sizes

---

[2] Or, as Ricky Gervais' Derek might ask, "Who would win if…?" *Spinosaurus* might have been the longest; *Giganotosaurus* and *Carcharodontosaurus* might be longer than *T. rex*; *T. rex* might have been heavier than all of them; but the definitive answer to this perennial unanswerable question is that *T. rex* has the mightiest name.

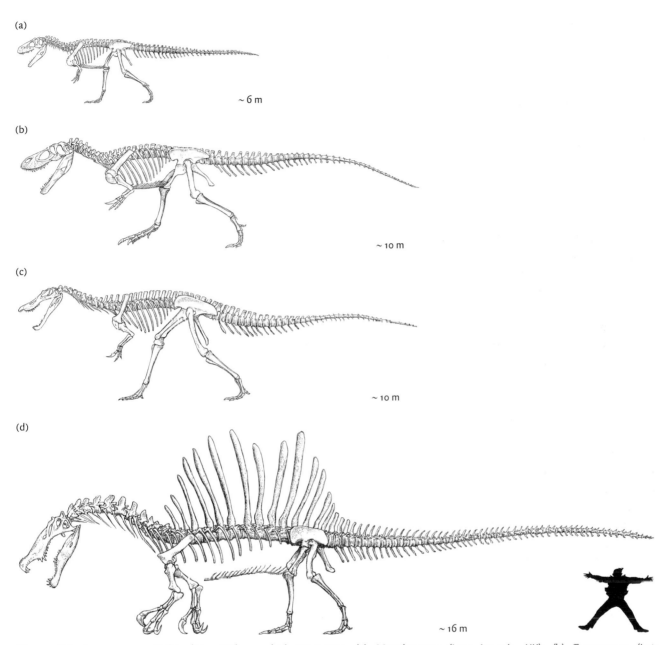

Figure 7.4. Some representative (non-avetheropodan) tetanurans. (a) *Megalosaurus* (Late Jurassic, UK); (b) *Torvosaurus* (Late Jurassic, western USA; Portugal); (c) *Baryonyx* (Early Cretaceous, western Europe); (d) *Spinosaurus* (mid-Cretaceous, northern Africa). Human for scale.

(tyrannosaurs to hummingbirds); many extraordinarily intelligent; some were hyper-carnivorous; and some… well, some still defy easy explanation. And of course out of this wellspring of genetic wealth, diversity and evolutionary ingenuity, come today's birds.

It was within Coelurosauria that the first obvious non-flying feathered dinosaurs were discovered, and by far the preponderance of feathered dinosaur specimens known to date are coelurosaurs. At this point, it would not be brash to state that the evidence suggests that *all* coelurosaurs, at a minimum, were feathered.

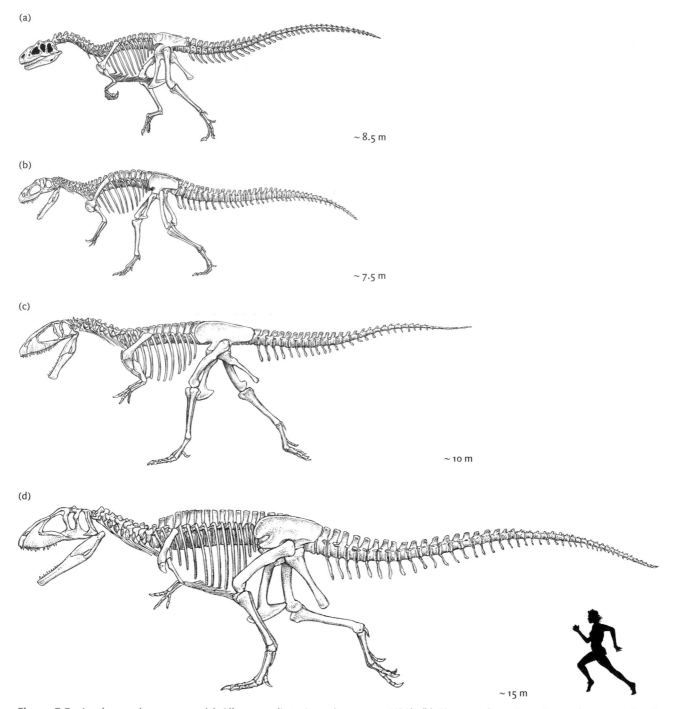

Figure 7.5. Avetheropod carnosaurs. (a) *Allosaurus* (Late Jurassic, western USA); (b) *Sinraptor* (Late Jurassic, northwestern China); (c) *Carcharodontosaurus* (mid-Cretaceous, North Africa); (d) *Giganotosaurus* (early Late Cretaceous, Argentina). Human for scale.

Because coelurosaurs are such an extraordinarily inclusive group, it has been tricky to diagnose them precisely. Paleontologist Stephen Brusatte, whose expertise is coelurosaur phylogeny, has proposed that the hallmark of all coelurosaurs in an enlarged brain.[3] An important character, and

---

[3] Other characters that he suggests include "an L-shaped quadratojugal without a prominent posterior process, an astragalus whose ascending process is separated from the condyles by a transverse groove or fossa, a jugal with a weakly concave orbital margin that is not deeply U-shaped, and a femoral fourth trochanter that is positioned near the center of the posterior

Figure 7.6. The early, primitive tyrannosaur *Raptorex* is dwarfed by the skeleton of *Tyrannosaurus rex*. Scale = 1.5 m.

one that as we shall see is critical in this group as evolution proceeded. The coelurosaurs we'll meet here include tyrannosaurs, diminutive compsognaths, alvarezsaurs; sloth-like therizinosaurs, oviraptors, deinonychosaurs, and avialans (Figure 7.7). Many of these we met in the preceding chapter as we discussed some of the things theropods do; now, we'll put all of them into their phylogenetic context.

Tyrannosauroidea

Tyrannosaurs are a rich clade, comprising ~20 known types. They first appeared in the Middle Jurassic, but stayed more or less medium-sized, under the semi-opposable thumbs, perhaps, of carnosaurs. With the disappearance of carnosaurs, tyrannosaurs flourished, and were around for the dinosaur extinction, 66 Ma.

Like so many dinosaurs, tyrannosauroids had an Asian and North American distribution. The earliest (and most primitive) forms are Asian, and throughout much of their time on Earth, they were largely Asian. At some point during the Late Cretaceous, however, they migrated (with many other kinds of dinosaurs (see Chapter 11), presumably across the Bering land bridge, to North America. There, life was evidently very congenial, and the classic huge North American tyrannosaurids evolved: *Albertosaurus*, *Dryptosaurus*, *Gorgosaurus*, and *Daspletosaurus*, sizeable (in the ~10 m range) creatures all. The largest tyrannosauroids were the very last: the end-Cretaceous, North American *T. rex*, and its smaller, but otherwise near-identical Asian twin, *Tarbosaurus*.

We would be remiss if we failed to mention the 2012 discovery of *Yutyrannus*, a ~9 m relatively primitive Early Cretaceous tyrannosauroid from China. *Yutyrannus* is coated in filamentous feathers (Figure 7.8); recall from Chapter 5 that these are an early stage in development of feathers. This is the basis for the suggestion that monofilamentous feather plumage likely graced that *über*-coelurosaur, *Tyrannosaurus*, whose image may require some readjustment for many of us.

surface of the shaft distally and extends proximomedially to become confluent with the posteromedial corner of the shaft proximally." Wow! These are a bit less accessible, so we'll stick with the enlarged brain. Brusatte, S. L., Lloyd, G. T., Wang, S. C., and Norell, M. A. 2014. Gradual assembly of avian body plan culminated in rapid rates of evolution across the dinosaur-bird transition. Current Biology, 24, 2386-2392.

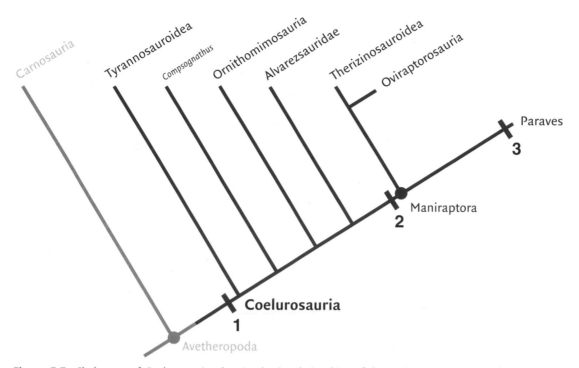

Figure 7.7. Cladogram of Coelurosauria, showing basic relationships of the major groups. At **1**, enlarged brain; narrow foot; at **2**, proportionately elongated forelimbs, bony sternum; at **3**, extremely long arms and hands, wings made up of multiple layers of quill-like (pennaceous) feathers, backwards facing pubis (**retroverted** pelvis), large retractable sickle claw on the second digit of the foot. Cladogram after Brusatte (2012).

Compsognathidae and Ornithomimosauria

Compsognathids were small (1–2 m), lightly built highly cursorial animals, also good candidates for the opening scene in *Jurassic Park II*. The bladed, recurved teeth suggest carnivory; however, insectivory is also a possibility.

Here, we'll highlight two compsognathids: the diminutive Late Jurassic *Compsognathus*, and *Sinosauropteryx*, from the Early Cretaceous of China. *Compsognathus* was found in Solnhofen, Bavaria (Germany), a very famous **Lagerstätte**, or locality of extraordinary preservation, which had been quarried for building stones since before the 1700s (Figure 7.9). *Sinosauropteryx* was found in another Lagerstätte, in Liaoning Province, in China. The fossil-rich rocks of the Early Cretaceous Jehol Group of Liaoning Province look superficially like those of Solnhofen. In both, preservation is spectacular; the specimens are generally complete *and* completely articulated, and the fine mudstones in which they are preserved show not only the impressions of the animals' coverings but also some darkened staining, possibly representing the original organic matter.

The 1997 discovery of the diminuitive *Sinosauropteryx* revealed a small compsognathid that obviously didn't fly. Yet it was covered with monofilamentous feathers, a thick coat covering a clearly non-flying theropod. The point was driven home: feathers were not originally about flight (Figure 7.10; Figure 6.30 shows another specimen of this dinosaur, highlighting color patterns in the plumage).

The highly cursorial ornithomimosaurs (*mimus* - mimics) were a group of ostrich-like dinosaurs that must have had a similar lifestyle. They include medium-sized runners like *Gallimimus*, *Struthiomimus*, and *Dromiceiomimus*, all of whom had long legs, small heads and large eyes, and were

Figure 7.8. Section of tail of *Yutyrannus*. Arrows indicate filamentous features interpreted to be (non-flight) feathers. From Liaoning Province, China.

toothless (Figure 7.11). As a group, they have been reasonably well understood for many years, although their precise diet had been questioned, because the large, powerful hands and claws (although these are not strongly curved, as one might expect in a carnivore) seem at odds with the small heads and toothless jaws; in at least one genus, straining though the beak was proposed. Most paleontologists now suspect a more or less herbivorous diet for these dinosaurs. Like the tyrannosaurids, they too have a Late Cretaceous Asian and North American distribution.

Figure 7.9. *Compsognathus*, the small Early Jurassic compsognathid from Solnhofen, Bavaria, Germany.

However, what we thought we knew about ornithomimds got upended in 1967 with *Deinocheirus* from the Gobi Desert of Mongolia. The late, legendary Polish paleontologist Zofia Kielen-Jawarowska describes her discovery of the first specimen (in a chapter fittingly titled "There ain't no such animal"), which consisted of only a pair of *huge* ornithomimid arms (and nothing else):

> While I was walking in the [Gobi] along the southern part of the outcropping, I suddenly noticed some fairly large bones showing on a small hill. There were more than a dozen of them...I began removing the sand from around the bones. Suddenly I saw there...an excellently preserved, powerful, strongly arched claw twelve inches long. It was undoubtedly a forelimb claw, but larger than any ever found before...the partly bare protruding bone looked like an arm bone; if this was actually the case it meant that I had come across a long forelimb bearing claws bigger than any ever seen before.
>
> Kielen-Jaworowska, Z. 1969. *Hunting For Dinosaurs*. MIT Press, Cambridge, MA, p. 141.

The arms, at longer than 2.5 m each, had massive hands with three elongate fingers, tipped with huge, powerful claws. If you blew up a conventionally sized ornithomimid to this size, the animal would be inexplicable. For the next 50 years, the arms hung on the walls of the Paleontological Museum in Ulan Baatar, Mongolia, a wake-up call for anybody who thought that we understand ornithomimid dinosaurs.

In 2009, tantalizing rumors began to circulate that much more of *Deinocheirus* had been found. But how much? And what did it look like? All rumors were put to rest in 2014, when a complete description was finally published. Instead of being a gargantuan fleet-footed monster ostrich mimic, *Deinocheirus* is hunch-backed, long-snouted, deep-jawed, with short hind legs: and absolutely *not*

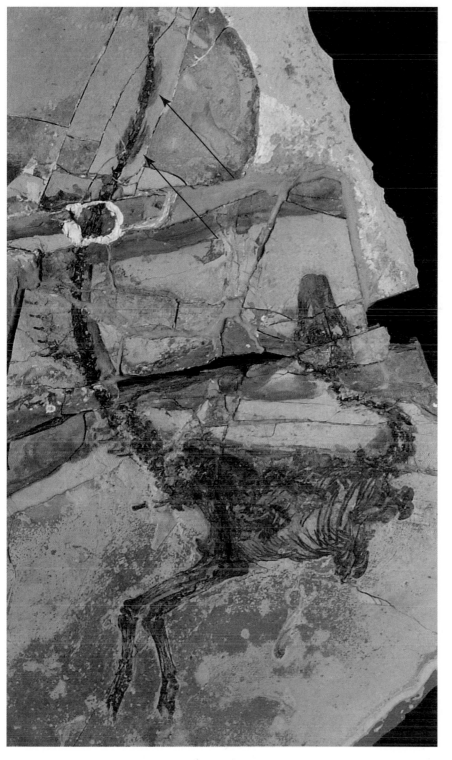

Figure 7.10. The feathered, non-flying compsognathid *Sinosauropteryx*. Filamentous features – interpreted to be early non-flying feathers – are preserved down the back of the animal, extending the length of the tail. From Liaoning Province, China.

fleetly cursorial! Moreover, >1000 gastroliths for grinding and fish remains were preserved in the stomach region – the thing ate fish?! The authors proposed that *Deinocheirus*, with its blunt-tipped feet and deep jaws, mucked around the rivers in which it lived, catching fish, and using those massive claws on the arms for digging plants, and its size for protection. They characterized it as a "megaomnivore." Figure 7.12 shows that you just can't make this stuff up.

Figure 7.11. Skeletons of an ornithomimid and an ostrich compared. (a) *Struthiomimus*; (b) ostrich. The similarity in design suggests similarities in lifestyle.

Figure 7.12. The skeleton of *Deinocheirus*, with a human drawn for scale.

Maniraptora

Maniraptora (*manus* – hand; *raptor* – stealer) includes everything else within Coelurosauria (see Figure 7.7). This is a grab bag of dinosaurs, from rare aberrant forms like therizinosaurs and alvarezsaurs, to oviraptorosaurs (we met these; see Figure 6.21) to the now-(in)famous and highly predaceous forms such as *Deinonychus* and *Velociraptor*, within Deinonychosauria. Chickens, penguins, sparrows, ostriches, and eagles (e.g., birds) as members of Avialae, are also found here. We've seen something of oviraptorosaurs and their nesting habits, as well as deinonychosaurs and their carnivorous habits in Chapter 6, so we'll devote more attention to the other groups, just touching on the origin of birds (which we'll address in greater detail in Chapter 8).

Maniraptorans as a group are identified by a number of characters (see Figure. 7.7), but a character than many share is the semi-lunate carpal, a wrist bone modification that increased manual dexterity and the ability to sever flesh from bone (Figure 7.13).

We begin with alvarezsaurs and therizinosaurs. Alvarezsaurs are a strange, uncommon group of dinosaurs, known from South and North America, and Asia. Epitomized by the *Mononykus* (Figure 7.14), they are not large (up to 2 m), but have strangely fused hands (convergent on those of a bird; see below), short arms, small teeth restricted to the front of the jaw, and were evidently feather-covered. What they did – and how they did it, remain enigmatic. The distinctive fused, short hands, however, were clearly for a particular function, and strategies from herbivory to termite-eating have been proposed.

Therizinosaurs are only slightly less enigmatic. Known from North America and Aisa, therizinosauroids represent yet another Cretaceous[4] theropod venture into – most likely – herbivory. Quite large and ponderous, these highly evolved and distinctive maniraptoran theropods have been compared to giant sloths, proposed as piscivorous, considered as tree climbers for wasp nests, and regarded as herbivores (Figure 7.15). Their behavior has thus been a matter of considerable confusion (as was, initially, their phylogenetic position), and the consensus was initially that they were an aberrant group of theropods; new finds, however, suggest that they were neither aberrant, nor sloth-like, nor wasp-eating tree climbers, but were likely – because stomach contents are to date unknown – herbivorous.

One thing that we do know about them is that they were feather covered – the kind of monofilamentous primitive feathers we saw in *Sinosauropteryx* (Figure 7.16).

Defying our old bugaboo "overall similarity" (see Chapter 3), the phylogenetically closest creatures to therizinosaurs were something that looked nothing like them: the very successful oviraptorisaurids (Figure 7.17; see also Figures 6.15b and 6.21). Oviraptorisaurids lived throughout the Cretaceous, started with teeth, but by the Late Cretaceous, were without teeth. They ranged in size from around ~1 m to ~8.5 m, and in estimated weight from ~5 kg to ~3000 kg (*Caenagnathasia* and *Gigantoraptor*, respectively). *Oviraptor*, who we met in Chapter 6, fits somewhere between these two extremes. These enigmatic (generally) toothless dinosaurs with their hollowed out skulls (Figure 6.21) and strong beaks were well equipped for running, as well as for doing something with powerful, grasping clawed hands.

Something... but what? There has been a lot of loose talk, some of it mutually exclusive, about invertebrate feeding in shallow waters, herbivory, scraping in crevices with the large claws, tree-climbing, prey catching, herbivory, omnivory, some of the above, all of the above, none of the

5 cm

Figure 7.13. Semi-lunate carpal (colored) in the left hand of *Ingenia*, an oviraptorosaurian maniraptoran.

---

[4] One form, *Eshanosaurus*, is from the Early Jurassic. The specimen consists of a left mandible only and some authors have referred it to basal Sauropodomorpha. Restudy by Paul Barrett in 2009 confirms it as a therizinosaur, albeit a very early one.

Figure 7.14. The skeleton of *Mononykus*, showing the unusual alvarezsaurid fused hands and short arms.

5 cm

1 m

Figure 7.15. The therizinosaur *Nothorhynchus* meets its skeleton. See also Figure 7.16.

above. . . for us, specialized functions for particular parts of the anatomy seem to be ruled out, given the size range of oviraptorisaurids – unless (obviously) different oviraptorsaurids did different things. Yet the unique skull implies a similar range of behaviors. At any rate, the slim gracile bones suggest a generally active, cursorial lifestyle.

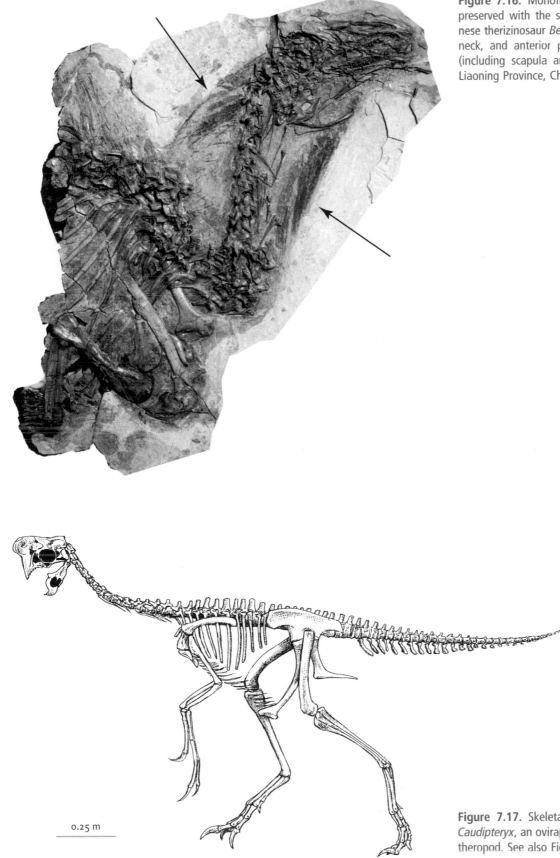

Figure 7.16. Monofilamentous feathers preserved with the skeleton of the Chinese therizinosaur *Beipiaosaurus* skull, neck, and anterior portion of rib cage (including scapula and humerus). From Liaoning Province, China.

0.25 m

Figure 7.17. Skeletal reconstruction of *Caudipteryx*, an oviraptorid maniraptoran theropod. See also Figure 7.18.

Figure 7.18. Feathers preserved in a specimen of the oviraptorid *Caudipteryx*. Arrows point to preserved feather fragments. From Liaoning Province, China.

Of one thing we can be confident: thanks to the basal oviraptorid *Caudipteryx*, from the Liaoning Lagerstätte, we can now be sure that they were covered with a colorful (?) monofilamentous feather coating (Figure 7.18).

Paraves

The last two groups of maniraptorans are Deinonychosauria, and a group at once more and less familiar: Avialae. By any measure, these constitute a startling clade of creatures – highly predaceous and crazily intelligent Paraves (Figure 7.19).

Deinonychosaurs rightly ought to evoke more fear than any large theropod, for they include the sickle-clawed troodontids and dromaeosaurids. Dromaeosaurs, as we saw in Chapter 6, were remarkably intelligent (see Box 13.4), hypercarnivorous, possibly social animals. Gracile bones, large claws, powerful hands, and a locked tail bespeak an active lifestyle and, potentially, "warm-bloodedness" (see Chapter 13). And with that kind of metabolism, it wouldn't be surprising to find some kind of insulation keeping the animals warm. And here, once again, the Early Cretaceous Liaoning lagerstäte

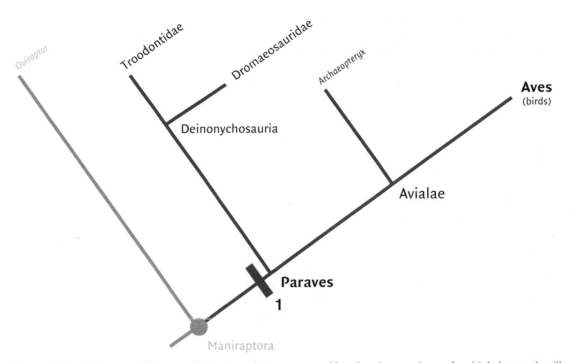

Figure 7.19. Cladogram of Paraves. At **1**, extremely long arms and hands, wings made up of multiple layers of quill-like (pennaceous) feathers, backwards facing pubis (retroverted pelvis), large retractable sickle claw on the second digit of the foot. Cladogram after Brusatte, 2012.

served up a feast of feisty, feathered carnivores. These include *Sinornithosaurus* (Figure 7.20), and *Microraptor* (Figure 7.21).

Microraptor is an extraordinary creature, with *flight* feathers (see Chapter 5) on the legs as well as the wings – not just the monofilamentous types of feathers we've seen previously – leading to considerable speculation about exactly how this "four-winged" dinosaur flew. Gliding was surely involved; and likely some kind of flapping flight. Regardless, of how it flew, however, it must have shimmered when it did: microscopic melanosomes preserved in the feathers suggest that it was iridescent.

Microraptor wasn't very large; it was less than 1 m long, and more than half of that was tail. By contrast, at least five other "four-winged" dromaeosaurids are known, including the very large *Changyuraptor*, whose tail feathers alone were up to 30 cm! These animals appear bird-like, but were not actually on the road to the evolution of birds. Rather, they seem to have been a unique radiation of gliding and/or flying dromaeosaurids.

Troodontids were another enigmatic group of animals, whose distinctive teeth (Figure 6.17c) caused them to be interpreted as ornithischians, as herbivores, and as *Deinonychus*-like carnivores. They too have the sickle-shaped claw seen in other dromaeosaurs; however, theirs is not as wickedly curved as those of deinonychosaurids, and may not have been held up off the ground as fully. Troodontids were nonetheless very large-brained, with stereoscopic vision, a rigid tail, and grasping hands. Troodontids, too, were feather covered, as clearly indicated by the Liaoning troodontid *Anchiornis* (Figure 7.22). But in this case, careful mapping of the distribution of color-indicating melanosomes preserved on the feathers offers us a startling glimpse of

Figure 7.20. Complete skeleton of the feathered dromaeosaur *Sinornithosaurus* from Liaoning Province, China. Arrows indicate monofilamentous feathers.

Figure 7.21. *Microraptor gui*, a "four-winged" dromaeosaur from Liaoning Province, China, with well-developed flight feathers present on all four limbs.

(a)   (b)

Figure 7.22. A beautifully preserved, complete specimen of the troodontid *Anchiornis* preserved between two slabs of claystone from Liaoning Province, China. Feathers are clearly visible. (a) Slab; feathers marked by white arrows; (b) counterslab.

Figure 7.23. A reconstruction of the feathers and coloration of the troodontid *Anchiornis*, based upon preserved melanosome size, shape, density, and distribution.

the original colors of the very extinct, very non-flying, feathered troodontid *Anchiornis* (Figure 7.23).

Now we're very close to the group Avialae (*aves* – birds), birds and their non-avian-theropod near relatives, many of whom evidently experimented with flight, as we have seen from *Microraptor* and its close kin. However, theropods continue to surprise us, and the 2015 discovery of *Yi qi* shows that learning to fly involved a little unexpected experimentation (Figure 7.24a, b) – as it perhaps did for humans as well! *Yi* is a theropod that has all the usual derived maniraptoran features that we've come to know: the large hands, the three fingers, feathers. But this one *also* has what appears to be a long,

Figure 7.24. The bat-winged theropod *Yi qi*. (a) Specimen; arrows labeled f point to feathery coat; arrow labeled m points to preserved membrane; arrow labeled b points to elongate bony support for wing membrane. (b) Reconstruction of *Yi qi*; and (c) two possible membrane reconstructions of *Yi qi*. Redrawn from *Nature*, doi:10.1038/nature14423.

thin extra bone in each hand. How that bone functioned remains a mystery; however, paleontologist Xu Xing and his team observed that *Yi* might developed a wing membrane – decidedly un-bird-like and more bat-like. Dinosaurs channeling bats? The remaining question is the extent of the membrane (Figure 7.24c), but *Yi* affords a glimpse into yet another unknown and completely unanticipated *cul de sac* of theropod evolution.

But at any rate, we've all seen birds, and so, with Avialae, things may get a little bit more familiar. These subjects are rightly deserving of their own chapter, so that's what we'll give them (Chapter 8).

SUMMARY

Theropods are a complex group, with a remarkable evolutionary history. The first major step in their evolution was the appearance of neotheropods; within these, *Coelophysis* and some related animals can be distinguished from tetanurans, whose zygapophysis-stiffened tails were used as dynamic counterbalances to grasping claws; it is among these that some of the most predatory dinosaurs reside. Tetanurans include a host of large theropods including *Megalosaurus*, *Spinosaurus*, and Avetheropoda, in which we see the venerable carnosaur–coelurosaur dichotomy. Carnosaurs include the North American *Allosaurus*, some even larger South American and African forms such as *Giganotosaurus* and *Carcharodontosaurus* (respectively); coelurosaurs include an astounding diversity of hypercarnivores, leggy herbivores, and possible omnivores. Coelurosaurs all bore feathers and colors and patterns are now known from some of them.

The many discoveries of feathered, non-flying theropods from the Early Cretaceous Liaoning Province of China blur the line between bird (such as we know them today) and non-bird. Clearly, although feathers are diagnostic for birds among living organisms, the Liaoning fossils demonstrate that presence of feathers does not guarantee that one is dealing with a bird. These discoveries reinforce the viewpoint that feathers had multiple uses, and were evidently invented as a form of insulation or perhaps display, only later to be co-opted for flight purposes.

SELECTED READINGS

Brusatte, S. L. 2012. *Dinosaur Paleobiology*. Wiley-Blackwell, New Jersey, 322 pp.

Holtz, T. 2012. Theropods. In Brett-Surman, M. K., Holtz, T., Jr., and Farlow, J. O. (eds.) *The Complete Dinosaur*. Indiana University Press, Bloomington, IN, pp. 347–378.

Hou, D., Hou, L., Zhang, L., and Xu, X. 2009. A pre-Archaeopteryx troodontid theropod with long feathers on the metatarsus. *Nature*, **461**, 640–643. doi:10.1038/nature08322.

Lamanna, M. C., Sues, H.-D., Schachner, E. R., and Lyson, T. R. 2014. A new large-bodied Oviraptorosaurian Theropod dinosaur from the Latest Cretaceous of western North America. *PLoS ONE*, **9**(3), e92022. doi:10.1371/journal.pone.0092022.

Li, Q., Gao, K-Q., Vinther, J., *et al.* 2010. Plumage color patterns of an extinct dinosaur. *Science Xpress*, www.sciencexpress.org / 5 February 2010 / Page 4 / 10.1126/science.1186290.

Xu, X., Tang, Z., and Wang, X. 1999. A therizinosaurid dinosaur with integumentary structures from China. *Nature*, **399**, 350–354.

Xu, X., Wang, X., and Wu, X. 2000. A dromaeosaurid dinosaur with a filamentous integument from the Yixian Formation of China. *Nature*, **401**, 262–266.

Xu, X., Zhou, Z., Wang, X., *et al.* 2003. Four-winged dinosaur from China. *Nature*, **421**, 335–340.

Zhou, Z.-H., Wang, X.-L., Zhang, F.-C., and Xu, X. 2000. Important features of Caudipteryx – evidence from two nearly complete new specimens. *Vertebrata Palasiatica*, **38**, 241–254.

TOPIC QUESTIONS

1. Describe some of the diversity of theropods. What are the different types, and what features characterize them?
2. What are the major formal groups of theropods; how are these distinct?
3. Given those groups, characterize the major events in theropod evolution.
3. In which groups do we have good evidence for feathers? How about the furcula?
4. Why is it that we no longer think feathers evolved for flight? Does this mean that they are unrelated to flight? Why?
5. The furcula and feathers are, in living animals, known only in birds. Does this mean that they evolved only once, in birds?
6. Which is more parsimonious – to say that feathers and the furcula evolved twice, once in modern birds and once in theropods; or to say that they evolved once, and thus birds must have a close relationship with theropods?
7. What are the possible original functions that feathers must have had?
8. If *Microraptor* and its close "four-winged" relatives are not directly on the main road to bird evolution – as we said – and if they evolved flight – as we said – *and* if birds are avialan theropods – as we said (!) – then how many times must flight have evolved in maniraptoran theropods? Please explain your answer.

Theropoda III
The origin and early evolution of birds

8

Q: Which came first, the chicken or the egg?
A: The egg. But it wasn't from a chicken.

CHAPTER OBJECTIVES

- Understand birds as theropod dinosaurs
- Learn about the origin of flight
- Learn about the evolutionary transition from maniraptoran dinosaurs to modern birds
- Gain familiarity with the diversity of Mesozoic avian theropods

Introduction

With Avialae, the step from non-avian dinosaurs to avian dinosaurs (birds) is a gentle one. As shown in Figure 8.1, we define Avialae as including *Passer* (again, representing living birds), *Archaeopteryx* (a derived maniraptoran theropod whom we shall meet next), and all the descendants of their most recent common ancestor. Figure 8.1 gives some diagnostic characters for Avialae.

Meet *Archaeopteryx*

Even with so many feathered theropod dinosaurs recently discovered, the road to the origin of birds still runs though the small, feathered, venerable paravian theropod *Archaeopteryx lithographica*. First discovered in the Upper Jurassic Solnhofen limestone quarry[1] as a feather in 1860 (Figure 8.2) and as bones shortly thereafter, the half-meter long dinosaur seemed chimeric for about 130 years, because it had "bird" features like feathers and a furcula, which co-existed with "reptilian" features, such as a tail and hands with claws (Figures 8.3 and 8.4).[2] Nothing else like it was known, and people were confused and got stuck in familiar, but unhelpful, Linnaean categories: was it a bird or a reptile? Because of its feathers and wings, most people considered it to be the world's oldest bird. It seemed somehow easy to derive familiar modern birds from the creature; less thought was devoted to *its* derivation.

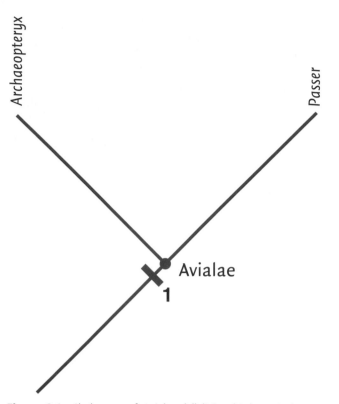

Figure 8.1. Cladogram of Avialae (all living birds + *Archaeopteryx*), again based upon the work of T. Holtz. Derived characters include: at **1**, fused scapula and coracoid; humerus longer than scapula; ulna longer than femur (again, a character indicating extremely long arms); ≤ caudal vertebrae. Multiple layers of pennaceous (quill-like) feathers *had been* a diagnostic character of avialans since the establishment of the group; however, it has become clear in the past seven years that these diagnose a group of theropods significantly more inclusive than Avialae; we put them at Paraves, but time may show them to be significantly more primitive than even this.

In the context of our long climb through Theropoda (Chapter 7), *Archaeopteryx* looks neither aberrant nor inexplicable; it doesn't even appear unusual. With its additional diagnostic avialan characters, it comfortably takes its place among the many feathered maniraptoran theropods that we've already seen. Indeed, of the eleven specimens of *Archaeopteryx* currently known, at least one was for many years believed to be a small theropod (which it is!) because the feather impressions were not well preserved on it and nobody *thought* to call it a bird!

Archaeopteryx – obviously – has well-preserved, unambiguous feather impressions (Figures 8.3 and 8.4). The wings show flight feathers that are indistinguishable from those of modern birds (see Figure 8.2), and the body was covered with symmetrical (non-flight) body feathers such as one might find on a modern bird. There were long feathers on the hind legs; these, however, shortened dramatically down the leg, and were not likely as much a key part of the flight

---

[1] The species name, *lithographica*, is a nod to the fine-grained limestones in which it was found; limestones of such extremely fine material that for hundreds of years they were used for engraving the stamps for detailed lithographic prints.

[2] Darwin had just published *On the Origin of Species* in 1859, proposing that species evolved into other species. Here, a mere two years later, was discovered an apparent "missing link" that mixed "reptilian" and avian features.

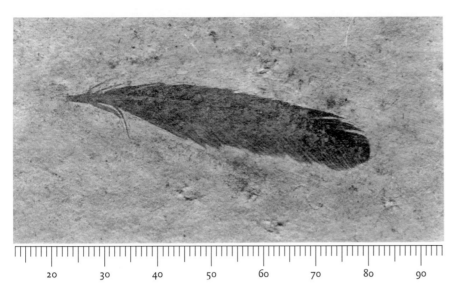

Figure 8.2. The first evidence for Jurassic-aged birds: the feather of *Archaeopteryx lithographica*, described in 1861, from the Solnhofen quarry in Bavaria (scale in centimeters).

apparatus as they were in the *Microraptor* group of "four winged" feathered theropods (Figure 7.20). The shape and distribution of melanosomes (see Chapter 7), suggests that the feathers covering *Archaeopteryx* were black.

Archaeopteryx isn't the only fossil non-Aves avialan known. There are a number of non-Aves avialans known, the relationships of which of which are shown in Figure 8.5. Along with these, some paleontologists have suggested that *Anchiornis* (see Figure 7.22a, b), although originally described to be a troodontid, is also an avialan; this underscores the subtlety of the differences among these clades of highly evolved theropods. Here, we showcase the undoubted avialan *Pedopenna*, which preserves beautiful feather impressions (Figure 8.6).

Although *Archaeopteryx* and other avialans are feathered and even possibly flew (see below), they definitely weren't quite modern birds (see Appendix 8.1 for a quick tune-up on the features of modern bird). Indeed, between basal avialans and modern birds, there are few significant morphological hurdles to traverse. Table 8.1 compares some of the differences and similarities between maniraptorans, basal avialans (like *Archaeotperyx*), and modern birds. It can be seen that for all their bird-like features, basal avialans maintain a host of primitive characters. They show a mix of maniraptoran and avian qualities that squarely place them as transitional between non-avian dinosaurs and modern birds.

Speed bumps

The story we've telling here – of birds as living theropod dinosaurs – is an impressive bit of evolutionary detective work and represents one of the best-known and most spectacular evolutionary transitions known. It didn't unfold without some hiccups along the way, and here we address some of the concerns that have been raised along the way, about the now unambiguous origin of birds.

Perhaps the most significant impediment to deriving birds from maniraptoran theropod dinosaurs was a dispute over the fingers involved in the carpometacarpus. Recall that the carpometacarpus of living birds is a fused structure composed of three fingers. Paleontology clearly tells us which three fingers these are: if Avialae is real, then the fingers must in fact be I, II, and III, since the fingers in avialan theropods, including *Archaeopteryx*, are I, II, and III.

But because the fingers fuse into the carpometacarpus as the hand forms during a living bird's development, embryologists since the 1870s have studied the hand of modern birds as the

Figure 8.3. The beautifully preserved, complete Berlin specimen of *Archaeopteryx*.

Figure 8.4. Berlin specimen of *Archaeopteryx* counterslab, preserving opposite side of specimen, primarily impressions. Note the exquisite feather impressions radiating out from the wings and tail.

carpometacarpus develops. They have repeatedly concluded that the fingers of the bird hand appear to be II, III, and IV. How could paleontology unambiguously identify the fingers as I, II, and III, when embryology identifies them as II, III, and IV? And, if the fingers are II, III, and IV as suggested by embryology, how could a bird come from a dinosaur that only had digits I, II, and III?

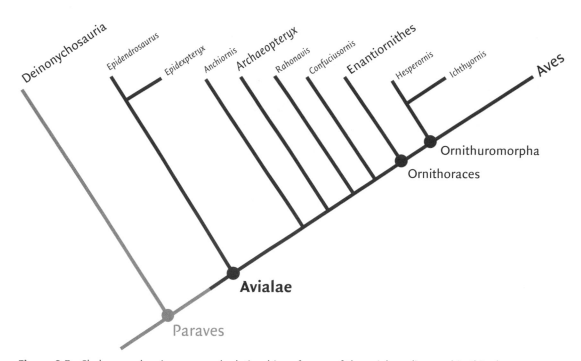

Figure 8.5. Cladogram showing proposed relationships of some of the avialans discussed in this chapter.

Moreover, at least one primitive theropod, a ceratosaur called *Limusaurus* (*Limus* – mud, or mire), may have reduced digits from both sides of the hand, in a *bilateral* reduction in digit size: both the "thumb" and "pinky" sides (digits I and V) become smaller. The remaining fingers of this animal thus seemed to point toward the embryological II, III, IV digit identification. More advanced theropods, of course, aren't built that way; the thumb is distinctive, with offset heads suggesting partial opposability, and the reduction and loss of digits is *unilateral*: from just the "pinky" side (digits IV and V), meaning that the remaining digits must be I, II, and III.

A possible solution to this conundrum was proposed by diapsid specialist J. A. Gauthier (see Figure 15.12) and developmental biologist G. P. Wagner. Embryologists identify the sequence of "*condensations*"; that is, early developed buds of material that later become fingers. As early growth occurs, embryologists thought they saw condensation I become digit I, condensation II become digit II, and so on. Gauthier and Wagner thought they saw something else, however, something that they called a "frameshift in the developmental identities" of the fingers. According to them, the bud considered to be embryological condensation II actually becomes adult digit I, condensation III actually becomes digit II, and condensation IV actually becomes digit III. More recent work now suggests that genes coding for the development of the original digit I cause its development in position II; and genes coding for the development of the original digit II cause its development in position III. The "frameshift" appears to be from the ancestral position to a new one.

And what about *Limusaurus*? The most probable explanation for its morphology is that its hand, rather than representing the ancestral condition for all theropods, was something of an evolutionary dead end. The future of theropod evolution – from Tetanurae onwards – appears to have been through a common ancestor that underwent a remarkable frameshift in the development of its hand.

With these new observations, the fingers identified in living birds and those in extinct theropods are the same. Wagner and Gauthier's work resolves the apparent discrepancy between embryology and paleontology, and reaffirms the avialan origin of birds.

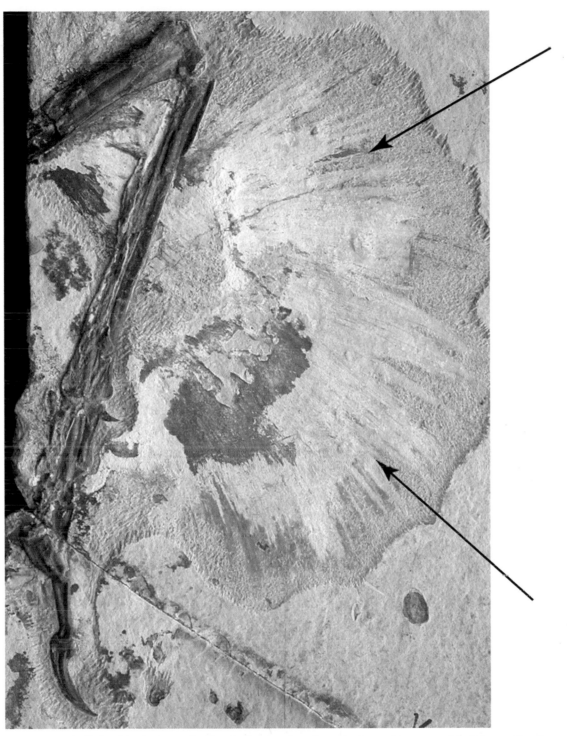

Figure 8.6. Hindlimb of *Pedopenna*, showing feather imprints. From the Lower Cretaceous Jehol Group, Liaoning Province, China. Arrows indicate feather impressions.

Another potential hiccup had to do with the feathers in theropods. Initially, it was uncertain whether or not the monofilmentous coverings seen, for example, in *Sinosauropteryx* and *Sinornithosaurus*, are anatomically the same thing as the vaned (pennaceous) feathers seen in

Table 8.1 Distribution of selected characters among maniraptoran theropods, *Archaeopteryx*, and modern birds (Neornithes).

Maniraptora	Avialae	Neornithes
Teeth (+)	Teeth (+)	Teeth (−)
Braincase slightly enlarged	Braincase slightly enlarged	Swollen braincase
Tail long, well developed	Tail long, well developed	Pygostyle (+)
Hand three-fingered; I, II, & III	Hand three-fingered; I, II, & III	Carpometacarpus (+); fused digits I, II, III
Legs: 1 Bipedal 2 Unfused foot	Legs: 1 Bipedal 2 Unfused foot	Legs: 1 Bipedal 2 Tarsometatarsus
Foot: 1 3 toes in front; 1 in back 2 Digit V (−) 3 Claws	Foot: 1 3 toes in front; 1 in back 2 Digit V (−) 3 Claws	Foot 1 3 toes in front; 1 in back 2 Digit V (−) 3 Claws
Hollow bones; some pneumatic	Pneumatic bones	Pneumatic bones
Furcula (wishbone)	Furcula (wishbone)	Furcula (wishbone)
Trunk not rigid 1 Sternum small; flat 2 Pelvis unfused 3 All vertebrae (+) 4 Flight adaptions (−)	Trunk not rigid 1 Sternum small; flat 2 Pelvis unfused 3 All vertebrae (+) 4 Partial flight adaptations	Rigidified trunk 1 Carinate sternum 2 Synsacrum 3 Some vertebrae (−) 4 Flight adaptations
Feathers (+)	Feathers (+)	Feathers (+)

The plus sign (+) indicates character present; the minus sign (−) indicates character absent.

fossils like *Microraptor*, *Anchiornis*, and *Archaeopteryx*. If they are some other anatomical structure, then their presence can hardly be used to evolutionarily link these non-flying theropods to birds. Now, however, the distinctive shared compound structure of both monofilamentous and vaned feathers and the recognition of identical melanosomes in both types of feather, as well as an improved understanding of the embryological of feather development, lead to the inescapable conclusion that both monofilamentous and vaned feathers are homologous.

Perhaps the least scientifically meaningful – but most frequently raised – concern about the theropod–bird hypothesis is one of timing. Most (but surely not all) of the known feathered non-avian theropods (many of which grace the pages of this book) come from the Jehol Group in Liaoning Province, and are Early Cretaceous in age. *Archaeopteryx*, on the other hand, is a Late Jurassic form. Because *Archaeopteryx* is older than the feathered Chinese dinosaurs, so the argument goes, the younger (Early Cretaceous) feathered fossils cannot have any bearing upon understanding the evolution that led to *Archaeopteryx*.

This, as we learned in Chapter 3, reflects a fundamental misunderstanding of cladograms. Because of the inconsistencies of preservation, the cladogram cannot incorporate time into its analysis in a way that is testable; there is no way to know whether an organism is not present in the fossil record because it didn't exist at that time, or because it simply was not preserved. As the old saying goes,

"Absence of evidence is not evidence of absence!"

The preservation of feathers is rare and requires most remarkable circumstances; it is a highly unlikely proposition – and certainly not testable – that the few times (and places) in the fossil record in which feathers are preserved are the first occurrence of such features. Recalling how we reconstruct phylogeny (again, Chapter 3), we look for the distributions of primitive and derived characters, and develop our phylogenetic hypotheses based upon these data. The Early Cretaceous age of the Liaoning feathered dinosaurs is thus all but irrelevant; the mix of primitive and derived characters that these dinosaurs possess is the key to understanding their relationships and reconstructing their evolutionary histories.

Lessons from history

In our modern world, many features of birds are unique; this gives the illusion that the evolution of Aves required a simultaneous suite of radical changes to go from "reptile" to "bird." However, when we look at theropods, it turns out that these apparently unique bird features are distributed throughout Theropoda – some even more primitive than this (see Chapter 6 and 7). So the point is that birds inherited a lot of features shared by many extinct theropods; the cladograms (Figures 7.1, 7.7, 7.19, and 8.1) show at which points during theropod evolution apparently "unique" features of modern birds actually appeared. The step from non-bird to bird is in fact so gradational, that it becomes difficult to exactly define when non-avian dinosaurs end and avian dinosaurs (birds) begin. Here are a few features of avialans (most unique to modern birds in our modern world); next to each, we've noted – *in blue* – the level in the hierarchy in which the character is diagnostic:

- Gastralia (belly ribs); found in *Archaeopteryx*, but lost in modern birds (Amniota);
- Bipedality (Dinosauromorpha);
- Mesotarsal ankle (Dinosauria);
- Feathers (Monofilamentous feathers may occur as early as Ornithodira; they certainly occur as early as Dinosauria, being present in Ornithischia (as we shall see in Part III) as well as Saurischia. Pennaceous (quill-shaped) feathers occur in maniraptoran theropods, long before flight was possible.);
- Thin-walled ("hollow") long bones (Saurischia);
- Blade-like, unserrated,[3] recurved teeth (Theropoda);
- Foot with three toes in front, and a fourth to the side with well-developed claws (Neotheropoda);
- Furcula (Neotheropoda);
- Semi-opposable thumb (Tetanurae; developing through Paraves);
- Hand with three, fully moveable, separate fingers, tipped with well-developed, recurved claws (Tetanurae);
- Elongate zygapophyses, meaning that the tail has little flexibility and has little potential for movement along its length (Tetanurae);
- Semi-lunate carpal (Maniraptora);
- Large tibia; small fibula thinning toward the ankle (Maniraptora; increasing dramatically through Paraves);

---

[3] The loss of serrations seems to have been an avialan character.

- Large brain (Coelurosauria);
- Rearward facing pubis (Paraves);
- Unidirectional breathing (now recognized in crocodylians; hollow bones go back as far as Saurischia, with elaborate pleurocoels, for air sacs, occurring in sauropods as well as theropods. In theropods, pneumatic bones with elaborate pleurocoels unquestionably go back to Avetheropoda; the inference is that the first avetheropod and all of its descendents, including birds, used avian-style unidirectional breathing).

Plainly, avialans inherited many features from their non-avian ancestors. Ironically, none of this is particularly new; the idea of birds-as-dinosaurs goes back as early as the discovery of *Archaeopteryx* in the mid 1800s (Box 8.1).

Box 8.1 Plus ça change...

The relationship of birds to dinosaurs as outlined here is not new. The famous early Darwinian advocate T. H. Huxley, as well as a variety of European natural scientists from the middle and late 1800s, recognized the connection between the two groups. Indeed, one did not have to be a Darwinian to recognize the important shared similarities, and Huxley's opinions were widely accepted at the time. As noted in 1986 by Yale's J. A. Gauthier (Figure 15.12), Huxley outlined 35 characters that he considered "evidence of the affinity between dinosaurian reptiles and birds," of which 17 are still considered valid today.

So what happened? Why is it news that birds are dinosaurs? During the very early part of the twentieth century, Huxley's ideas fell into some disfavor, as it was proposed that many of the features shared between birds and dinosaurs were due to convergent evolution.

What evidence was there to argue for convergence in the case of dinosaurs and birds? Really, not terribly much. But in light of the limited knowledge of dinosaurs at the time, the group just seemed too specialized to have given rise to birds. Moreover, clavicles were not known from theropods (then, as now, the leading contender as the most likely dinosaurian ancestor of birds). Thus, fused clavicles (furcula) in birds had to have originated outside Dinosauria. What was needed was a more primitive group of archosaurs that did not seem to be as specialized as the dinosaurs.

In the early part of the twentieth century, such a group of archosaurs, the ill-defined "Thecodontia" (see Chapters 5 and 15), was established by Danish anatomist G. Heilmann as the group from which all other archosaurs evolved. Since this was, by definition, the group that gave rise to all archosaurs, and since birds are clearly archosaurs, it was concluded that birds must have come from "thecodonts." Heilmann had in mind an ancestor such as *Ornithosuchus* (note the name: *ornitho* – bird; *suchus* – crocodile), a 1.5 m long carnivorous bipedal archosaur that, among living archosaurs, looks a bit like a long-legged crocodile. For over 50 years, Heilmann's detailed and well-argued analysis held sway over ideas about the origin of birds.

Several events caused the thecodont ancestry hypothesis to fall into general disfavor (see also Chapter 15). The first was that clavicles were found in coelurosaurians among theropods. Moreover, it came to be recognized that "Thecodontia" is not monophyletic; that is, it is diagnosed by no unique, diagnostic characters pertaining to all its members and no others. How could one derive birds (or anything else) from a group that had no diagnostic characters?

The renaissance of the dinosaur–bird connection must be credited to J. H. Ostrom of Yale University (Figure 15.9). In the early 1970s, through a series of painstakingly researched studies, he spectacularly documented the relationship between *Archaeopteryx* and dinosaurs, in particular coelurosaurian theropods. His ideas inspired R. T. Bakker and P. Galton (Figure 15.10), who in 1974 published a remarkable paper suggesting that birds should be included within a new vertebrate class: Dinosauria. The idea didn't catch on, in part because it involved controversial assumptions about dinosaur physiology and because the anatomical arguments on which it was constructed were not completely convincing. It may also have been just a bit too radical for its time.

In 1986, however, Gauthier applied cladistic methods to the origin of birds, and with well over a hundred characters demonstrated that *Archaeopteryx* (and hence, birds) is indeed a coelurosaurian dinosaur.

Stuff n' feathers!

A look at the modern biota, without understanding its evolutionary history (preserved in the fossil record) leads to some very erroneous ideas about the "why" of many features. For example, pneumatic bones and feathers are commonly singled out as marvelous adaptations to maintain lightness and permit flight. Well, they surely are marvelous, they clearly maintain avian lightness, and there is no doubt that feathers work well for flight. But did pneumatic bones and feathers evolve for lightness and flight, respectively?

In both cases, now that we have a sense of bird ancestry, the answer is a resounding "No!". Hollow bones are a saurischian character (even the names *Coelophysis* and *Coelurosauria* contain a reference to the hollow bones in these non-flying dinosaurs), and pneumaticity is likely related to efficient breathing (as implied above, we will see this again in immense, ponderous sauropods – Chapter 9 – which you can safely assume didn't fly). Avian pneumaticity likely developed long before there were things we would call birds. Modern birds inherited pneumaticity, unidirectional breathing, feathers, and a host of other "bird" features from non-avian theropods, even though these features were further refined in modern birds.

Most of us would say that the purpose of feathers is flight. We agree; feathers are used for flight, but we doubt that they evolved for that purpose, particularly as many non-flying, non-avian dinosaurs had them. This, in turn, provides a real insight into what the origin of feathers was all about. One striking feature of all feathered dinosaurs – flying or no – are skeletal designs suggesting very active, commonly predatory, lifestyles. With feathers most primitively appearing on non-flying theropods, we infer that feathers likely provided the insulation that is a prerequisite for warm-bloodedness, allowing theropods to maintain high levels of activity for extended periods of time; an attribute that was eventually used for flight (see Chapter 13). Feathers undoubtedly were useful in a cursorial, predatory lifestyle. It is perhaps not coincidence that the most highly evolved feathers are flight feathers; other feathers, such as monofilamentous feathers and down, likely represent earlier stages in feather development, and graced the coats of more primitive non-avian theropods.

But insulation for warm-bloodedness as the real reason for the origin of feathers doesn't completely satisfy, either, because many of the early feathers that we've seen are monofilamentous, and may not have insulated very well. But as we all know, modern birds are brightly colored, highly social, and visually acute. Moreover, we know that at least some non-avian theropods were highly social creatures (Chapter 6). We also know that ancient feathered theropods – non-flying and flying – had distinctive patterns and colored markings. Could feathers – at first – have been at least as much for display as for insulation?

Recall that in Chapter 5 we discussed the sensory importance of feathers. Here, then, is a hypothetical, but plausible, scenario for the evolution of feathers: widely distributed (across the skin), monofilamentous feathers could have first arisen for tactile sensation. Color might have appeared as a byproduct – or a even a driving force – of evolving theropod sociality, associated with denser packing of the feathers. Still more densely packed, multifilamentous (downy) feathers might then (or concurrently) have evolved as a means of insulation. And finally, as in Figure 5.16, pinnate, flight feathers may have ultimately evolved, associated with small size and a very active lifestyle. That is simply a scenario; however, what we know for sure, now, is that early feathers weren't about flight!

So what is a bird?

Clearly, the old equation (feathers = bird) won't fly; there are now many examples of feathered, non-flying dinosaurs *way* more primitive than Avialae on the cladogram. Likewise, the equation (warm-blooded = bird) also doesn't work; feathered, non-flying maniraptoran dinosaurs were very likely warm-blooded. After all, why insulate yourself from your heat source if you are cold-blooded? Should Aves – traditionally living birds – be restricted to all those organisms bearing the distinctive suite of characters of living birds? That would, of course, exclude *Archaeopteryx*, which certainly has a plausible claim on the designation "bird." Should the equation be bird = flight? Certainly, it is clear that flight is a quality that, at least primitively, characterizes all modern birds. Some have argued that the loss of teeth might be another strong diagnostic character for modern birds.

Origin of avian flight

Somewhere in paravans, flight evolved. But how? Two opposing hypothetical endpoints exist as regards the origin of bird flight (Figure 8.7). The first is the so-called **arboreal** (or "trees down") **hypothesis**: that bird flight originated by birds gliding down from trees (Figure 8.7a). In this hypothesis, gliding is a precursor to flapping (powered) flight; as birds became more and more skillful gliders, they extended their range and capability by developing powered flight. Perhaps flapping developed as a modification of the motions used in controlling flight paths.

Antithetical to the arboreal hypothesis is the **cursorial** (or "ground up") **hypothesis** for the origin of flight (Figure 8.7b). The cursorial hypothesis states that bird flight originated by an ancestral bird running along the ground. In this scenario, perhaps as obstacles were avoided, the animal became briefly airborne. Flapping (powered) flight appeared early on, as the animal strove to overcome more fully the force of gravity. This idea obviously requires a highly cursorial, feathered ancestor – the

Figure 8.7. The arboreal and cursorial hypotheses for the origin of bird flight. (a) The arboreal hypothesis, which suggests that bird flight evolved by "birds" gliding down from trees. (b) The cursorial hypothesis, which suggests that bird flight evolved by "birds" running along the ground until the animals became airborne.

evidence for which is abundant. In this hypothesis, the legs, feet, and hands of birds are viewed as an inheritance from a cursorial maniraptoran ancestry.

Towards a model for the origin of flight in birds

The arboreal hypothesis is intuitively appealing, and getting airborne is easy. On the other hand, the cursorial hypothesis is strongly supported because ultimately the ancestor of birds had to have been a cursorial creature.

A problem with the cursorial hypothesis is that it has so far proven nearly insurmountable to model a cursorial theropod that developed flight by running along the ground. For this reason, an arboreal stage intermediate in the development of flight has been attractive to many scientists. Yet indications of a cursorial heritage are present in all living birds as, indeed, their limbs are little changed from the non-flying coelurosaurian condition. However, could a group of small, once cursorial maniraptorans have developed an arboreal lifestyle?

Recently an interesting compromise position has been proposed. Perhaps flapping wings helped early cursorial theropods to get a purchase on steep slopes, overhangs, or even tree trunks. From this it would not have been a big leap, as it were, to flapping flight.

Paleontologists Q. Li, M. J. Benton, M. S. Y. Lee, and a variety of colleagues have proposed that the process of becoming a bird must have involved several steps: the decoupling of the forelimbs from the hindlimbs (as the forelimbs became flight-dedicated structures), the appearance of flight feathers, a startling plunge in overall size, and the assumption of an arboreal lifestyle (likely made possible by the diminution in size). Certainly, the decoupling of the forelimbs and hindlimbs was well underway in early theropods, all of whom had inherited the primitive bipedal dinosaurian morphology. As regards some of these other events, several teams of paleontologists have attempted to gauge their rapidity and nature.

Some of it, unfortunately, is shrouded in obscurity; the oldest avialans, *Archaeopteryx*, *Epidendrosaurus*, and *Epidexipteryx*, were likely already arboreal by Late Jurassic time, suggesting that much of the key evolutionary steps outlined above took place earlier – perhaps beginning as early as the Early Jurassic. The drop in size was certainly precipitous: Benton and colleagues demonstrated that paravans shrank dramatically before the advent of true birds, the size decrease picking up dramatically around 178 Ma.

Important changes in body proportions accompanied the decreases in size. The most obvious of these is visible from the cladogram: the increase in forelimb length relative to body length. The rapid changes in size and proportion of the bird lineage are shown in Figure 8.8.

In the case of feathers, P. Godefroit and colleagues estimated that feathers must first have evolved at least 50 million years earlier than *Archaeopteryx*; with *Archaeopteryx* dated at ~150 Ma, that would bring the origin of feathers into the Early Jurassic or even Late Triassic!

The story of the origin of flight reminds us of a fundamental property of evolution. Structures are not commonly invented wholesale in evolution. Evolution modifies existing structures. Here, the feathers and grasping arms of warm-blooded, colorful, non-avialan paravans were modified – remarkably little – to permit flight. It was a breathtaking evolutionary achievement.

Mesozoic birds

Archaeopteryx, if one would term it a "bird" at all, was very primitive and conveniently among the oldest known as well; it may serve to give a sense of what primitive Mesozoic birds were like. *Archaeopteryx* had many features that are far from the condition found in living birds, including

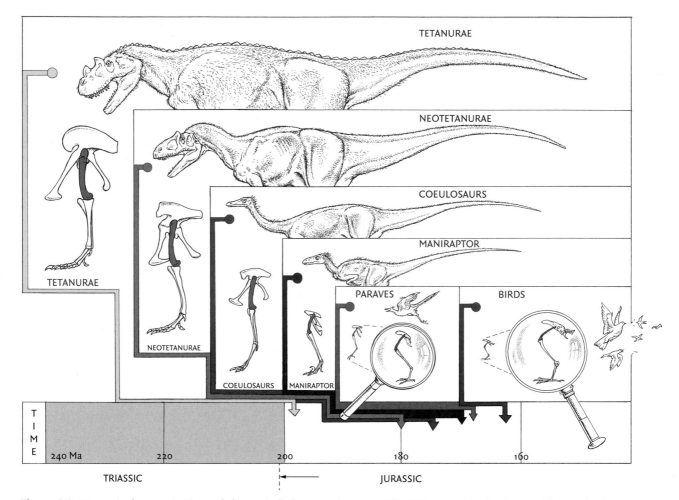

Figure 8.8. Dramatic decrease in size and change in limb proportions over 50 milllion years, in the theropod lineage leading to Aves. Here, the measured length of the femur (thigh) represents body size (here also represented by inferred weight). These are plotted against time, to show the dramatic decrease in body size and proportion that took place during in the theropod lineages leading to birds, during the first part of the Jurassic. From Lee, M. S. Y., Cau, A., Naish, D., and Dyke, G. J. 2014. Sustained minaturization and anatomical innovation in the dinosaurian ancestors of birds. *Science*, **345**, 562–566.

teeth, an unfused hand, a bony tail, no synsacrum, and gastralia. And as it turns out, there is much about its physiology as well that would also be unfamiliar to a modern ornithologist. Indeed, recent work by G. Erickson and colleagues suggests that it grew much more slowly than modern birds grow, although its growth rates appear to have been comparable to similarly sized maniraptoran theropods (Figure 8.9). So physiologically, compared to its near relatives, it was not unusual, although next to a modern bird it was quite primitive.

Archaeopteryx and its close theropod relatives had another unexpected feature as well: a well-developed sense of smell. Paleontologist D. Zelenitsky and colleagues used CT-scans to study the relative size of the olfactory bulbs in 157 non-avian theropods, fossil birds, and living birds. The relative size of the bulb was used to infer smelling capability: the proportionately larger the bulb, the better the sense of smell. While modern birds are characterized by relatively weak olfaction, their antecedents were not: the relative size of the olfactory bulbs – and the inferred olfactory acuity – increased from primitive maniraptorans through *Archaeopteryx* and its near relatives, and into early birds. Only later, when modern groups of birds became ascendent, did the sense of smell begin to diminish.

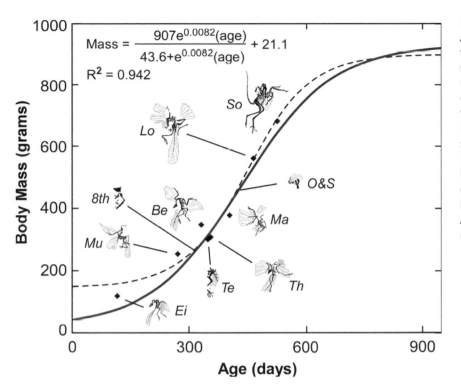

$$\text{Mass} = \frac{907e^{0.0082}(\text{age})}{43.6+e^{0.0082}(\text{age})} + 21.1$$

$$R^2 = 0.942$$

Figure 8.9. Inferred growth rates for *Archaeopteryx*. On the horizontal axis is time in days; on the vertical axis is presumed body mass. The curve is based upon eight specimens; growth was calculated from femur length. Growth rates were calculated to be between those of living birds (which themselves span a significant range of growth rates) and living reptiles. From Erickson, G. M., Rauhut, O. W. M., Zhou, Z., *et al.* 2009. Was dinosaurian physiology inherited by birds? Reconciling slow growth rates in *Archaeopteryx*. *PLoS ONE*, **4**(10), e7390. doi:10.371/journal.pone.0007390.

How and when did the other changes take place that distinguish living birds from *Archaeopteryx*? Here our interest will be within Avialae, the clade that includes *Archaeopteryx*, Aves (living birds), and everything in between.

The Mesozoic avialary

Within Avialae, very close to *Archaeopteryx* is *Rahonavis* from the Late Cretaceous of Madagascar. We've opted to emphasize its avian features in tentatively considering it more derived than *Archaeopteryx*. Slightly larger than *Archaeopteryx* (the size of a crow; Figure 8.10), recent work places *Rahonavis* as a dromaeosaurid theropod, but its position above or below *Archaeopteryx* remains uncertain. It possessed an enlarged sickle-shaped claw on its feet (similar to that of dromaeosaurids and troodontids), and a long, *Archaeopteryx*-like tail. Younger than *Archaeopteryx* by 25 million years, it had forward-looking features such as pneumatic foramina leading into pleurocoels in its thoracic vertebrae, which as we've seen implies unidirectional breathing and possibly a more efficient metabolism (see Appendix 8.1 and Chapter 12), along with a series of other bird-like characters (Figure 8.11). Time and further analyses will eventually consolidate its position.

Flight proficiency seems to have been a driving force in avialan evolution. Subsequent events included the formation of a pygostyle as well as the development of the synsacrum and other features for a rigid trunk, all of which contribute to the efficient flight that characterizes modern birds. Confuciusornithidae, a group containing *Confuciusornis* (Figure 8.12) and *Changchengornis* from the Early Cretaceous of China, is characterized by these relatively modern features (see Figure 8.11).

Thereafter, birds reduced the number of trunk vertebrae, altered the shoulder joint, and fused the digits of the hand into a carpometacarpus, among other flight-related features. All birds with these transformations are united within **Ornithothoraces** (*ornith* – bird; *thora* – thorax, or chest).

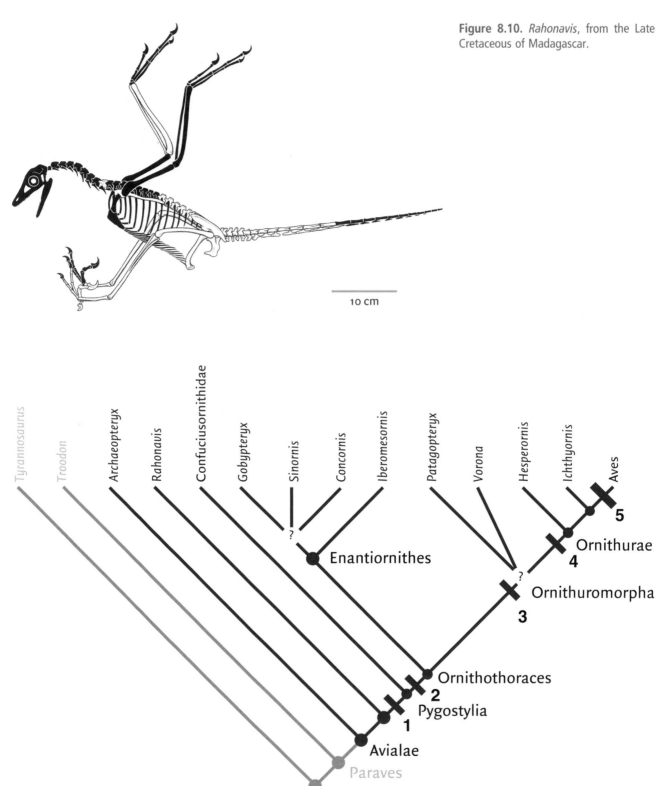

Figure 8.10. *Rahonavis*, from the Late Cretaceous of Madagascar.

10 cm

Figure 8.11. Cladogram of Mesozoic birds, depicting some of the steps in early avialan evolution. Derived characters include: at **1**, pygostyle; at **2**, reduction in number of trunk vertebrae, flexible furcula, strut-like coracoid, alula, carpometacarpus, fully folding wings; at **3**, further reduction in number of trunk vertebrae, loss of gastralia, final rotation of pubis to lie parallel with ilium and ischium; at **4**, reorientation of pubis to lie parallel to ilium and ischium, reduction of number of trunk vertebrae, decrease in size of acetabulum, patellar groove on femur; at **5**, loss of teeth.

Ornithothoracans have two main divisions: on the one hand, **Enantiornithes** (*enant* – opposite), and on the other **Ornithuromorpha** (*ornith* – bird; *uro* – tail; *morpha* – form), the lineage leading to Aves (see Figure 8.11).

Enantiornithes. Enantiornithes ("opposite bird") are all sparrow-sized, small, arboreal birds (Figure 8.13) that had well-developed flight capabilities. They were the most diverse clade of Mesozoic birds, yet all went extinct before the close of the Era.

Enantiornithes modified the wrist joint to allow, for the first time, the wing to fold tightly against the body, and developed adaptations indicative of perching: the first toe is positioned fully opposite to the others. The perching foot is a clue that these Mesozoic birds lived in trees, reinforcing other evidence that flight was an integral part of their life habits.

Still, one would hardly call Enantiornithes modern birds – they had gastralia across their belly (a carry-over from the primitive archosaurian condition), relatively numerous back vertebrae (a number intermediate between the 13 or 14 found in *Archaeopteryx* and the 4–6 found in living birds), an unfused tarsometatarsus, and an unfused pelvis.

Enantiornithes apparently had a worldwide distribution. *Nanantius* is from Australia, *Iberomesornis* hails from Spain, and *Sinornis*

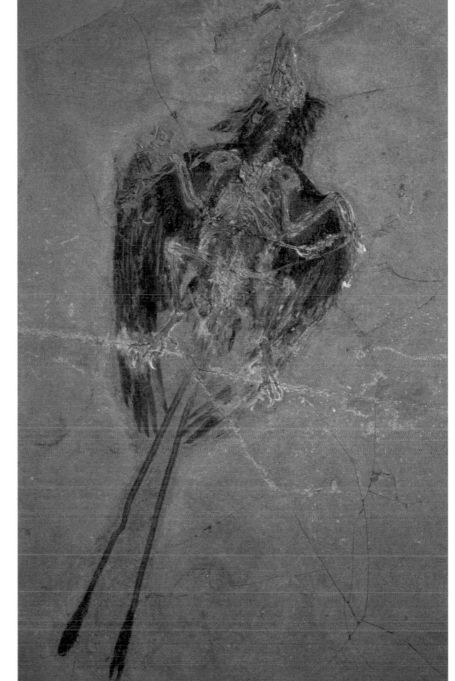

Figure 8.12. *Confuciusornis,* from the Early Cretaceous of Liaoning Province, China.

comes from China. Others include *Kizylkumavis* and *Sazavis* from Uzbekistan (Asia), *Alexornis* from Mexico, *Enantiornis* itself, and *Avisaurus*, known from both Argentina and the USA.

Ornithomorpha. Returning to the line leading to Aves, Ornithomorpha is represented by *Vorona* from the Late Cretaceous of Madagascar, *Patagopteryx* (Figure 8.14) from the Late Cretaceous of Argentina, and all remaining birds, the clade known as **Ornithurae** (*ornith* – bird; *uro* - tail).

Ornithurae is one of the most robust of all avialan clades, united by at least 15 unambiguous characters (see Figure 8.11). Not surprisingly, ornithurans include not only the closest relatives to living birds (Hesperorniformes and Ichthyorniformes), but also Aves (modern birds) as well.

Hesperornithiform (*hesper* – western) birds were a monophyletic clade of large, long-necked, flightless, diving birds that used their feet to propel themselves, much like modern loons or cormorants (Figure 8.15). They had highly reduced arms and developed powerful hindlimbs for propulsion. The hindlimbs were oriented to the side of the creature, and could not be brought under the body. For this reason, locomotion on land must have been a kind of seal-like waddling (at best).

In the water, on the other hand, hesperornithiform birds were supremely adapted for marine life. The long, flexible neck must have been useful in catching fish, a behavior indicated from coprolites preserved with their skeletons. In many respects, the group is quite close to modern birds, yet they retained teeth in their jaws. Like modern diving birds, some of the pneumaticity in the bones was lost. Presumably because strength in the chest and shoulder regions for flight was no longer needed, the furcula, coracoid, and forelimbs were highly reduced.

All of these adaptions indicate that this group had a long evolutionary history prior to their appearance in the Late Cretaceous. Hesperornithiforms include *Enaliornis* from the Early Cretaceous of England, *Hesperornis* (Figure 8.15) and its smaller relative *Baptornis*, and *Parahesperornis*, all from North America.

Closer related yet to Aves were the toothed Ichthyornithiformes (Figure 8.16; see Figure 8.11). Unlike hersperornithiforms, ichthyornithiforms were excellent flyers. *Ichthyornis*, from the Late Cretaceous of North America, had a massive keeled sternum and an extremely large deltoid crest on the humerus that was probably an adaptation for powerful flight musculature. In other respects, it shared many of the adaptations of modern birds, including a shortened,

3 cm

Figure 8.13. *Iberomesornis*, a member of Enantiornithes from the Early Cretaceous of Spain.

Figure 8.14. *Patagopteryx*, from the Late Cretaceous of Argentina.

fused trunk, a carpometacarpus, a pygostyle, a completely fused tarsometatarsus, and a synsacrum formed of 10 or more fused vertebrae. Found exclusively in marine deposits, ichthyornithiforms must have been rather like Mesozoic seagulls – but with teeth.

Figure 8.15. *Hesperornis*, the diving bird from the Late Cretaceous of the USA.

Figure 8.16. *Ichthyornis*, a gull-like bird from the Late Cretaceous of the USA.

Evolution of Aves

Getting to be a modern bird

For all of its limitations, the Mesozoic record provides for us insights into the transition from primitive theropods, through eumaniraptorans, to Avialae (including *Archaeopteryx*), and finally to Aves. The first part of the long evolutionary sequence – the part that ran from primitive theropods to avialans – was detailed in Chapter 7. The second part, the evolution of Aves from the primitive avialan condition, represented by *Archaeopteryx*, through the remarkable avian discoveries recently made in China, can be read from the cladogram in Figure 8.11.

Our best understanding of the sequence of evolutionary events is as follows:

(1) Development of the perching adaptation in the foot, in combination with limited flapping flight capabilities.
(2) Development of a pygostyle.
(3) Reduction in number of trunk vertebrae; and development of a flexible furcula, strut-like coracoid, carpometacarpus, and fully folding wings.
(4) Further reduction in the number of trunk vertebrae, loss of gastralia, final rotation of pubis to lie parallel with ilium and ischium.
(5) Reduction of number of trunk vertebrae, decrease in size of acetabulum, formation of **patellar groove**, a groove at the distal end of the femur to accommodate the patella (knee cap).
(6) Finally, loss of teeth (Aves).

Box 8.2   Molecular evolution and the origin of Aves

In this book, we've emphasized the fossil record as the means of telling when events occurred, mainly because this book deals with extinct organisms, the only record of which has historically been the fossil record. But in those cases in which we are dealing with living organisms, a whole different type of technique is available for study: **molecular evolution**.

Molecular evolution involves measuring the timing of molecular changes. So, for example, take two somewhat closely related living organisms, A and B. Now, choose a particular protein that they share, say, serum albumin (a protein in their blood). Their serum albumins might be quite closely related, but if some time has elapsed since A and B last shared a common ancestor, the exact molecules may have evolved and now differ slightly, in either form or composition. If we knew the rate at which the molecules diverged, we would know how distantly in the past A and B shared a single common ancestor (whose serum albumin composition and form were the ancestral ones). This very technique (and indeed this very molecule) was used in the case of humans and chimpanzees to show that the two shared a most recent common ancestor only ~5 million years ago, instead of the 15 million that had been inferred from the geological record.

More recently, molecular biologists have been using a technique called **DNA hybridization**. This technique works similarly to the one described above for proteins, except that it compares two strands of DNA (instead of proteins). In the same species, the strands of DNA should be virtually identical. DNA hybridization allows molecular biologists to measure the differences between the two strands. Knowing what rate **substitutions** (or changes) occur in the DNA allows us to calculate how long ago two different creatures shared identical DNA. That number should equal the time of divergence from a common ancestor.

And what of birds? Molecular estimates of the earliest Aves have consistently been somewhat earlier than has been inferred from the fossil record. Until very recently (see Figure 8.17), the fossil record seemed to suggest that the major radiation of birds took place *after* the extinction of non-avian dinosaurs. Yet, the fossil record of birds is, as we have seen, rather spotty, and perhaps most trustworthy only in its broadest outlines.

Estimates of the radiation of Aves, based on molecular data – primarily DNA hybridization – have put the time of the radiation well within the Cretaceous, *before* the boundary. How to resolve this contradiction?

Recently, the fossil record has begun to support the molecular record. . . a little bit. The fossil record of modern bird groups in the Cretaceous now includes the ancestral relatives of ducks, chickens, and large, flightless birds such as ostriches and emus (see text). Even with these newfound discoveries, however, whether or not the major radiations of birds took place before or after the K/T boundary remains unclear.

Molecular evidence has also been used in an entirely different context. A study published in April 2008 compared proteins from 21 different living creatures, including an alligator, an ostrich, a chicken, and two extinct creatures *T. rex* and a mammoth. The *Tyrannosaurus* proteins were types of collagen extracted from an unaltered femur (see Box 6.3). The results were unequivocal: the proteins showed that the mammoth and an elephant were phylogenetically close and, more important for our story, that the *Tyrannosaurus* and the birds were close – closer to each other than any is to an alligator.

Steps (1)–(5) occurred in the Mesozoic; step (6) may have occurred after the Mesozoic was over, because all Mesozoic avialans for which a skull is known, had teeth. Skulls are unfortunately not well known for Mesozoic Aves. Likewise, no Cenozoic toothed bird is known, although the fossil record of birds in the earliest Cenozoic is also very poor. Teeth – that final step in getting to be a modern-style bird – somehow got lost during the Mesozoic–Cenozoic transition (Box 8.2).

Aves

Aves (modern birds) is a well-supported clade, involving as many as 11 characters of the skull, pelvis, and ankle. Birds clearly suffered a major extinction when the non-avian dinosaurs went

Figure 8.17. Bones of the anseriform (modern bird clade including ducks, geese, and swans) *Vegavis*, from the Cretaceous of Antarctica. This fossil was the first body fossil evidence of modern a modern bird group co-existing with non-avian dinosaurs.

5 cm

extinct at the end of the Cretaceous (see Chapter 16). Certainly the toothed birds of the Mesozoic, some of which are described in this chapter, appear not to have made it into the new Era. But the data are beginning to suggest that modern groups birds – such as we've described here, in Appendix 8.1 – actually began their evolutionary explosion before the non-avian dinosaurs went extinct; that is, some familiar modern families of birds co-existed with non-avian dinosaurs.

Recent data – some biogeographic and some molecular (see Box 8.2) – suggest that the origins of modern bird families lie in the Cretaceous. Screamers and water-fowl (Anceriformes), loons (Gavii-formes), and possibly shorebirds such as sandpipers, gulls, and auks (Charadriiformes), landfowl (Galliformes), wing-propelled divers such as modern petrels (Procellariiformes), and parrots (Psitta-ciformes) all may have had Mesozoic origins.

The discovery of the bones of a small anseriform (water fowl), *Vegavis*, from the Cretaceous of Antarctica is physical evidence that modern groups of birds were alive and perhaps thriving along with non-avian dinosaurs (Figure 8.17). University of Texas avian paleontologist Julia Clarke (see Figure B15.9.7), and her colleagues, have suggested that ducks, chickens, and ratites (including emus and ostriches) were likely present during the latter days of non-avian dinosaurs. Clearly, these early records of modern birds speak, however incompletely, to the origin, initial radiation, and establishment of Aves in the closing moments of the Cretaceous.

SUMMARY

Birds *are* dinosaurs. We don't mean that they are *related* to dinosaurs – although, if they are dinosaurs, they must be related to them. We don't mean that they *come from* dinosaurs – although they obviously evolved from something that was itself a dinosaur. We mean that birds *are* dinosaurs, a statement that is no more radical than saying that humans are mammals.

Living birds have a suite of highly derived anatomical features that at first glance appear to be unique. In fact, a close look at Theropoda shows that most of the supposedly uniquely avian features of living birds are actually distributed thoughout theropods: cladistic analysis demonstrates that many presumed "bird" characteristics are distributed in non-bird theropods, particularly within Tetanurae and Eumaniraptora.

The discovery in Bavaria of small, Late Jurassic theropod, *Archaeopteryx lithographica*, and the many discoveries of feathered, non-flying theropods from the Early Cretaceous of Liaoning Province of China blur the line between bird and non-bird. Clearly, although feathers are diagnostic for birds among living organisms, these fossils all demonstrate that presence of feathers does not guarantee that one is dealing with a bird. These discoveries reinforce the viewpoint that feathers had multiple uses, and were evidently invented as a form of insulation or perhaps display, only later to be co-opted for flight purposes.

An important quality of Avialae appears to be flight. The origin of flight is shrouded in mystery; birds may have evolved flight by leaping down from trees (the "arboreal hypothesis") or alternatively, may have evolved flight by fast running (the "cursorial hypothesis"). For a variety of reasons, neither hypothesis is completely satisfactory, and the evolution of flight – and thus the final step to Avialae – likely was a multi-million-year process involving small size, arboreality, and the complete separation of the forelimbs and hindlimbs.

While *Archaeopteryx* was clearly important in identifying the relationship between Aves and dinosaurs, there were still a number of evolutionary steps to take before the anatomical condition seen in Aves was achieved. Despite their rarity as fossils, the fossil record of birds indicates the general order in which these evolutionary events occurred.

The improvement of flight capability was a driving force in post-*Archaeopteryx* bird evolution. Pneumatic foramina became better developed, along with, sequentially, the pygostyle, a reduction in the number of trunk vertebrae, modifications of the shoulder, and the development of the carpometacarpus. Still, some primitive characters such as gastralia, were retained. All these features were present in Cretaceous ornithothoracan birds, including the small, comparatively common Enantiornithes, and the line leading to Aves, Ornithuromorpha.

Within Ornithuromorpha, several highly derived birds evolved, notably the diving hesperornithiformes and the seagull-like ichthyornithiformes. These birds, for all their advancement, were not exactly like living birds, lacking a number of features diagnostic for Aves, including loss of teeth. The earliest fossil record of modern Aves is very fragmentary, but goes back into the Late Cretaceous.

SELECTED READINGS

Benton, M. J. 2014. How birds came to be birds. *Science*, **345**, 508–509.

Brusatte, S. L., Lloyd, G. T., Wang, S. C., and Norell, M. A. 2014. Gradual assembly of avian body plan culminated in rapid rates of evolution across the dinosaur–bird transition. *Current Biology*, **24**, 2386–2392.

Carney, R. M., Vinther, J., Shawkey, M. D., D'Alba, L., and Ackermann, J. 2012. New evidence on the colour and nature of the isolated *Archaeopteryx* feather. *Nature Communications*, DOI: 10.1038/ncomms1642.

Chiappe, L. M. 1995. The first 85 million years of avian evolution. *Nature*, **378**, 349–355.

Chiappe, L. M. and Dyke, G. J. 2002. The Mesozoic radiation of birds. *Annual Review of Ecology and Systematics*, **33**, 91–124.

Chiappe, L. M. and Dyke, G. J. 2007. The beginnings of birds: recent discoveries, ongoing arguments, and new directions. In Anderson, J. S. and Sues, H.-D. (eds.) *Major Transitions in Vertebrate Evolution*. Indiana University Press, Bloomington, IN, pp. 303–336.

Chiappe, L. M. and Witmer, L. M. (eds.) 2002. *Mesozoic Birds*. University of California Press, Berkeley, 520 pp.

Clarke, J., Tambussi, C. P., Noriega, J. I., Erickson J. M., and Ketcham, R. A. 2005. Definitive fossil evidence for the extant radiation of Aves in the Cretaceous. *Nature*, **433**, 305–308.

Cracraft, J. and Clarke, J. 2001. The basal clades of modern birds. In Gauthier, J. and Gall, L. F. (eds.) *New perspectives on the origin and early evolution of birds*. Proceedings of the International Symposium in Honor of John H. Ostrom, pp. 143–156.

Dial, K. 2003. Wing-assisted incline running and the evolution of flight. *Science*, **299**, 402–404.

Dingus, L. and Rowe, T. 1997. *The Mistaken Extinction*. W. H. Freeman and Company, New York, 332 pp.

Erickson, P. G. P., Anderson, C. L., Britton, T., *et al.* 2006. Diversification of Neoaves: integration of molecular sequence data and fossils. *Biology Letters*, **2**, 543–547.

Gauthier, J. A. 1986. Saurischian monophyly and the origin of birds. In Padian, K. (ed.), *The Origin of Birds and the Evolution of Flight*. California Academy of Sciences Memoir no. 8, pp. 1–56.

Gauthier, J. A. and Gall, L. F. (eds.) (2001). *New Perspectives on the Origin and Early Evolution of Birds*. Peabody Museum of Natural History, Yale University Press, New Haven, CT, 613 pp.

Foth, C., Tischlinger, H., and Rauhut, O. W. M. 2014. New specimen of *Archaeopteryx* provides insights into the evolution of pennaceous feathers. *Nature*, **511**, 79–82.

Hecht, M. K., Ostrom, J. H., Viohl, G., and Wellenhofer, P. (eds.) 1984. *The Beginnings of Birds*. Proceedings of the International Archaeopteryx Conference Eichstatt, Freunde des Jura–Museums Eichstätt, Willibaldsburg, 382 pp.

Ji, Q., Currie, P. J., Ji, S., and Norell, M. A. 1998. Two feathered dinosaurs from northeastern China. *Nature*, **399**, 350–354.

Ostrom, J. H. 1974. Archaeopteryx and the origin of flight. *Quarterly Review of Biology*, **49**, 27–47.

Ostrom, J. H. 1976. Archaeopteryx and the origin of birds. *Biological Journal of the Linnaean Society*, **8**, 91–182.

Prum, R. O. and Brush, A. H. 2003. Which came first, the feather or the bird? *Scientific American*, **288**, 84–93.

Schweitzer, M. H., Suo, Z., Avci, R., *et al.* 2007. Analyses of soft tissue from *Tyrannosaurus rex* suggest the presence of protein. *Science*, **316**, 277–280.

Shipman, P. 1998. *Taking Wing*. Simon and Schuster, New York, 336 pp.

Wagner, G. P. and Gauthier, J. A. 1999. 1, 2, 3 = 2, 3, 4: a solution to the problem of the homology of the digits in the avian hand. *Proceedings of the National Academy of Sciences*, **96**, 5111–5116.

Zhang, F., Kearns, S. L., Orr, P. J., *et al.* 2010. Fossilized melanosomes and the colour of Cretaceous dinosaurs and birds. *Nature*, **436**, 1075–1078.

TOPIC QUESTIONS

1. What are the diagnostic characters of modern birds that might be preserved in the fossil record?
2. Compare these characters to those present in non-avian theropods.
3. What characters point to maniraptoran theropods as good candidates for the ancestry of modern birds?
4. Why did the ancestor of birds have to have been cursorial if one goes back far enough?
5. Why is it that we no longer think feathers evolved for flight? Does this mean that they are unrelated to flight?
6. How did bird flight evolve?
7. What were the evolutionary steps from *Archaeopteryx* to Aves?
8. What were the evolutionary steps from basal tetanurans to *Archaeopteryx*?
9. If we were unable to resolve the discrepancy between the paleontological and the embryological identifications of the fingers in a bird's carpometacarpus, would this in any way affect the hypothesis that birds are dinosaurs? Why?
10. During bird evolution, how did the wing evolve? Was a whole new wing structure required, or were all the pieces – as well as the correct proportions – there already?
11. If *Archaeopteryx* had never been found, would we be able to tell that birds are dinosaurs? Support your answer.
12. What characters suggest that considerable evolution occurred before the appearance of hesperornithiform birds?
13. What are the features of ichthyornithiform birds that indicate powerful, efficient flight?
14. Formulate a thoughtful answer to the question "What is a bird?".

APPENDIX 8.1 WHAT MAKES A MODERN BIRD A MODERN BIRD?

Among living vertebrates, birds possess a remarkable and largely unique suite of diagnostic features (Figure A8.1.1). There are many; most of those discussed here are likely to be preserved in a fossil.

Feathers

All living birds have feathers – complex, distinctive structures that consist of a hollow, central shaft that decreases in diameter toward the tip. Radiating from the shaft are **barbs**, feather material that, when linked together along the length of the shaft by small hooks called **barbules**, form the sheet of feather material called the **vane** (Figure A8.1.1a). Feathers with well-developed, asymmetrical vanes are usually used for flight and are therefore called **flight feathers**. Feathers in which the barbules are not well developed tend to be puffy, with poorly developed vanes, and are called **down**; as we know from sleeping bags, comforters, and ski parkas, they are superb insulation.

Loss of teeth

No living bird has teeth. The jaws of birds are covered with a rhamphotheca (beak).

Large brains

Living birds have well-developed brains protected by a large braincase.

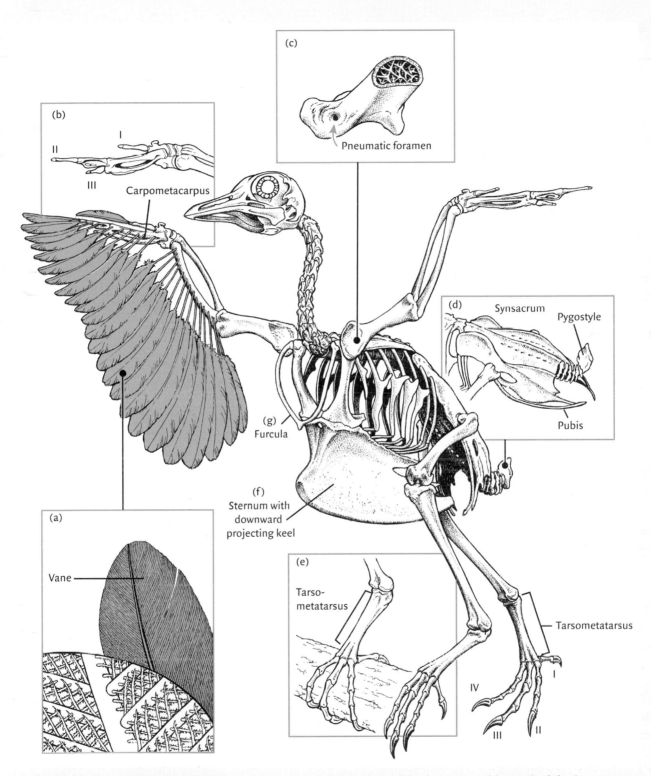

Figure A8.1.1. The skeleton of a pigeon, showing major features of its skeletal anatomy. (a) Detail of feather structure; (b) carpometacarpus with digits labeled; (c) hollow bone with pneumatic foramina; (d) synsacrum (fused pelvic bones) with pygostyle; (e) fused foot bones (tarsometatarsus); (f) sternum with large downward-projecting keel; and (g) furcula (fused clavicles).

Carpometacarpus

The wrist and hand bones in the hand of modern birds are fused into a unique structure called the **carpometacarpus**[4] (Figure A8.1.1b). The carpometacarpus is composed of three fused fingers, now generally thought to be digits I (the thumb), II, and III.

Legs and feet

Birds are fully bipedal, and have an erect stance (see Chapter 4). The twin shin bones (tibia and fibula; together, the "drumstick" on the dinner table) are unequal: the tibia is large, but the fibula thins to a sliver close to the ankle.

The feet of all living birds are clawed and have three toes in front (digits II, III, and IV), and a smaller toe (digit I) at the back. The three central metatarsals (foot bones, to which the toes attach; in this case II, III, and IV) are fused together and with some of the ankle bones, to form a unique structure called a **tarsometatarsus** (Figure A8.1.1e).

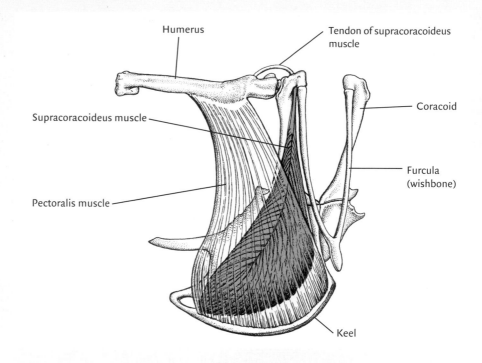

Figure A8.1.2. The two major muscles for flight: the pectoralis and the supracoracoideus. The pectoralis is the muscle used in the downward (power) stroke, while the supracoracoideus is used in the recovery stroke.

Labels: Humerus; Tendon of supracoracoideus muscle; Supracoracoideus muscle; Coracoid; Pectoralis muscle; Furcula (wishbone); Keel

Pygostyle

No living bird has a long tail skeleton. Instead, in most cases, the bones are fused into a compact, vestigial structure called a **pygostyle** (*pygo* – rump; *stylus* – stake; Figure A8.1.1d).

Pneumatic bones

Living birds breathe unidirectionally with a complex system of air sacs (see Box 7.1). Their bones are thus pneumatic and have pneumatic foramina (Figure A8.1.1c).

Rigid skeleton

Bird skeletons have undergone a series of bone reductions and fusions to produce a light, rigid platform to which the wings and the muscles that power them attach. Fused vertebrae in the back are connected with a well-developed breastbone, or sternum, by ribs with upper and lower segments. The sternum is large and, in flapping flyers, has a broad, deep **keel**, or downward-protruding bony sheet, for the attachment of flight muscles (Figure A8.1.1f). The pelvic region of the vertebral column is fused together into a **synsacrum**, a single structure consisting of

---

[4] Spiced and served with beer, we call them "buffalo wings"!

many sacral vertebrae fused together (Figure A8.1.1d). The pubis is very slender and points posteriorly.

In the shoulder, pillar-like coracoid bones buttress against the front of the sternum, the shoulder blade (scapula), and against the furcula (Figure A8.1.1g). No *living* organism except birds has a furcula, although of course they are present in all neotheropods (e.g., most!).

Flight musculature

In modern flying birds, the downward stroke of the wing is obtained by the **pectoralis** muscle, which attaches to the front of the coracoid and sternum, and to the furcula and humerus. The recovery stroke is carried out by the **supracoracoideus** muscle. The supracoracoideus attaches at the keel of the sternum, runs up along the side of the coracoid bone, and attaches via a tendon at the top of the upper arm bone through a hole (the **trioseal foramen**) formed by the coracoid, furcula, and scapula (Figure A8.1.2). This is an adaptation unique to living birds.

CHAPTER OBJECTIVES

- Introduce Sauropodomorpha
- Develop familiarity with current thinking about lifestyles and behaviors of sauropodomorphs
- Develop an understanding of sauropodomorph evolution using cladograms, and an understanding of the place of Sauropodomorpha within Dinosauria

Figure 9.1. *Diplodocus*, one of the best-known sauropodomorphs, from the Late Jurassic of the Western Interior of the USA.

Sauropodomorpha
The big, the bizarre, and the majestic

9

Sauropodomorpha
Living large

Brontosaurs were *extremely* large, not too bright, and extinct. Isn't that what dinosaurs are supposed to be all about?

But what about mighty, or majestic? Some of these dinosaurs pushed the extremes of terrestrial body size – to the tune of 75 000 kg and possibly more (Figure 9.1). And some of them were considerably smaller. As growing gargantuans, however, they taxed biomechanical and physiological design – weight support, neural circuitry, respiration, digestion, *everything* – to the limit. Viewed from that (biomechanical) perspective, sauropods were some of the most sophisticated animals that ever walked the face of the Earth.

Sauropodomorphs (*saurus* – lizard; *pod* – foot; *morpho* – form) lived for >160 million years, from the beginning of dinosaur history until its close. Over this long interval, sauropodomorphs managed to walk or be carried to every continent (Figure 9.2), and spawned well over a hundred different species. They definitely did *something* right.

Who are sauropodomorphs?

Sauropodomorpha is a well-diagnosed group of saurischian dinosaurs (Figure 9.3). The dinosaurs that look like *Brontosaurus* – Sauropoda – are but one part of Sauropodomorpha; the other consists of a relatively short-lived group of dinosaurs: basal sauropodomorphs called "prosauropods" (*pro* – before; see Figures 4.5 and 9.4).

Specialists have had some trouble figuring out what to do with "prosauropods." Originally, they were thought to be the ancestors of sauropods; then it was thought that they were a monophyletic sister group, but not ancestral; current thinking has them as a series of successively more derived, early sauropodomorphs (Figure 9.3a). Here we use quotation marks around the term to denote the fact that although we refer to basal sauropodomorphs as "prosauropods," the group is not monophyletic.

Figure 9.2. Global distribution of Sauropodomorpha. "Prosauropods" indicated by solid circles, Sauropoda indicated by open squares.

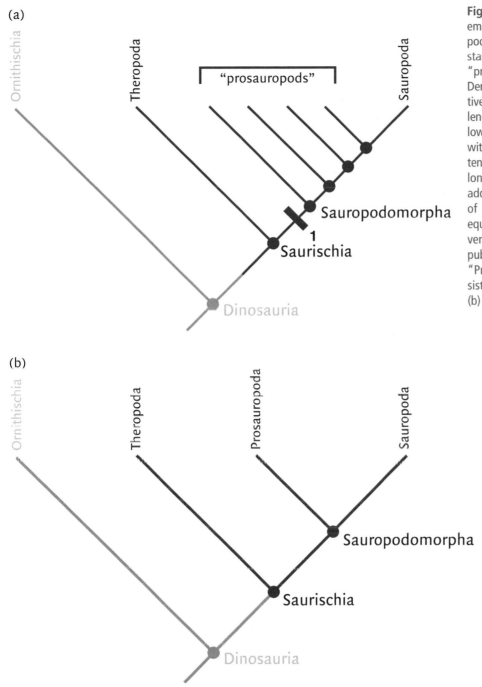

Figure 9.3. Cladogram of Dinosauria emphasizing the monophyly of Sauropodomorpha. (a) Our best understanding of the relationships of "prosauropods" and Sauropoda. Derived characters include: at **1**, relatively small skull (about 5 per cent body length), deflected front end of the lower jaw, elongate lanceolate teeth with coarsely serrated crowns, at least ten neck vertebrae that form a very long neck, dorsal and caudal vertebrae added to the front and hind ends of the sacrum, enormous thumb equipped with an enlarged claw, a very large obturator foramen in the pubis, and an elongate femur. (b) "Prosauropods" as a monophyletic sister group to sauropods. This view (b) is currently rejected.

Thinking about sauropodomorphs in terms of a formal definition, we might call Sauropodomorpha the group specified by the most recent common ancestor of *Plateosaurus* and *Saltosaurus*, and all of its descendants.[1]

---

[1] Another way to think about this definition might be: those sauropodomorphs closer in relationship to *Plateosaurus* than to *Saltosaurus*.

Figure 9.4. Left lateral view of the skull and skeleton of *Plateosaurus*.

Sauropodomorphs: among the very first dinosaurs

Sauropodomorphs are well represented among the very earliest dinosaurs (see also Chapter 5). They appeared early, and quickly diversified. The changes that they underwent were almost progressive: starting with a bipedal ancestor, heads became smaller, necks became longer, and, as we have seen, hind legs became closer in length to that of the front legs: the animals were becoming quadrupedal. Paralleling this were increasing steps towards an unquestionably herbivorous lifestyle.

The link between herbivory and "prosauropod" evolution is underlined by the fact that the history of basal sauropodomorphs parallels the rise of **gymnosperms** – seed-bearing plants (see Figure 14.10). That is, as gymnosperms – particularly tall ones – became an important component of the land plant biota, prosauropods became an increasingly important component of the terrestrial vertebrate fauna. "Prosauropods" were the first land creatures ever to take advantage of tall-growing plants.

"Prosauropods"

"Prosauropods" were a collection of relatively primitive dinosaurs with small heads, long necks, barrel-shaped bodies, and long tails, known from the Late Triassic through Early Jurassic, from all continents except Australia (see Figure 9.2). In general, the front limbs were somewhat shorter than the hindlimbs, and all had five digits. "Prosauropod" hands were equipped with a large,

(a)

(b)

Figure 9.5. Left hand of the "prosauropod" dinosaur *Plateosaurus*, showing its well-developed thumb claw: (a) reconstructed hand; (b) thumb showing amount of movement permitted by skeleton.

half-moon-shaped thumb claw (Figure 9.5). Whether for food procurement, defense, and/or some unspecified social activity, the function of this claw remains unknown.

"Prosauropod" lives and lifestyles

Feeding

In the mood for food, sure, but which? The skulls show almost none of the advanced design features associated with chewing (see Part III: Ornithischia); however, the jaw joint is slightly lower than the tooth row (Figure 9.6), a characteristic sometimes associated with chewing. The teeth are generally separated, leaf-shaped (Figure 9.7), and reveal few grinding marks, suggesting puncturing as the dominant tooth function.

Although paleontologists have traditionally considered sauropodomorphs to be herbivores, carnivory has been suggested in some of the basal forms because primitive sauropodomorph skulls, teeth, and general body proportions lack herbivore specializations. Yet, supporting herbivory in "prosauropods," the skull is proportionately smaller than that generally seen in carnivores. Recent treatments of the group split the difference, calling them predominantly herbivores that might have enjoyed an occasional meaty snack.

Once the food was past the mouth, grinding likely took place via gastroliths – which have been found in association with some "prosauropod" skeletons – and perhaps by stomach fermention, to judge from their barrel-shaped torsos (see Chapter 10).

Need for speed?

In the most primitive of "prosauropods," the forelimbs are shorter than the hindlimbs and the trunk region is relatively short, suggesting that these animals walked principally on their hindlimbs rather than on all fours. However, the largest and most derived "prosauropods" (among them *Riojasaurus* and *Melanorosaurus*; see Figure 9.21) appear to have become fully quadrupedal. The early history of locomotion in Sauropodomorpha is consistent with the primitive condition for all dinosaurs: bipedality (see Chapters 4 and 5). This interpretation is supported by evidence from trackways as well; the trackway genus *Otozoum*, generally attributed to "prosauropods," is all but exclusively bipedal.

"Prosauropods" appear to have been quite slow. Calculations from trackways (see Box 13.3) suggest speeds of no more than 5 kmh, about the average walking speed of humans. Of course,

(a)

(b)

10 cm

5 cm

(c)

(d)

10 cm

5 cm

Figure 9.6. Left lateral view of the skull of (a) *Anchisaurus*, (b) *Coloradisaurus*, (c) *Lufengosaurus*, and (d) *Yunnanosaurus*.

(a)   (b)   (c)

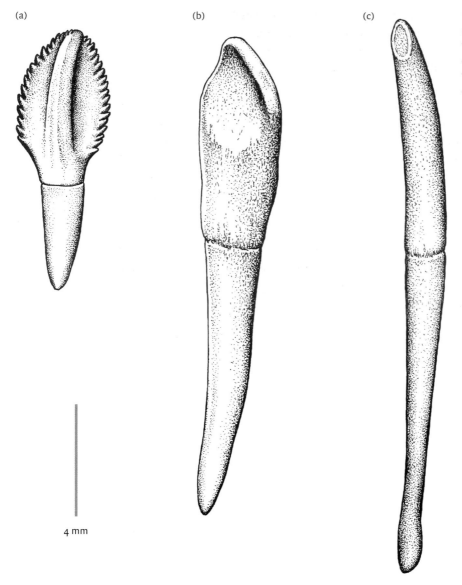

4 mm

Figure 9.7. Teeth in selected sauropodomorphs. (a) Leaf-shaped "prosauropod" tooth of *Plateosaurus*; (b) spatulate tooth of sauropod *Camarasaurus*; (c) pencil-like tooth of *Diplodocus*. The lower part of each tooth is the root.

those are just the tracks that they left one Sunday (?) back in the Late Triassic; in fact, their top speeds were likely much faster.

Socializing

Very little is known of "prosauropod" social habits. The famous *Plateosaurus* bonebeds in Germany and Switzerland, as well as others elsewhere, however, hint at herds; indeed, herds of prosauropods migrating across the European continent were proposed as early as 1915.

Detailed analyses of *Plateosaurus*, *Thecodontosaurus*, and *Melanorosaurus* (prosauropods for which large numbers of individuals are known) reveal sexual dimorphism in skull dimensions and in thigh bone (femur) size. Sexual dimorphism tends to be more pronounced in highly social animals, and thus there may be a connection between the likelihood of herding and sexual dimorphism.

Eggs, nests, and babies

Eggs and nests are known for the "prosauropods" *Mussasaurus* (Argentina) and *Massospondylus* (South Africa). The eggs were very thin-shelled; clutches tended to be small by dinosaur standards (something like ten eggs), and the hatchlings proportionately large-headed, toothless, and quite small. Adult "prosauropods" are roughly 500–1000 times larger than the hatchlings, so the formative years of basal sauropodomorphs must have been times of radical growth and reorganization. How all this occurred metabolically is unclear, but rapid growth rates are indicated by the disparity in bone sizes between juvenile and adult (see Chapter 13).

"Prosauropods" were not particularly common beasts and did not hang around on Earth for a very long time, all dutifully going extinct by the Early Jurassic. Still, as the first tall-browsing herbivores, they represent the first appearance on Earth of the modern ecosystem that is with us today. Our hope is that, with more attention and finds, the future will bring more insights into this enigmatic and basal group of dinosaurs.

Sauropoda

Sauropods are the familiar *Brontosaurus*-type dinosaurs: quadrupedal, herbivorous, small-headed, long-necked and long tailed. Although there has been considerable discussion about relationships within sauropods, most specialists would be comfortable defining Sauropoda itself as the monophyletic group comprised of the most recent ancestor of *Vulcanodon*, *Saltosaurus*, and all of its descendants (Figure 9.8).

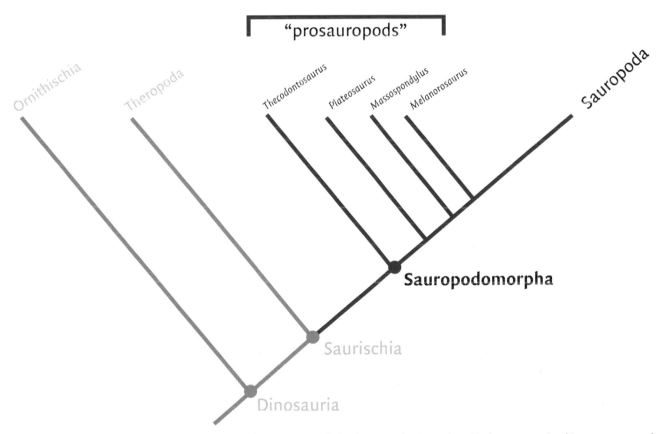

Figure 9.8. Cladogram of Sauropodomorpha, emphasizing monophyly of Sauropoda. More than 40 characters unite this group, some of which are listed in Figure 9.21. Many of these characters are about the relatively large size, elongate necks, herbivory, and the obligate quadrupedal stance of all sauropods.

1 m

Figure 9.9. Left lateral view of the skull and skeleton of *Apatosaurus*.

Design

Getting *really* big takes some serious evolution, and many sauropods were really huge dinosaurs. Yet, the sophisticated basic sauropod design, once it appeared, remained unique and little changed during their ~140 million years on Earth (Figure 9.9). It obviously worked!

The skull itself was distinctive: the tooth row was not inset, as one sees in mammalian and ornithischian herbivores. The teeth, depending upon the sauropod, had simple crowns and were triangular, spatulate, or slender and pencil-like (see Figure 9.7). In some derived forms, there is a tendency to limit the teeth to the front of the jaws (Figure 9.10). Yet, like their ornithischian cousins, advanced sauropods inclined towards some chewing (as we will see in Part III, Ornithischia, chewing is a highly specialized behavior which most vertebrates don't do – even if mammals happen to!). Some sauropods likely had cheeks; there is evidence for tooth occlusion in many, especially primitive forms, and in at least one case (*Nigersaurus*) there is even evidence for the development of a distinctive block of teeth: a dental battery (see introduction to Part III: Ornithischia, and Chapter 12).

Sauropod skulls tended to be delicately built, with large openings (Figure 9.10). The external nares, instead of residing at the tip of the snout, had an as-yet unexplained phylogenetic tendency to migrate upward, toward the top of the head (Figure 9.11). The skulls appear absurdly tiny – until you realize that only an idiot would design a large, heavy skull at the end of an extremely long neck.

That "extremely long neck" turns out to have been made up of a complex and very sophisticated system of girders and air pockets that maximized lightness and strength (Figure 9.12a). Unlike the vertebrae in mammals, most sauropod veretebrae have an internal system of pleurocoels, as well as pneumatic foramina (see Chapters 7 and 8). These same features are found in modern birds, which augment the volume of air in their lungs by a complicated series of interconnected auxillary air sacs (see Box 7.1). By analogy with modern birds, auxiliary air sacs in sauropods must have reached (and expanded into) the internal cavities via the pneumatic foramina (Figure 9.12b).

Also distinctive in sauropods were the Y-shaped neural arches on the vertebrae. These held the **nuchal ligament**, an elastic rope of connective tissue that ran down the back of the animal and supported the head and neck, so that it was not held up exclusively by muscles (Figure 9.13).

Sauropods were quadrupeds, having, as we have seen, secondarily evolved their stance from their bipedal, basal saurischian ancestors. The limbs were pillar-like, and would have done a Greek temple proud (Figure 9.14). The bones are composed of denser material than that found in the upper parts of

(a)

(d)

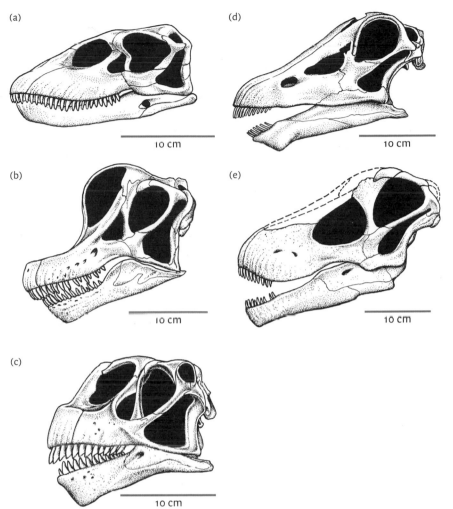

Figure 9.10. Left lateral view of the skull of (a) *Shunosaurus*, (b) *Brachiosaurus*, (c) *Camarasaurus*, (d) *Diplodocus*, and (e) *Nemegtosaurus*.

10 cm

10 cm

(b)

(e)

10 cm

10 cm

(c)

10 cm

the skeleton, an adaptation locating the weight and strength in the skeleton where it was most needed. The hindlimbs articulated with an immense, robust pelvis.

The forefeet (the "hands," as it were, on the forelimb) were **digitigrade**, which means that the animal stood on its fingertips. The fingers were arranged in a nearly symmetrical horseshoe-shaped semicircle, and the first digit (the thumb) carried a large claw (Figure 9.15a). By contrast, the hindfeet were semi-**plantigrade**, which means that the animal's weight was supported along the lengths of its toe bones.[2] The foot was asymmetrical, and generally had three large claws (on digits I, II, and III; Figure 9.15b). Sometimes the trackways reveal the impression of a heel pad that nestled behind the claws of both the fore- and hindfeet, and supported the body as well. Sauropod footprints are immense, a single print not uncommonly spanning as much as 1 m!

Despite the fact that the trunk of sauropods was relatively broad, most sauropod trackways tend to be quite narrow, with the feet aligned toward the midline of the body. Most significantly, relatively few trackways include a tail-drag mark, providing strong evidence that many sauropods carried their immense, whip-like tails – as long as 15 m in the longest cases – entirely off the ground (Figure 9.16).

---

[2] For reference, humans are supported along the lengths of their toe and foot bones, and are thus fully plantigrade.

Figure 9.11. Dorsal view of the skull of (a) *Brachiosaurus* and (b) *Diplodocus*. Note the dorsally placed external nares (arrows), especially in *Diplodocus*.

For many years, reconstructions commonly showed sauropods as swamp-dwellers, their great bulk buoyed up by water. In this way, so the story went, they could have remained deeply submerged, breathing with only their high nostrils poking out of the water, à la crocodile. But sauropods don't exactly look like crocodiles. With four powerful pillar-like limbs, what is it about sauropods that says "amphibious?" In fact, close study of sauropod anatomy gives no evidence that sauropods whiled away their palmiest days in Mesozoic swamps, their great bulk buoyed up by the water. Our best evidence, supported by biomechanical studies of sauropod limbs, is that the dense-boned, massive, and pillar-like limbs were designed for fully terrestrial locomotion.

Thoughts of a sauropod

This section will necessarily be quite short because, if brain size alone is considered, sauropods did not have obvious pretensions to deep thought. The head is proportionally small on the body, and the brain is proportionally small within the head; the fact is that sauropods had the smallest brains for their body size (and thus the lowest EQs; see Box 13.4) of any dinosaur. Yet their long, successful record of survival speaks volumes; as we shall see, their behavioral repertoire may have been more sophisticated than one might expect for an animal with proportionally so small a brain.

Lifestyles of the huge and ancient

A place to roam

When we find the remains of these magnificent animals, they come from a myriad different environments, from river floodplains to sandy deserts. Some environments, such as those of the Upper Jurassic Morrison Formation in the American West, evidently required sauropods to cope with long dry seasons during the year. Annual droughts may have been severe enough to have forced sauropods to migrate, a point to which we will return when considering sauropod herding.

Figure 9.12. The neck of the *Apatosaurus*. (a) The complicated girder-like structure of the zygapophyses was used as a framework for muscle and ligament attachment, used to support the neck. (b) In-pocketings in two vertebrae, marking the location of the pneumatic foramina and pleurocoels. White blobs are plaster filling pleurocoel entrances; not all pleurocoels, however, are plaster-filled.

Yet, at Tendaguru (see Box 15.7) in southeastern Tanzania, as well as in the USA in northern Texas at the famous Glen Rose trackway sites and in Maryland where the remains of the brachiosaurid *Astrodon* have been uncovered, there is strong evidence that these environments were once close to the sea and quite humid. Perhaps these were some of the conditions that sauropods found most congenial.

(a)

Figure 9.13. Anterior neck vertebrae in *Diplodocus*. The neural spines are bifurcated, and are thought to have held a ligament supporting the neck, the nuchal ligament. (a) Nuchal ligament (shown in solid blue) running from the head, down the neck, and beyond. (b) Bifurcated vertebral spines (indicated by arrows) in *Apatosaurus*, showing the groove where the nuchal ligament would have been located along the length of the neck.

(b)

Beating hearts and necking

Sauropod necks have been likened to those of giraffes, inviting the inference that they fed in tall trees. Recent reconstructions, however, indicate that the head in most sauropods was generally held at or near the height of the shoulder, and that a giraffe-like, vertically oriented neck was, at best, a strain, and, at worst, not physically possible. It is very likely that most sauropods held their heads at, or very near, the level of the body and browsed downward.

For brachiosaurs, however, things may have been different. Not only was the neck very long, the front limbs were longer than those in the rear (Figure 9.17). With this extra boost, its head could potentially have been raised to as high as 13 m, providing the opportunity to feed on foliage to which virtually no one else had access. But animals like *Brachiosaurus* would have had to pay a price for such posture. If the head was truly perched so high, its brain (the relatively smallest among dinosaurs) would have towered about 8 m above its heart.

To push blood through the arteries up its 8.5 m long neck, the heart of a *Brachiosaurus* must have pumped with a pressure exceeding that known in any living animal – indeed, double that of a giraffe. It would indeed take a very powerful heart – some estimate one weighing as much as 400 kg – to do the pumping (Figure 9.18). How fine capillaries in the brain might have withstood such pressures is again a matter for speculation as, unfortunately, is the basic question of how high *Brachiosaurus* could actually reach in life.

Some paleontologists have suggested *Diplodocus* and other long-necked sauropods may have gained access to foliage at high levels in the trees by adopting a tripodal posture, rearing up on their hindlimbs and using their tails as a "third leg" (Figure 9.19). In tripodal posture, these dinosaurs would have had to pay the same price as *Brachiosaurus* (which, with its elongated forelimbs, was

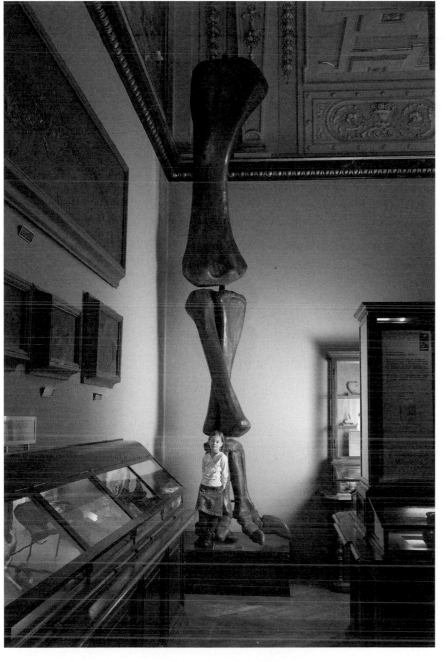

Figure 9.14. The right forelimb of *Apatosaurus* on display in the imperial surroundings of the Naturhistorisches Museum Wien.

probably not able to rear up): elevated blood pressure and a large, powerful heart to produce it. Interestingly, however, a number of modern studies have suggested that the shape of the pelvis in most sauropods precluded rearing back on the hind legs. This would leave dramatic images,[3] such as a mother *Barosaurus* rearing back and protecting its young from a ravenous *Allosaurus* (see Figure 1.12), more in the realm of science fiction than science.

None of these considerations addresses yet another challenge posed by long-necked life: the extraordinary amount of unused, wasted, air contained in the neck if sauropods simply breathed

---

[3] The root of the word "imagination" is, after all, image!

Figure 9.15. Sauropod left (a) forelimb and (b) hindlimb. The "hand" is far more digitigrade than the foot, which is nearly plantigrade (as shown in the limb cross-sections).

Figure 9.16. Five parallel trackways of Late Jurassic age, Morrison Formation, Colorado, USA. Tracks are thought to have been made by diplodocids walking alongside each other. Notice how close to each animal's midline each of the tracks is and the absence of any mark made by the tail.

in and out, bidirectionally, as do mammals and most tetrapods. If, however, sauropods used a unidirectional, avian style of respiration, in which air was pumped in a single direction through the lungs by auxilliary air sacs, almost all of the oxygen could be extracted from the inhaled air, and the problem of the amount of air contained within the neck would be diminished (Box 7.1; Figure 9.12).

Figure 9.17. Left lateral view of the skull and skeleton of *Brachiosaurus*.

Pneumatic bone and pleurocoels (see above) are convincing evidence that sauropods must have breathed unidirectionally (Box 7.1). With sauropods as well as theropods possessing air sacs and unidirectional breathing, this apparently highly derived character might actually be a basal saurischian characteristic – reinforcing our point that birds, so apparently derived in the modern fauna, contain a lot of characters unexpectedly shared by their long extinct, "primitive" relatives.

Considerations of blood pressure, heart size, lung capacity, and breathing style leave us unsure of how sauropods really functioned, but remind us that, in these respects at least, sauropods were a highly evolved group of animals.

Feeding

Tooth form and especially tooth wear indicate that sauropods nipped and stripped foliage, unceremoniously delivering a succulent vegetative bolus to the gullet, variably chewed, depending upon the sauropod. Some

Figure 9.18. Systolic blood pressures compared: (a) a sauropod (approximately 630 mm), (b) a giraffe (320 mm), and (c) a human (150 mm).

sauropods, such as *Camarasaurus* and brachiosaurs, had short snouts and proportionately deeper skulls, suggesting perhaps a more powerful bite. This idea is reinforced by considerable tooth wear that is preserved in these forms. By contrast, the longer-snouted diplodocoids, with their thin, pencil-like teeth restricted to the front of the jaws, may have stripped plant material, and left the heavy food processing to gastroliths and bacteria in the stomach (see below). Diplodocoids evidently had some ability to move the lower jaw fore and aft relative to the skull (humans can do this too!) which may have contributed to the stripping process. Here, too, the well-developed thumb claw may have played a role, stripping vegetation off plants into bite-sized veggie filets. Swallowing with powerful neck musculature sped the bolus of food down its long journey through the **esophagous** (tube leading from the throat to the stomach) whereupon it entered the stomach.

The sauropod cross-section is wide and surely accommodated a capacious gut, even considering the forward-projecting pubis (in contrast to all ornithischians, which rotated the pubis rearward to accommodate an enlarged gut; see introduction to Part III: Ornithischia). Sauropods likely had an exceptionally large fermentation chamber (or chambers) that would have housed **endosymbionts**; that is, bacteria that lived within the gut of the dinosaur. The endosymbionts would have chemically broken down the cell walls of the plant food, thereby liberating whatever nutrition was to be gotten. Considering the size of the abdominal cavity in sauropodomorphs, these animals probably fed on foliage with high fiber content (see Chapter 14); perhaps they also had slow rates of passage of food through the gut in order to extract a maximum of nutrients from such low-quality food. We can only conclude that these huge animals, with their comparatively small mouths, must have fed constantly to acquire enough nutrition to maintain themselves. The digestive tract of a sauropod had to have been a non-stop, if low-speed, conveyor belt.

Locomotion

The top speeds of *Brachiosaurus*, *Diplodocus*, and *Apatosaurus* have been calculated to be between 20 and 30 kmh, a reasonable clip for animals the size of a house and weighing in excess of three to ten elephants. They undoubtedly walked a good deal more slowly most of the time, perhaps at rates of 20–40 km per day, as calculated from sauropod trackways (see Box 13.3).

Hanging with the big dogs

The many now-famous mass accumulations in the Morrison Formation in the USA (see Figure 9.16), the Tendaguru bonebeds of Tanzania (see Box 15.7), the Lower Jurassic sauropod sites of India, and most recently the Middle Jurassic of Sichuan, China, together with the vast sauropod footprint assemblages, all speak loudly to the existence of gregariousness in sauropods, including *Shunosaurus*, *Diplodocus*, and *Camarasaurus*.

Sauropods living in large groups must have been capable of wreaking severe damage on local vegetation, both by stripping away all the foliage they could reach and by trampling into the ground all of the shrubs, brush, and trees that might have got in the way. So while herds of sauropods likely depleted their food sources and had to move on for more, other sauropods, including *Brachiosaurus* and *Haplocanthosaurus* from the Morrison Formation of the western USA, and *Opisthocoelicaudia* from Mongolia, are not so numerous and may thus have lived a more solitary existence, not needing to keep traveling far for sustenance. Other than these generalities, we know little about how sauropods communicated within herds, who ran the show, or what might have caught a sauropod's wayward eye.

Defense

In sauropods, size must have been the best deterrent against an attack; since, depending upon the sauropod, they were between 50 per cent and 300 per cent larger than known, co-existing predatory dinosaurs. Living in herds, healthy animals must have been nearly invulnerable. The tail was flexible and whip-like, and could have been employed to brush aside would-be predators. In the case of an attack from a pack of predators, both size and the tail may have been brought into play (but see Chapter 6 for how predators might have handled sauropod prey). Sauropod claws were not sharp like those of carnivorous theropods, but they could doubtless inflict significant damage, particularly when backed by the full weight and power of an adult sauropod. Like many anatomical features, sauropod claws likely served multiple functions.

Growth and development

Until recently, we knew next to nothing about sauropod nesting, and indeed, as recently as the early 1990s, it was seriously proposed that sauropods gave birth to live young (which, if it were true, would have made them unique among dinosaurs).

Sauropod eggs began to pop up as paleontologists looked for them and, in 1997, a sauropod nesting ground was discovered in Patagonia. This site, known as Auca Mahuevo ("Auca more eggs"), consists of a massive nesting ground covering more than a square kilometer and littered with tens of thousands of large, unhatched eggs. Upon further investigation, four (or six, depending upon how you count them) layers of eggs were uncovered and, in each layer, the eggs were organized into clusters of 15–34 linearly paired eggs, thought to represent individual nests or clutches. Most spectacularly, a high proportion of these eggs contained embryonic skeletons, some with impressions of embryonic skin (Figure 9.20)!

The geographical extent of the nesting horizons confirms herding and/or social behavior in sauropods. Clearly several enormous colonies were preserved, to which mothers may have regularly returned. Because there is no fossil evidence of adults preserved at Auca Mahuevo, the females may have left the site after laying their eggs, although it is possible that they may have communally guarded the whole nesting area from its periphery. If so, the eggs may also have been covered by mounds of vegetation to keep them at optimal temperature and humidity.

Figure 9.20. Titanosaurian remains from the Auca Mahuevo locality of Patagonia, Argentina. (a) Titanosaur skull (fossil); (b) reconstructed skull; (c) titanosaur skin (fossil) impressions; (d) reconstructed egg/embryo; and (e) schematic field of nests.

Since the find at Auca Mahuevo, eggs have been associated with particular sauropods in the Upper Cretaceous of southern France, Mongolia, and India, where, in 2007, a second, massive sauropod nesting ground was also uncovered. These vast clutches of eggs give us insights into dinosaurian life history strategies; that is, the ways in which particular organisms grow, reproduce, and die. One strategy, called the *r*-strategy, is to produce enormous numbers of eggs that result in thousands of offspring, the vast majority of which do not survive to reproduce during their relatively short lifespans. Think mayflies. Such offspring, born as near-adults, are called **precocial**. Not so much parental care here – too many children for serious parental investment.

This contrasts with the *K*-strategy, involving fewer offspring, lots of parental care, and longer lifespans. Think whales. Such offspring, born with a longer trek toward adulthood and requiring parental investment to get there, are called **altricial**.

Sauropods clearly laid a lot of eggs; thus, the resources expended by the parents on the offspring, as well as infant survival, may not have been so great. Sauropods evidently were *r*-strategists.

Beyond these few details, what do we know about the general aspects of sauropod reproduction, growth, and life histories? Sex in these animals assuredly involved coupling between a tripodal male and a quadrupedal female; however, beyond this most elemental of positions all else remains speculative. For example, could the trenchant thumb claw have been used in this aspect of sauropod behavior as well? The question of the existence of a penis in male sauropods also looms large.

Once hatched, sauropodomorphs apparently grew at very high rates. New studies of the microscopic structure of sauropod bone indicate rapid and continuous rates in both prosauropods and sauropods. Studies of bone tissue (see Chapter 13) suggest that, during the time of its peak growth, *Apatosaurus* grew at the astounding rate of $\sim 5.5 \times 10^3$ kg per year (see Chapter 13). The food consumption required to support such growth rates beggars description. Moreover, rather than imagining animals taking a leisurely 60 years to reach sexual maturity and having a longevity of perhaps 100–150 years, as had once been thought, the best current estimates are that it took about 30 years or less for a sauropod (and perhaps for a prosauropod as well) to become sexually mature. Similarly, lifespans for these animals were probably on the order of not much more than 100 years. Life for a sauropod may well have been in the fast lane.

The evolution of Sauropodomorpha

Sauropodomorpha is easily diagnosed by more than a dozen derived features (Figure 9.21; see also Figure 9.3). With "prosauropods" as a grade of primitive sauropodomorphs, we move directly to Sauropoda. Sauropoda is supported by more than a dozen unique features, many of which relate to the attainment of great size and weight on land (Figure 9.21).

Sauropoda

There appears to be consensus on at least the larger contours of sauropod evolution. As currently understood, sauropods consist of several primitive taxa (among them *Blikanasaurus*, *Vulcanodon*, and *Kotasaurus*) on the one hand, and the more derived clade Eusauropoda on the other. Eusauropods are diagnosed by many features (Figure 9.22). The most primitive known member of the group, *Shunosaurus*, was a 9 m long sauropod from the Middle Jurassic of China (Figure 9.23). Its

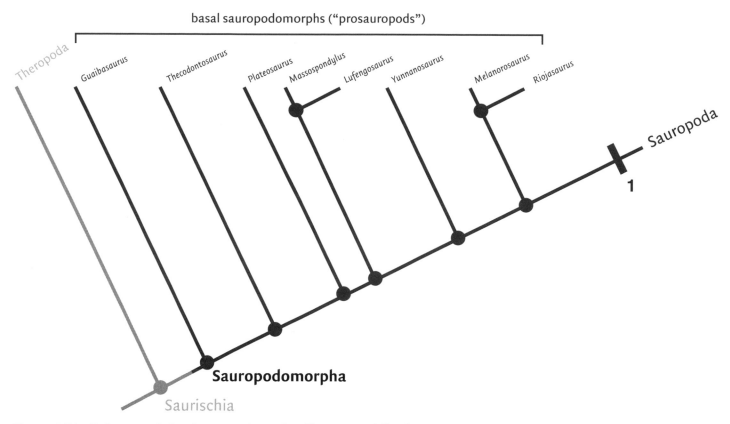

Figure 9.21. Cladogram of basal sauropodomorphs ("prosauropods") showing successive phylogenetic proximity to the clade Sauropoda. Derived characters of Sauropoda include, at **1**, special laminar system on forward cervical vertebrae, forelimb length greater than 60 per cent hindlimb length, triradiate proximal end of ulna, subrectangular distal end of radius, length of metacarpal V greater than 90 per cent that of metacarpal III, compressed distal end of ischial shaft, reduced anterior trochanter on femur, femoral shaft elliptical in horizontal cross-section, tibia length less than 70 per cent femur length, metatarsal III length less than 40 per cent tibia length, proximal end surfaces of metatarsals I and V are larger than those of metatarsals II, III, and IV, metatarsal III length is >85 per cent metatarsal V length.

skull is relatively long and low, and vaguely reminiscent of the primitive sauropodomorph condition, with nostrils near the front of the snout and a mouth filled with many small and spatulate teeth.

As sauropod evolution proceeded, various aspects of jaw mechanics and body form appear to have been linked. Sauropods evidently inherited a fully quadrupedal stance from their basal sauropod ("prosauropod") ancestors. At the same time, there was a significant increase in body size; in fact, body sizes evidently increased throughout sauropod evolution, until the highly derived titanosaurids (see below), which in some cases appear to have been somewhat smaller.

With the advent of the great clade Neosauropoda, the snout broadened, the lower jaw strengthened, and wear facets from grinding appear on the teeth, indicating front and rearward movement of the jaws. The great division within Neosauropoda is between the stocky, pumped Macronaria, and the elongate, ectomorphic Diplodocoidea (see Figure 9.22). These differences in body type match differences in the skulls: macronarians (including camarasaurs (Figure 9.24) and brachiosaurs (Figure 9.17), generally show a shortening and elevation of the skull, indicating a more

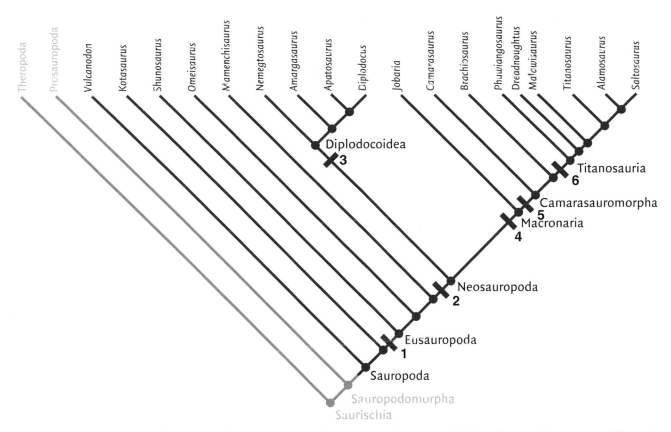

Figure 9.22. Cladogram of relationships of the major sauropod groups. At **1** (Eusauropoda), broadly rounded snout, caudal margin of external nares that extends behind the posterior margin of the antorbital fenestra, lateral plate on premaxillae, maxillae, and dentaries, loss of the anterior process of the prefrontal, frontals wider than length, wrinkled tooth crown enamel, most posterior tooth positioned beneath antorbital fenestra, at least 12 cervical vertebrae, neural spines of the cervical vertebrae that slope strongly forward, dorsal surface of sacral plate at the level of dorsal margin of ilium, block-like carpals, metacarpals arranged in U-shaped colonnade, manual phalanges wider transversely than proximodistally, two or fewer phalanges for manual digits II–IV, strongly convex dorsal margin of ilium, loss of the anterior trochanter of femur, lateral muscle scar at mid length of fibula, distally divergent metatarsals II–IV, three phalanges on pedal digit IV, ungual length greater than 100 per cent metatarsal length for pedal digit I; at **2** (Neosauropoda), subnarial foramen on premaxilla–maxilla suture, preantorbital fenestra in base of ascending process of maxilla, quadratojugal in contact with maxilla, pedal digit IV with two or fewer phalanges; at **3** (Diplodocoidea), subrectangular snout, fully retracted external nares, elongate subnarial foramen, reduction of angle between midline and premaxilla–maxilla suture to 20° or less, most posterior tooth rostral to antorbital fenestra; at **4** (Macronaria), greatest diameter of external nares greater than that of orbit, subnarial foramen found within the external narial fossa; enlarged jaw muscles; at **5** (Camarasauromorpha), nearly vertical dorsal premaxillary process, splenial extending to mandibular symphysis, acute posterior ends of pleurocoels in anterior dorsal vertebrae, metacarpal I longer than metacarpal IV; at **6** (Titanosauria), prominent expansion of rear end of sternal plate, very robust radius and ulna.

powerful bite; by contrast, diplodocoids had lower, more elongate skulls, with the nostrils fully migrated away from the snout, and the teeth generally pencil-like and restricted to the front of the jaws (see Figure 9.10 for differences in the skulls).

Our quick survey of Macronaria would not be complete without a stop at titanosaurs (see Figure 9.22), an extremely diverse group of sauropods, known from all continents, and among the last sauropods to grace the planet Earth. Titanosaurs achieved incredible sizes: the heaviest animals ever known were titanosaurs; some, such as *Argentinosaurus* (from Argentina) and *Paralititan* (from

2 m

Figure 9.23. The primitive eusauropod *Shunosaurus*, from the Middle Jurassic of Sichuan Province, China.

3 m

Figure 9.24. The skeleton of *Camarasaurus*, a well-known macronarian from the Late Jurassic of the western United States. This somewhat dated drawing dates back to its original description in the 1880s by legendary paleontologist O.C. Marsh (see Box 14.3)

Egypt) may have exceeded 60 metric tons,[4] however, their fragmentary remains, along with the problems of estimating weight, make such estimates very crude approximations (Box 9.1). The mighty Argentine *Dreadnoughtus* (Figure 9.25), first described in 2014, is relatively complete (some fragmentary jaw material plus ~70 per cent of the post-cranial skeleton is represented) and, using a new computer algorithm for weight estimation based upon the circumference of the humerus and femur, the thing weighed an unbelievable 59.3 metric tons.[5] Yet sauropod evolutionary history is not entirely one of getting bigger. In Transylvania is found the titanosaur *Magyarosaurus*, a creature 5–6 m in length, much smaller than contemporary sauropods elsewhere in the world. These smaller forms were living on islands, a common phenomenon in the Mesozoic as now. This dwarf sauropod was a "mere" 6 m long.

Several titanosaur forms, including *Saltasaurus*, from the Late Cretaceous of Argentina, and *Malawisaurus* (obviously from Malawi), are known to have been covered with special bones, embedded in the skin, called **osteoderms**[6] (*osteo* – bone; *derm* – skin); in this case they are scattered, immense (up to 22 cm in length), and hollow (Figure 9.26). Why scattered? Palentologist Kristina Curry Rogers proposed that in juveniles, the osteoderms were close together and served, as we will see in ankylosaurs (Chapter 10), as a kind of armor pavement. When the animals grew, the osteoderms, she proposed, grew apart, and then their hollow interiors potentially served as reservoirs for minerals and other nutrients during times of environmental stress.

Among the most famous of titanosaurs is, ironically, among the most poorly known: *Alamosaurus*, from the Late Cretaceous of the southwest of the United States (Figure 9.27). While Late Cretaceous sauropods are very well known from Europe, Asia, and South America, North American ones are quite rare; and it appears that they never got farther north than the American Southwest. What *that* is about is a story for another day – or at least for Chapter 14.

By contrast to macronarians, diplodocoids were elongate and slender animals with extraordinarily long necks and tails; the longest, though not the heaviest animals that ever lived on land. The Late Jurassic *Diplodocus* extended ~30 m, yet its weight is supposed to have been ~15 metric tons (again, see Box 9.1 for the perils of dinosaur weight estimates).

As we have seen, the long-necked, long-tailed, gracile diplodocoids (can an animal whose weight is in double-digit tons *ever* be called "gracile"?) developed a distinctive skull: they restricted their dentition to just a series of highly evolved, pencil-like teeth to the front of the jaws. Tooth wear is apparent at the apices of the teeth (instead of along the side of the tooth), and front–back jaw movement is thought to have characterized the way these animals managed food in the mouth.

Some remarkable dinosaurs are among the diplodocoids, including *Amargosaurus*, with its extraordinary neural arches, from the Early Cretaceous of Argentina (Figure 9.28). The 21 m long *Apatosaurus* (Late Jurassic, Western Interior, USA) is best known by its incorrect name "*Brontosaurus*" (Box 9.2). *Diplodocus*, renowned for its long neck and tail, evidently carried dragon-like dermal spines along the length of its back (Figure 9.29).

---

[4] 1 metric ton = 1000 kg; 1 kg = 2.20462 lbs, or 0.001 tons.

[5] But is this number right? More recent (and reasonable?) estimates of the weight of *Dreadnoughtus*, using different computer algorithms, put its heaviest weight at 30–40 metric tons (see Box 9.1). Nobody doubts that it was *very* big and *very* heavy, and in the end, that may have to be enough. Bragging rights have little relationship to science.

[6] Although very rare in mammals, osteoderms are relatively common in tetrapods; we are familiar with them, for example, as the bony plates in crocodile skin. We'll see them in many different dinosaurs.

Box 9.1   Weighing in

Back in the day, when people really didn't understand these things very well, it used to be easy to estimate the weight of a dinosaur. Just as American paleontologist E. H. Colbert did in the early 1960s, you made a scale model of the dinosaur, and then calculated its displacement in water (Figure B9.1.1).

That displacement was then (a) multiplied by the size of the scale model (for example, if the model were 1/32 of the original, the weight of the displaced water would have to be multiplied by 32) and (b) further modified by some amount using a fudge factor corresponding to the density of tetrapod bodies.

But what is the density of a tetrapod? Based upon studies with a baby crocodile, Colbert determined that baby crocodiles, at least, have a specific gravity of 0.89 (e.g., slightly less dense than the water they displaced). Studies with a large lizard (*Heloderma*) showed that the specific gravity of that lizard was 0.81. Among mammals, it would not be surprising to find the density of a whale differing from that of a cheetah, which might in turn differ from that of a gazelle. Full disclosure: there is no uniform density shared by all tetrapods.

What if you had a fossil represented by a big thigh bone, say, but not much else? Your plans are not quite DOA, because there turns out to be a relationship between limb cross-sectional area and weight. This relationship makes intuitive sense, because obviously, as a terrestrial beast becomes larger, the size (including cross-sectional area) of its limbs, to support its ever-increasing weight, must also increase. The question is, does it increase in the same manner for all tetrapods? If so, a single equation could apply to all. No surprise: it does *not*!

As noted by J. O. Farlow, weight is dependent upon muscle mass and muscle mass is really a consequence of behavior. Therefore weight estimates of dinosaurs are in part dependent upon presumed behavior. For example, reconstructing the weight of a bear would involve assumptions of muscle bulk and gut mass very different from those used in reconstructing the weight of a deer (Figure B9.1.2).

Indeed, the cross-sectional area of their limb bones may be identical, but they may weigh very different amounts; likewise, two animals might have the same weight, but be very differently sized. Moreover, our knowledge of dinosaurian muscles and muscle mass is incomplete. So the cross-sectional area method, although convenient and used by a number of workers, has the potential for serious misestimations of dinosaur weights.

In these days of CT-scans, computers, and sophisticated algorithms, can we do better? Of course we can, as demonstrated by the University of Liverpool's K. T. Bates and colleagues. These scientists begin with digitized, complete skeletons and have digitally reconstructed the fleshed-out bodies using laser-scanning techniques (Figure B9.1.3).

Among the most extraordinary features of these models were reconstruction of real internal anatomy, such air sacs, features that doubtless took up considerable space in non-avian dinosaurs but would cause the density of their owners to diverge significantly from that hopeful (but meaningless) quantity, the "specific gravity of tetrapods." The use of computers allowed the researchers to make innumerable other decisions as well; generally, how much "meat" to put on any given series of bones. Different parts of the body were accorded different densities, in accordance with their biology.

They began with work in the mid 2000s with five dinosaurs: three large theropods, an ornithomimid, and a duck-billed dinosaur (hadrosaur). Ultimately, the mass estimates for these dinosaurs varied significantly, "emphasizing the high levels of uncertainty inevitable in such reconstructions," as Bates and colleagues put it. Still, some

Figure B9.1.1. Estimating the weight of a dinosaur using displacement. For explanation of (a) and (b), see the text

Figure B9.1.2. Estimating the weight of a dinosaur by comparing the cross-sectional areas of bones.

(a)

(b)

(c)   (d)

Figure B9.1.3. The "best estimate," fleshed-out reconstruction of the ornithomimid *Struthiomimus* in (A) right lateral, (B) dorsal, (C) cranial and (D) oblique right craniolateral views (not to scale), using laser-scanning technology.

"best estimates" were produced: 7655, 6071, and 6177 kg (= 8.44, 6.69, and 6.81 tons, respectively) for the three large-bodied theropods, 423 kg (= 932 lbs) for the ornithomimid, and 813 kg (= 1792 lbs) for the duckbill. Science is supposed to be testable; how could these scientists figure out whether or not their final numbers had any meaning?

One test was to use the method on the skeleton of a living ostrich. If the method had any validity, it should predict from the skeleton the correct weight of the modern bird, in which the actual mass is known. In fact, the ostrich's predicted weight closely matched that of the estimate calculated from the skeleton, suggesting that the method has some validity.

Ultimately, the most striking thing about the research was what it said about the weight distribution in these creatures. In all cases, the center of mass was consistently located low on the body, in front of the hip. This reinforces the idea, based upon the design of the bones, that these animals all tended to keep the backbone parallel, or nearly so, with the ground (as we saw in Chapter 4).

In 2015 Bates and some other colleagues turned their attention to *Dreadnoughtus* and two other sauropods (*Apatosaurus* and *Giraffatitan*). *Dreadnoughtus*, they concluded, likely weighed between 30 and 40 metric tons, still in their view (and ours!) an extraordinary amount for a land-dwelling creature. But as for sauropod weight estimates of upwards of 60 tons, they concluded, "We find that 59 300 kg (59.3 metric tons) for *Dreadnoughtus* is highly implausible and demonstrate that masses above 40 000 kg require high body densities and expansions of soft tissue volume outside the skeleton several times greater than found in living quadrupedal mammals."[a]

[a] Quote from Bates, K. T., Falkingham, P. L., Macaulay, S., Brassey, C., and Maidment, S.C.R. 2015. Downsizing a giant: re-evaluating *Dreadnoughtus* body mass. *Biol. Lett.*, 11, 2015. dx.doi.org/10.1098/rsbl.2015.0215. See also, Bates, K. T., Manning, P. L., Hodgetts, D., and Sellers, W. I. 2009. Estimating mass properties of dinosaurs using laser imaging and 3D computer modeling. *PLoS ONE*, 4(2), e4532.

Figure 9.25. *Dreadnoughtus schrani*, an immense titanosaur from the Late Cretaceous of Argentina. *Dreadnoughtus* is among the largest known sauropods, as well as among the best preserved. Despite its immense size, it was likely still growing when it died. Human for scale.

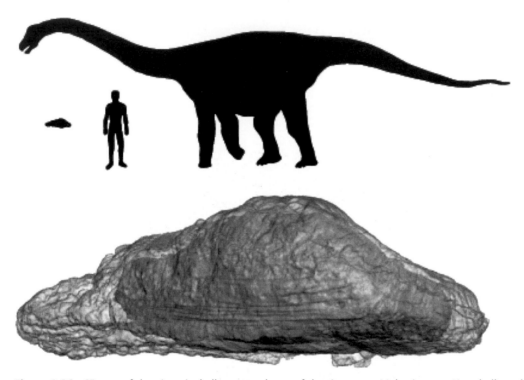

Figure 9.26. CT-scan of the gigantic, hollow osteoderms of the titanosaur *Malawisaurus*. Note hollowed, central region. Silhouetted dinosaur and human for scale. Inset: reconstruction of *Malawisaurus*.

Sauropodomorphs were the largest terrestrial vertebrate life forms of their time and indeed of all time. We often think of *Brachiosaurus*, from the Late Jurassic of the western USA, as well as *Giraffatitan* from Tanzania (see Figure 9.17), which captured several decades' worth of people's imaginations as the largest land-living animals of all time (measuring 23 m long and weighing in excess of 50 000 to 60 000 kg). Now supplanted by the likes of *Argentinasaurus* and *Dreadnoughtus*, *Brachiosaurus* nevertheless is still by far the best known of all of these earthly giants.

1 m

Figure 9.27. *Alamosaurus*, a poorly known Late Cretaceous sauropod from the southwest of the United States.

Figure 9.28. The diplodocoid sauropod *Amargasaurus*, from the Early Cretaceous of Argentina.

Sauropods were the fast-growing, yet slow-paced high-browsing giants of the Mesozoic. Today we view them as evolutionary marvels, as they continue to baffle, surprise, and inspire with biomechanical and evolutionary consequences of "living large."

Box 9.2   On again; off again: the checkered career of *Brontosaurus*

With the discoveries of dinosaurs in the Western Interior of the USA during the late nineteenth century, box-car-loads of brand-new, but often incomplete, sauropod skeletons were shipped back east. Yale's O. C. Marsh (Box 15.3)described one of these new sauropods as *Apatosaurus* in 1877. With further shipments of specimens and more studies, Marsh found another "new" sauropod in 1879, which he named *Brontosaurus*.

Years went by and – thanks to the burgeoning popularity of many kinds of dinosaurs – *Brontosaurus* became as famous as any dinosaur ever discovered, while *Apatosaurus* languished in semi-obscurity. Nevertheless, there was the

suspicion by many sauropod researchers that *Apatosaurus* and *Brontosaurus* were the same. In fact, this case was made as early as 1903 by E. S. Riggs, of the University of Kansas. Since then, most sauropod workers have regarded *Brontosaurus* as synonymous with *Apatosaurus*. If *Apatosaurus* and *Brontosaurus* are the same sauropod, the older name has priority, and thus *Apatosaurus* should be applied to this Late Jurassic giant.

Meanwhile, O. C. Marsh, lamenting as early as 1883 that no head was known for his *Brontosaurus* (now *Apatosaurus*), made his best guess as to the kind of skull this animal would

Figure B9.2.1. (a) "*Brontosaurus*" on display at the American Museum of Natural History in the early 1970s. This dinosaur was a chimera with a camarasaur-type head on an *Apatosaurus* body. (b) *Apatosaurus* being readied for display at the American Museum of Natural History in the mid 1990s.

Figure 9.29. Dermal spines reconstructed along the back of the Late Jurassic North American diplodocoid, *Diplodocus*.

have had; he concluded that it ought to have one like yet another sauropod, *Camarasaurus*. And so it was that *Brontosaurus/Apatosaurus* got the short-snouted type of skull of *Camarasaurus*.

Early in the twentieth century, H. F. Osborn, curator of vertebrate paleontology and powerbroker of the American Museum of Natural History in New York, and W. J. Holland, curator of fossil vertebrates at the Carnegie Museum of Natural History in Pittsburgh, each equally confident of his interpretations, skirmished over the issue of whose head should reside on the neck of *Apatosaurus* (for it was by then acknowledged to be such). Osborn followed Marsh and had his skeleton of *Apatosaurus* – in New York – topped with a camarasaur head (Figure B9.2.1a), while Holland, in Pittsburgh, was equally certain that *Apatosaurus* had a more *Diplodocus*-like head (based on a somewhat removed yet associated skull found near an otherwise quite complete skeleton at what is now Dinosaur National Monument in Colorado). But Holland gained no adherents and his mount of *Apatosaurus* in the Carnegie Museum remained headless in defiance of Osborn's dogma. After Holland's death, however, the skeleton was fitted with a camarasaur skull, almost as if commanded from Beyond, by Osborn himself.

Which head belongs to which dinosaur was finally resolved in 1978 by Carnegie Museum Curator of Paleontology D. S. Berman and the late sauropod authority J. S. McIntosh. Through some fascinating detective work on the collection of sauropod specimens at the Carnegie Museum, these two researchers were able to establish that *Apatosaurus* had a rather *Diplodocus*-like skull – long and sleek, not blunt and stout as had previously been suggested. As a consequence, a number of museums that display *Apatosaurus* skeletons celebrated the work of Berman and McIntosh (and Holland) by conducting painless head transplants (Figure B9.2.1b).

Still, people were attached to the name "*Brontosaurus*," and nobody gets points for raining on parades. So an unofficial compromise was reached: the word "brontosaur" could be a common name like "dog" – you call a sauropod a brontosaur, and everybody knows what you're talking about. In any event, the scientific upshot of all this was "*Brontosaurus*" is actually *Apatosaurus*; *Apatosaurus* had an elongate, *Diplodocus*-like skull. Case closed, right?

Uh, no. In April 2015, three European specialists did an immense, exhaustive analysis of dinosaurs very closely related to *Diplodocus* (for example *Apatosaurus* and even *Brontosaurus*). Three hundred pages later, they concluded that *Brontosaurus* is very real (if not alive and well). In fact, there are actually three species of the thing: *B. parvus*, *B. excelsior*, and *B. yahnapin*. And everybody had *Diplodocus*-like elongate skulls.

This discussion has gone on for 135 years; do we really think we've heard the last word on these iconic sauropods? We hope so.

SUMMARY

Sauropodomorpha consists of the great herbivorous saurischian quadrupeds Sauropoda, and an early non-monophyletic (e.g., paraphyletic) offshoot, called "prosauropods." "Prosauropods" were primitive large dinosaurs, appearing in the Late Triassic at the dawn of Dinosauria, and surviving through the Early Jurassic. Initially thought to be the ancestors of Sauropoda, they are now considered to represent an early saurischian radiation, representing the world's first high-browsing herbivores.

Sauropoda were the largest land animals ever to walk the Earth, reaching 40 m from the tip of the snout to the tip of the tail. These obligate herbivorous quadrupeds were highly evolved, with many biomechanical adaptations for large size and weight, including four pillar-like limbs and a massive pelvis, a tendency to lighten bones not immediately involved in bodily support functions, and a complex girder-like neck design, tipped by a small skull, to maximize leverage and lightness. Among the many striking features in the design of sauropod bones are the presence of pleurocoels, hollow spaces suggestive of avian-style air sacs. The likely presence of air sacs, as well as the inefficiency, in an animal of such great size, of mammalian-style bellows breathing, suggest that sauropods likely enjoyed avian-style unidirectional breathing.

The skulls of sauropods were relatively small, and the group showed a general tendency toward migrating the nostrils or nares to the top of the skull, which is emphasized in diplodocoids. Dentition varied, from simple leaf-shaped teeth to pencil-like teeth restricted to the front of the mouth. Sauropods in general lack the chewing adaptations present in ornithischians (see Part III), and we may be confident that the stomach and gastroliths did much of the heavy lifting in the breakdown of food. Nonetheless, for all that there appear to have been a variety of strategies for handling food at the mouth, reflected in jaw musculature, skull design, and tooth wear.

Sauropods appear to have been social animals, particularly as reflected in sauropod bonebeds and in trackways. The trackways clearly indicate that sauropods did not drag their tails. Sauropod gregariousness is also reflected in the recent discoveries of extremely large sauropod nesting grounds. Sauropod lifespans, once thought to be in the hundreds of years, are now thought to have been around 100 years, with extremely rapid growth of juveniles. The large number of eggs and babies associated with sauropod nesting grounds implies that these dinosaurs may have been *r*-strategists.

Sauropod defense was likely accomplished mainly by their huge body size, with perhaps an assist from the whip-like tail and the broad, trenchant claw on the forefoot.

The immense size of sauropods makes analogizing them with living terrestrial vertebrates extremely difficult. Although not likely possessed of a "warm-blooded" metabolism – at least as adults (see Chapter 12) – they nonetheless must have consumed copious quantities of food, and must have required virtually incessant food consumption through the comparatively small mouth. If the number of individuals was large enough, a healthy herd might have have defoliated landscapes. This in turn leads to at least the possibility that sauropods were often on the go, searching out new foliage to consume.

SELECTED READINGS

Curry Rogers, K. and Wilson, J. A. (eds.) 2005. *The Sauropods: Evolution and Paleobiology.* University of California Press, London, 349 pp.

Curry Rogers, K., D'Emic, M., Rogers, R., Vickaryous, M., and Cagan, A. 2011. Sauropod dinosaur osteoderms from the Late Cretaceous of Madagascar. *Nature Communications*, **2** (564), 1–4. doi: 10.1038/ncomms1578.

Galton, P. M. and Upchurch, P. 2004. Prosauropoda. In Weishampel, D. B., Dodson, P., and Osmólska, H. (eds.) *The Dinosauria*, 2nd edn. University of California Press, Berkeley, pp. 232–258.

Henderson, D. M. 2003. Typsy punters: sauropod dinosaur pneumaticity, buoyancy, and aquatic habits. *Proceedings of the Royal Society of London, Series B (suppl.)*, **271**, S180–S183.

Klein, N., Remes, K., Gee, C.T., and Sander, P.M. (eds.) 2011 *Biology of the Sauropod Dinosaurs – Understanding the Life of Giants.* Indiana University Press, Bloomington, IN, 331 pp.

Seymour, R. S. and Lillywhite, H. B. 2000. Hearts, neck posture and metabolic intensity of sauropod dinosaurs. *Proceedings of the Royal Society of London, Series B*, **267**, 1883–1887.

Tidwell, V. and Carpenter, K. (eds.) 2005. *Thunder-Lizards – The Sauropodomorph Dinosaurs.* Indiana University Press, Bloomington, IN, 495 pp.

Upchurch, P., Barrett, P. M., and Dodson, P. 2004. Sauropoda. In Weishampel, D. B., Dodson, P., and Osmólska, H. (eds.) *The Dinosauria*, 2nd edn. University of California Press, Berkeley, pp. 259–321.

Upchurch, P., Barrett, P. M., Zhao, X., and Xu, X. 2007. A re-evaluation of *Chinshakiangosaurus chunghoensis* Ye vide Dong 1992 (Dinosauria, Sauropodomorpha): implications for cranial evolution in basal sauropod dinosaurs. *Geological Magazine*, **44**, 247–262.

Wedel, M. J. 2003a. The evolution of vertebral pneumaticity in sauropod dinosaurs. *Journal of Vertebrate Paleontology*, **23**, 344–357. doi:10.1671/0272-4634(2003)023[0344:TEOVPI]2.0.CO;2.

Wedel, M. J. 2003b. Vertebral pneumaticity, air sacs, and the physiology of sauropod dinosaurs. *Paleobiology*, **29**, 243–255. doi:10.1666/0094-8373(2003)029,0243: VPASAT.2.0.CO;2.

Yates, A. M. 2012. Basal Sauropodomorpha: the "Prosauropods." In Brett-Surman, M. K., Holtz, T. M., and Farlow, J. O. (eds.) *The Complete Dinosaur*, 2nd edn. Indiana University Press, Bloomington, pp. 424–443.

Yates, A. M., Bonnan, M. F., Nwevelling, J., Chinsamy, A., and Blackbeard, M. G. 2009. A new transitional sauropodomorph dinosaur from the Early Jurassic of South Africa and the evolution of sauropod feeding and quadrupedalism. *Proceedings of the Royal Society of London, Series B*, **277**, 787–794. doi:10.1098/rspb.2009.1440

TOPIC QUESTIONS

1. What is Sauropodomorpha? How does it relate to Saurischia?
2. What are the diagnostic characters of Sauropodomorpha? Why are the quotation marks always around the word "prosauropods"?
3. Outline what is known of sauropod reproductive strategies. Would you classify them as *r*-selected, or *K*-selected? Why?
4. Describe some features that are associated with large size in sauropodomorphs.
5. What is the evidence that sauropods didn't really chew their food? How can sauropods have been herbivores, but not have developed much chewing ability?
6. Why are sexual dimorphism and gregariousness often linked?
7. What kinds of design constraints are associated with having an extremely long neck?
8. What is the evidence that indicates that sauropods did not dwell in swamps? Where, then, and in what kinds of environment did they live?
9. In Box 7.1, we wrote: "isn't it possible that unidirectional breathing should be described as *saurischian* and not "avian?"." Can you think of some reasons why this statement might be true?
10. Why do we think that the quadrupedal stance of sauropods evolved secondarily?

PART III
ORNITHISCHIA: ARMORED, HORNED, AND DUCK-BILLED DINOSAURS

Figure III.1. Right lateral view of the ornithischian pelvis as exemplified by *Stegosaurus*. Note that the pubis has a rearward-facing splint of bone, lying underneath the ischium in what is known as the opisthopubic condition (see also Figures 5.9 and 5.10).

A long long time ago, and many saurischians away (Chapter 5), we met Ornithischia, one of the two great branches of dinosaurs. Although Ornithischia was identified as early as 1887, nobody at that time had an inkling how diverse the group really was. Now we know a bit better. We'll check out individual ornithischians in Chapters 10–12. But first, let's introduce Ornithischia as a group.

What makes an ornithischian an ornithischian?

Unlike saurischians, ornithischians show obvious family resemblance, and diagnostic characters abound. Among these, two stand out:

- In all ornithsichians, at least a part of the pubis has rotated backward to lie close to and parallel with the ischium;[1] this orientation is called **opisthopubic** (Figure III.1).
- All ornithischians had a unique bone, the **predentary**, an unpaired, scoop-shaped element that capped the front of the lower jaws (Figure III.2).

Both of these adaptations were associated with food consumption and processing. The evolution of the rearward-directed pubis (recall that, primitively, the pubis points forward; see Figure 5.9) is believed to be associated with the development of a large stomach (or stomachs?) and intestinal region (gut), the better for extracting nutrients from plants. Accommodating the large gut was a barrel-shaped torso, recognizable from the shape of the ribs. The predentary supported the lower portion of a beak, a characteristic feature of all ornithischians (see below). Other ornithischian diagnostic characters are listed in the cladogram shown in Figure III.3. You can see that the monophyly of Ornithischia is *extremely* well supported.

---

[1] In birds, the pubis is rotated backwards as well, and so, confusingly, ornithischian dinosaurs are called "bird-hipped," while birds, as we have seen, are undoubted saurischians. This means that birds belong to the "lizard-hipped" branch of Dinosauria while the "bird-hipped" branch – Ornithischia – seems to have given rise to nothing still alive; certainly not birds.

Figure III.2. Left lateral view of the skull of the ceratopsid *Triceratops*. The predentary caps the front of the lower jaw, and is outlined in white.

The discovery of one entirely unexpected attribute of ornithischians came in 2009, with the unearthing of a primitive (basal) ornithischian dinosaur called *Tianyulong* (Figure III.4). *Tianyulong* bears a row of long, monofilamentous structures – Stage 1 proto-feathers? – down its neck, back, and tail. An ornithischian with some kind of monofilamentous feather coating is hard to imagine, since at most we've only associated feathers with theropods; however, feathers or feather-like plumage are shared by other ornithischians as well (see Chapter 11). Do feathers – or their precursors – go all the way back to basal Dinosauria? The short answer is "quite probably," pointing the way to very primitive feathers as a basal dinosaur – or even older – character (see also Chapter 5).[2]

Chew chew chew-boogie!

One essential quality of ornithischians is that, to a greater or lesser extent, all apparently chewed their food. Because humans and many other mammals chew, we naturally assume that everybody

---

[2] The question really hinges on whether or not feathers evolved more than once. In the past, received wisdom suggested that feathers only evolved one time; that being the case, feathers would be primitive for all ornithodirans, since they are present in at least pterosaurs, theropods, and ornithischians. If, however, they evolved more than once, then perhaps they occurred independently in these groups, and it cannot be said the Ornithodira is characterized by the possession of feathers. The jury is still out on this question, but we can't wait to learn the answer!

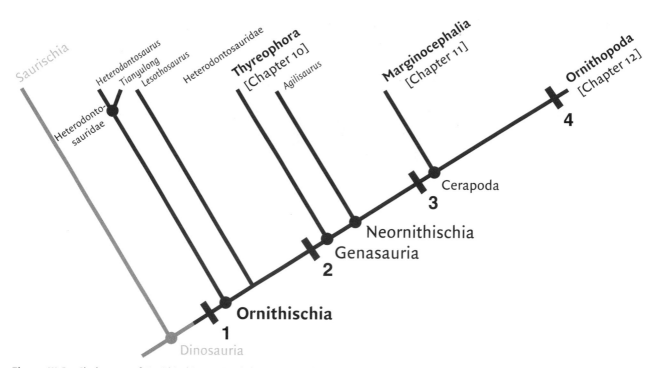

Figure III.3. Cladogram of Ornithischia. Derived characters include: at **1** (Ornithischia), opisthopubic pelvis, predentary bone, toothless and roughened tip of snout, reduced antorbital opening, palpebral bone, jaw joint set below level of the upper tooth row, cheek teeth with low subtriangular crowns, at least five sacral vertebrae, ossified tendons above the sacral region, small prepubic process along the pubis, long and thin preacetabular process on the ilium; at **2** (Genasauria), emarginated dentition (indicating large cheek cavities, and reduction in the size of the opening on the outside of the lower jaw (the external mandibular foramen); at **3**, gap between the teeth of the premaxilla and maxilla, five or fewer premaxillary teeth, finger-like anterior trochanter; at **4**, high-crowned cheek teeth, denticles on the margins restricted to the terminal third of the tooth crown, canine-like tooth in both the premaxilla and dentary.

does, and it can be surprising to learn that (as we noted earlier) most vertebrates don't chew; that teeth are really most commonly used only for biting off chunks of whatever is being eaten. The fundamental act of *chewing* – of grinding food down into a paste that can be digested relatively efficiently – is done by other organs in most vertebrates; generally, as in sauropods, a gastric mill; commonly with gastroliths. But ornithischians got into the chewing game in a big way, so a look at what chewing's all about is useful for understanding these dinosaurs.

Chewing in mammals

We start by looking at the basics of chewing in a familiar living group: mammals. Among living herbivorous mammals, no matter what the size, the skull is generally divided into three sections (Figure III.5): at the front is the **cropping** part, where blade-like teeth (generally incisors) bite off chunks of food. Behind the cropping section is the **diastema**, or gap, a region that is toothless (or nearly so), likely used for the manipulation of the food by the tongue. Finally, further back in the mouth are the **cheek teeth** (molars in mammals) – in herbivores especially (which tend to chew their food more than carnivores), a block of teeth, relatively tightly fitted against one another, which are used to grind or shear plant material. In mammals the upper and lower cheek teeth **occlude** – or fit tightly against each other when the jaw is shut – which ensures that, as the chewing takes place, the grinding is efficient.

Two other features are associated with chewing in mammals. Firstly, toward the back of the lower jaw is a large expansion of bone, the **coronoid** process, which serves as an attachment site for strong

Figure III.4. The tail and left leg of *Tianyulong*, a basal ornithischian evidently bearing monofilamentous feathers (indicated by arrow). Head is to right of photograph. Because feathers are found in both primitive ornithischians as well as many saurischians (so far, theropods), the most parsimonious conclusion is that feathers must also have characterized the common ancestor of both groups: the earliest dinosaur.

jaw-closing muscles (Figure III.5). Secondly, the tooth row is deeply inset toward the midline of the skull. This makes room for **cheeks**, muscular tissues that play the obviously essential role of keeping food in the mouth while it is being chewed.

In mammals, then, chewing leaves a recognizable imprint on the design of the skull, the lower jaw, and the teeth. It's striking that, in almost all ornithischian dinosaurs, many of these same adaptations for chewing can be found, where they doubtless arose independently.

Chewing in dinosaurs

The primitive tetrapod condition, exemplified in theropods, is that in which the jaw joint is at the same level as the tooth row (Figure III.6b): the jaw thus functions like a scissors: the bite slices sequentially, from the back of the jaw forward; we saw this in theropods (see also Figure 6.16).

By contrast, when the jaw joint is below the level of the tooth row, as it is in all ornithischians, the jaw functions a bit like water-pump pliers, in which the jaws close simultaneously along their entire length (Figure III.6a). The blocks of cheek teeth grind against each other simultaneously instead of shearing past each other sequentially. As it turns out, even the most primitive ornithsichians had many of the required goods for serious chewing. And the many ornithischians that followed never kicked the habit.

In many dinosaurs, including all ornithischians, the cropping function of the mouth was carried out not by teeth, but by a beak, or **rhamphotheca**. Rhamphothecae were made of **keratin**, a protein-based substance that makes up horns, nails, hooves, and claws. It is also a key ingredient of hair, although you'll never find *that* in a dinosaur!

Figure III.5. Selected herbivorous mammal skulls (not drawn to scale). (a) Horse (*Equus*), (b) llama (*Lama*), (c) rabbit (*Lepus*), and (d) rat (*Rattus*). Divisions of skulls indicate: anterior cropping section (yellow), diastem (orange), block of grinding cheek teeth (light purple), and coronoid process (light blue). Despite the range of sizes and herbivorous behaviors, all skulls show the same basic organization.

Ornithischia: the big picture

Specialists agree that fundamental split within Ornithischia is between the very primitive hetero-dontosaurids and *Lesothosaurus*[3] on the one hand, and everything else on the other (see Figure III.3).

"Everything else" is within the clade **Genasauria** (*gena* – cheek, a reference to their all having well-developed cheeks). Genasaurs include two great groups of ornithischians, **Thyreophora** (Chapter 10), and **Cerapoda** (Marginocephalia + Orthinopoda; Chapters 11 and 12, respectively), as well as a few assorted forms with which we won't concern ourselves here. They all share the derived characters of muscular cheeks, indicated by the deep-set position of the tooth rows, away from the sides of the face, a spout-shaped front to the mandibles, and reduction in the size of the opening on the outside of the

---

[3] *Lesothosaurus* has not gone quietly into the phylogenetic night. A major reanalysis of Ornithischia (Butler *et al.*, 2008) removed *Lesothosaurus* from its conventional basal position in Ornithischia and made it a basal thyreophoran (see Chapter 10), closely related to stegosaurs and ankylosaurs. This idea, however, does not appear to have been widely adopted; we therefore return it to its conventional spot at the base of Ornithischia pending a broader scientific consensus on this idea. The relationships of basal neornithischians are very much in a state of flux (see Box 12.1).

(a)

(b)

Figure III.6. Position of jaw joints. (a) In ornithischians, the jaw joint is below the level of the tooth row, and the blocks of cheek teeth *grind* against each other simultaneously, much like the water-pump pliers; (b) by contrast, the theropod condition (as we have seen) has the jaw joint at the same level as the tooth row; the dominant movement, therefore, is scissor-like *cutting*, back to front.

50 cm

Figure III.7. Left lateral view of the skeleton of *Heterodontosaurus*.

(a)

10 mm

(b)

15 cm

Figure III.8. (a) Upper tooth of the heterodontosaurid *Lycorhinus*. (b) Left lateral view of the skull of *Heterodontosaurus*.

Figure III.9. Jaw mechanics in Heterodontosauridae, showing mobility of the lower jaws. As the lower jaw closed, each side rotated slightly along its long axis.

lower jaw (the external **mandibular foramen**), among others. Because it's hard to understand the point of cheeks without chewing, chewing should be thought of as a fundamental genasaur behavior. Then it only becomes a matter of how efficiently the various groups of genasaurs chewed.

Moving up the cladogram is the slightly-more-derived **Neornithischia** (*neo* – new; see Figure III.3). Primitive neornithischians include a variety of small, bipedal forms – we'll call them *basal neornithischians* – whose relationships within Ornithischia are still very uncertain. Rather than getting mired here among the basal neornithischians, we revisit them again in Chapter 12.

Heterodontosauridae and *Lesothosaurus*: the most primitive ornithischians

Heterodontosaurids were small, bipedal ornithischians (Figure III.7) of uncertain phylogenetic position; with a mix of primitive and advanced characters, some authors consider them to have been basal ornithischians, while others have them as the sister group Marginocephalia (Chapter 11), and still others as the sister group to Euornithopoda (Chapter 12). Because heterodontosaurids bear the

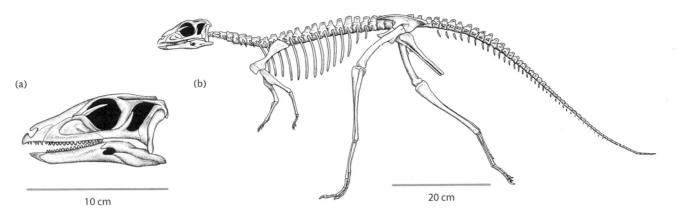

(a)

(b)

10 cm

20 cm

Figure III.10. Left lateral view of the skull (a) and skeleton (b) of the basal ornithischian *Lesothosaurus.*

inset tooth rows (and cheeks) diagnostic of genasaurs, some consider them to be basal genasaurs. We tentatively consider them to be basal ornithischians (Fig. III.3).

In many respects heterodontosaurids are not exactly primitive; for example, they evolved teeth bearing a high, chisel-shaped crown ornamented with tiny bumps (Figure III.8a; correctly termed **denticles**) as well as a large canine-like tooth on both upper and lower jaws (the basis for the "heterodonto-" part of the name; Figure III.8b). Moreover, skull design and tooth-wear striations suggest that they chewed distinctively: they amplified the familiar vertical movement of the lower jaws with slight outward rotations of each side of the mandible about its long axis (Figure III.9). This allowed them to get a bit more grind out of each bite. Powerful forelimbs and clawed hands likely were used to grab vegetation or to dig up roots and tubers.

Like many ornithsichians, the evolution of canine-like teeth of heterodontosaurids likely was related to combat between animals of the same species (males?), ritualized display, social ranking, and possibly even courtship (see Chapters 10 and 11). A modern analog is found in tusked tragulids, living mammals from southeastern Asia and Africa, closely related to deer. In these mammals, tusk development is tied to sexual maturity, and is a dimorphic (differently expressed in the sexes) feature that is, as we propose for heterodontosaurids, used for intraspecific combat, ritualized display, and social ranking. Similarly, the development of a large knob of bone, a **boss**, in the cheek region (therefore called the jugal boss[4]) in heterodontosaurids might also be interpreted as a form of visual display.

Display in ornithischians may have come in another form as well. We've already met *Tianyulong*, the feathered heterodontosaur from western Liaoning Province, China (see Figure III.4). Depending upon the colors and patterns, these primitive feathers could have served a display function and may have been intimately associated with behaviors ranging from choosing mates (see Chapter 6) to marking territory, to camouflage, to keeping warm (Chapter 13).

Lesothosaurus was a small, leggy Early Jurassic herbivore from South Africa (Figure III.10). It had a typical suite of diagnostic ornithischian characters including a jaw joint lower than the tooth row (see Figure III.6). That character hints at chewing; we'll have far more than hints for the rest of Ornithischia.

The rest of Ornithischia

Thyreophora (*thyreo* – shield; *phora* – bearer; a reference to the fact that these animals have dermal armor) consists of genasaurs in which there are parallel rows of keeled dermal armor **scutes** (or bony

---

[4] Named for the jugal, the skull bone on which it appears.

plates) on the back surface of the body. The most familiar thyreophorans are stegosaurs and ankylosaurs, but we'll encounter others along the way (Chapter 10).

Cerapoda (*cera* – horn) are those genasaurs that share a pronounced diastem between the teeth of the premaxilla and maxilla among other derived characters (see Figure II.3). Marginocephalians (*margin* – margin; *cephal* – head), a group united by having, primitively, a distinctive, narrow shelf that extended over the back of the skull (see Figure II.3), consists of two well-known ornithischian taxa, the dome-headed pachycephalosaurs and ceratopsians (Chapter 11), the latter being the horned dinosaurs most famously known from the Late Cretaceous of North America.

Ornithopods, especially euornithopods, took chewing to new levels, possibly unmatched in the history of life. Within this group are found the familiar duck-billed dinosaurs, as well as one of the very first dinosaurs ever recorded, *Iguanodon*. We'll visit these animals in Chapter 12.

SELECTED READINGS

Boyd, C. A. 2015. The systematic relationships and biogeographic history of ornithischian dinosaurs. *PeerJ*, 3:e1523; DOI 10.7717/peerj.1523.

Butler, R. J., Upchurch, P., and Norman, D. B. 2008. The phylogeny of the ornithsichian dinosaurs. *Journal of Systematic Paleontology*, **6**, 1–40.

Norman, D. B., Witmer, L. M., and Weishampel, D. B. 2004. Basal Ornithischia. In Weishampel, D. B. and Dodson, P. (eds.) *The Dinosauria*, 2nd edn. University of California Press, Berkeley, pp. 325–334.

Sereno, P. C. 1986. Phylogeny of the bird-hipped dinosaurs (Order Ornithischia). *National Geographical Research*, **2**, 234–256.

Figure 10.1. Representatives of the two great groups of thyreophorans: *Gastonia*, an ankylosaur from the Early Cretaceous of Utah, USA (left), and *Stegosaurus*, a stegosaur from the Late Jurassic of the United States.

Thyreophorans
The armor-bearers

10

CHAPTER OBJECTIVES

- Introduce Thyreophora, particularly its two large constituent groups, Stegosauria and Ankylosauria
- Develop familiarity with current thinking about the lifestyles and behaviors of thyreophorans
- Develop an understanding of thyreophoran evolution using cladograms, and an understanding of the place of Thyreophora within Dinosauria

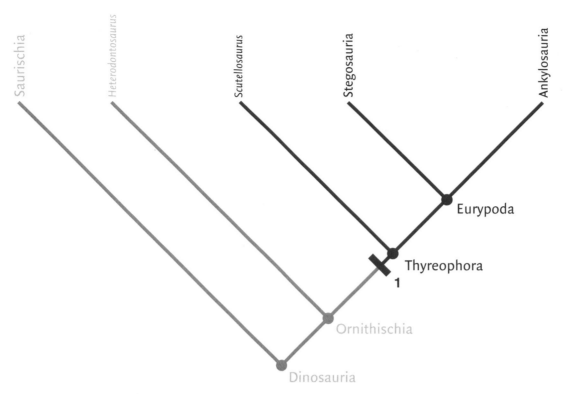

Figure 10.2. Cladogram of Thyreophora, emphasizing relationships within Ornithischia. Derived characters include: at **1**, transversely broad process of the jugal, parallel rows of keeled scutes on the back surface of the body.

Thyreophora

In life as in games, offense and defense are strategies, each with its own advantages. **Thyreophora** (*thyr* – shield; *phora* – bearing or carrying; literally, "armor-bearers") went with defense, evolutionarily opting for fortress-like protection and armor. And the strategy paid off: these dinosaurs did very well during their approximately 100 million years on Earth, spawning upward of 50 species currently known.

Who are thyreophorans?

All thyreophorans are characterized by parallel rows of osteoderms, which run down the necks, backs, and tails. The group is dominated by two great clades: **Stegosauria** (*stego* – roof); and **Ankylosauria** (*ankylo* – fused). Together, stegosaurs and ankylosaurs make up a monophyletic clade known as **Eurypoda** (*eury* – broad; *poda* – feet). Along with these two big groups, a few other miscellaneous, primitive Early Jurassic thyreophorans round out our story. Figure 10.2 lays out basic thyreophoran relationships.

Primitive Thyreophora

Primitive thyreophorans, outside of Ankylosauria and Eurypoda, are represented by three forms: *Scutellosaurus*, *Emausaurus*, and *Scelidosaurus* (Figure 10.3).[1] Although primitive ornithischians in most respects, all have the diagnostic thyreophoran character of rows of osteoderms down the back

---

[1] And possibly a fourth: *Lesothosaurus*, if the proposal of Butler and colleagues (2008) is correct.

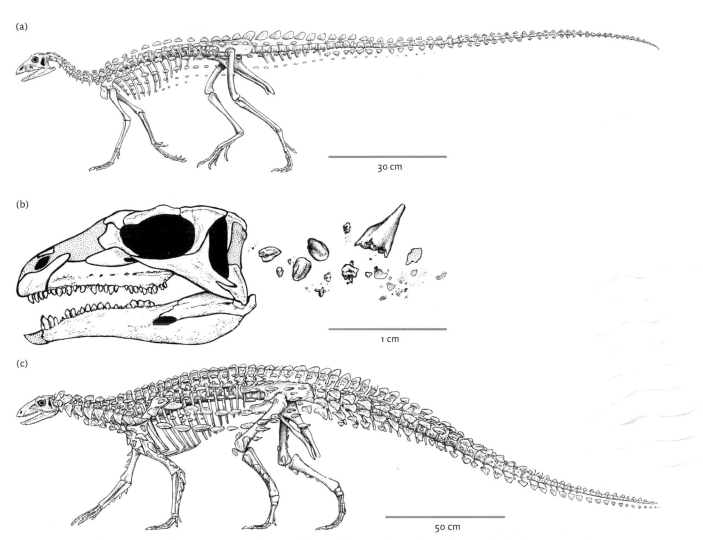

(a)

30 cm

(b)

1 cm

(c)

50 cm

Figure 10.3. Left lateral view of primitive thyreophorans. (a) *Scutellosaurus*, from the Early Jurassic of Arizona, USA; (b) *Emausaurus*, from the Early Jurassic of Germany; and (c) *Scelidosaurus*, from the Early Jurassic of England.

and tail. Two were bipedal, reflecting the primitive condition in Dinosauria; *Scelidosaurus* was quadrupedal, foreshadowing the trend in the rest of Thyreophora.

Eurypoda: Stegosauria

Stegosaurs were medium-sized dinosaurs, 3–9 m in length and weighing 300–1500 kg, characterized by bearing osteoderms that developed into spines and plates, as well as by their quadrupedal stance (Figure 10.4). Their profiles sloped strongly forward and downward toward the ground as a result of the hindlimbs being substantially longer than the forelimbs (Figure 10.5). All toes had broad hooves. They seem to have been relatively uncommon dinosaurs, yet clearly had a global distribution (Figure 10.6).

Stegosaurian lives and lifestyles

Locomotion

The general body plan of stegosaurs does not suggest life in the fast lane (Figure 10.7). Indeed, with their long back legs and short front legs, stegosaurs must have had a locomotor conundrum: at the same **cadence** (the rate of feet hitting the ground), the hindlimbs would have covered much more

Figure 10.4. *Tuojiangosaurus*, a stegosaur from the Late Jurassic of Sichuan Province, China.

distance than the forelimbs. At high speeds, therefore, the rear of the animal would have overtaken its head! This problem could be avoided in two ways: (1) by drawing the forelimbs up from the ground (that is, temporarily being bipedal while running) or (2) by limiting movement to a slow walking gait. Because of the mass distribution of stegosaurs, the first option is unlikely. Our best guess is that the pace of stegosaur life was leisurely, on the order of 6.5 to 7.0 kmh maximum speed (see Box 12.3). Chasing fleet prey was not too important to a hungry herbivore.

Neither, it would seem, was growing up quickly. Paleontologists can gain insights about the rate at which a dinosaur grew from juvenile to adult by the type of bone tissue it produces (Chapter 13). Recent studies on stegosaur development, based upon the type of bone tissue they have, suggest that by the standards of most other dinosaur groups, stegosaurs generally took their time to grow, not getting too large too quickly. This has implications for how stegosaurs used their distinctive plates (see "Spines and plates" below).

Dealin' with mealin'

The business end of feeding in stegosaurs began at the rhamphothecae, similar to those seen in modern turtles and birds, which covered the fronts of both the upper and lower jaws (Figure 10.8). The rhamphothecae were probably sharp-edged, and were used to crop and strip foliage.

Like all genasaurs, stegosaurs had an inset tooth row, implying cheeks, which in turn suggest chewing; however, exactly how that must have worked is baffling. The cheek teeth of stegosaurs were relatively small, simple, and triangular (Figure 10.9), and not tightly pressed together in a block for efficient grinding. Moreover, the teeth lack regularly placed, well-developed worn surfaces,

(a)

Figure 10.5. The best-known of all plated dinosaurs, the North American *Stegosaurus* from the Late Jurassic. (a) Front three-quarter view; (b) rear three-quarter view. Note that the tail is now reconstructed as having been held straight out and back (rather than dragging in the familiar hump-backed reconstruction), a position supported by the anatomy of the ligaments and bases of the plates, both of which supported the tail.

features present in herbivores that chew by grinding. Furthermore, the coronoid process was low, lending little mechanical advantage to the jaw musculature. Chewing? Perhaps, but not particularly efficient when compared with other chewing vertebrates.

So how else might stegosaurs have ground their food? Birds use gastroliths (see Chapter 6) to grind food. The problem is that gastroliths have never been found with stegosaur remains, as they have with other dinosaurs (prosauropods, sauropods, psittacosaurs, and ornithomimosaurs). Ultimately, however, the co-existence in stegosaurs of a small skull (in proportion to the size of the animal), simple, irregularly worn teeth, large gut capacity, cropping rhamphothecae, weak jaw musculature, and cheeks all conspire to make the business of dealin' with mealin' – beyond knowing that they were herbivorous – poorly understood in these dinosaurs.

What might stegosaurs have eaten? In most, the head was held near the 1 m level. Thus stegosaurs were likely low-browsers, consuming ground-level plants such as ferns, cycads, and other herbaceous gymnosperms (see Chapter 14).

(b)

Figure 10.5. (*cont.*)

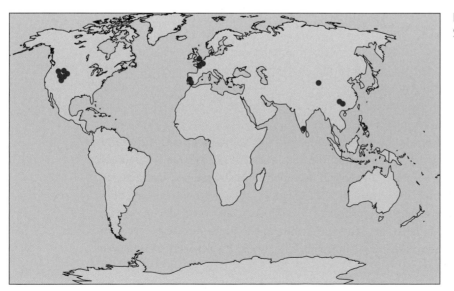

Figure 10.6. Global distribution of Stegosauria.

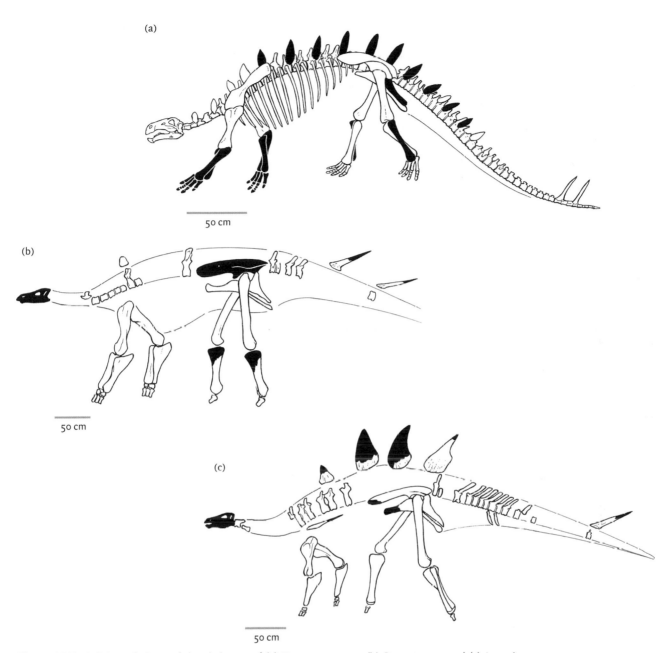

Figure 10.7. Left lateral views of the skeletons of (a) *Huayangosaurus*, (b) *Dacentrurus*, and (c) *Lexovisaurus*.

Of course, the Mesozoic world of the low-browsers was not filled only with stegosaurs. It is very likely that stegosaurs competed with a variety of other dinosaurs, many of which appear to have been more efficient chewers. Could stegosaurs have used their narrow skulls to select only the most nutritious parts of the plant, while everybody else dined less discriminately?

On the other hand, maybe stegosaurs weren't confined to low-browsing. Some paleontologists have argued that stegosaurs could have reared up on their hindlimbs in order to forage. Then the strong, flexible tail might have acted as a third "leg" to form a tripod. If so, these animals could have reached as high as 6 m in the largest forms.

(a)

(b)

(c)

(d)

(e)

(f)

(g)

10 cm

Figure 10.8. Left lateral views of the skulls of (a) *Stegosaurus*, (b) *Huayangosaurus*, (c) *Tuojiangosaurus*, and (d) *Chungkingosaurus*. Dorsal views of the skulls of (e) *Stegosaurus*, (f) *Huayangosaurus*, and (g) *Tuojiangosaurus*.

No brains, one brain, or two brains?

It is as clear as most anything can be at a distance of 150 million years that stegosaurs were just not all that bright. Their brains were an estimated 0.001 per cent of the adult stegosaur body weight, putting them near the bottom of the dinosaur – for that matter, vertebrate – gray-matter scale (Figure 10.10). Brainy-ness must not have been part of the stegosaur largely defensive life strategy, as indeed they were so small-brained that early workers felt compelled to assign them an extra brain: based upon an enlargement of the canal in the centra of the vertebrae (see inset to Figure 4.5) in the hip region, in which the spinal cord rests. Here began the legend of the dinosaur with two brains: a small one in the head and another in the pelvis, presumably to pick up the slack left by the first. All of this inspired literary effulgence, two samples of which we offer in Box 10.1.

The enlargement of the stegosaur spinal canal in the pelvic region, the putative "hind brain," is yet another stegosaur mystery. Many vertebrates have enlargements in the sacrum for nerves going to the hind legs,

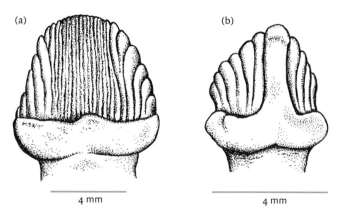

(a)   (b)

4 mm   4 mm

Figure 10.9. Inner views of an upper tooth of (a) *Stegosaurus* and (b) *Paranthodon*.

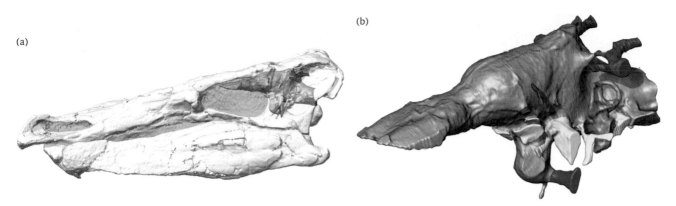

Figure 10.10. CT-scanned reconstruction of skull and brain of *Stegosaurus*. (a) cross-section through skull showing location of brain; (b) brain. Images courtesy of Larry Witmer and Witmerlab, Ohio University, USA.

but the neural canal at the front of the stegosaur pelvis is upward of 20 times the volume of the brain. Some living birds have a similar enlargement that houses an organ whose function is thought to supply glycogen (a complex sugar-based molecule which the body stores, but can break down to obtain energy) to the nervous system. In the early 1990s, evolutionary biologist and anatomist Emily Bucholtz proposed that the enlargement in the stegosaur sacrum could have housed a glycogen body, an idea that still is perhaps the best explanation for the high degree of sacral enlargement.

Speculation aside, the two stegosaurs in which the brain cavities are known – *Kentrosaurus* and *Stegosaurus* – suggest that stegosaur brains were relatively long, slightly flexed, and *small* (Figure 10.10). Only the **olfactory bulbs**, the portions of the brain that provide the animal with its sense of smell, are somewhat enlarged. Clearly stegosaurs, animals that had an unhurried lifestyle and possibly a relatively uncomplicated range of behaviors, would have had time to stop and smell the roses. . . had there *been* any roses![2]

Social lives of the enigmatic

We don't have much of an idea about the social behavior of stegosaurs or know much about their life histories. No nests, isolated eggs, eggshell fragments, or hatchling material is yet known for any stegosaur. In fact, only a few juvenile and adolescent stegosaur specimens can tell us anything about the lives of subadult stegosaurs.

Among fully adult individuals, some workers have proposed that there is some **sexual dimorphism**; that is, differences between the sexes reflected in the morphology. It has been claimed that this shows up in, almost biblically, the number of ribs that contribute to the formation of the pelvis, as well as in the size of the thigh. While sexual dimorphism was never definitively demonstrated using these features, recently, differences in the size and shape of the plates was clearly shown for *Stegosaurus*. All of the appropriate criteria for the differences being related to gender were observed: there were only two varieties and no intermediate forms, one form was 45 per cent larger and wider than the other; there were approximately equal numbers of each kind, and these differences occur at every plate position on the animal.

Little is also known about the degree of sociality among stegosaurs. A mass accumulation of disarticulated, yet associated, *Kentrosaurus* material from Tendaguru in Tanzania (see Chapter 15)

[2] Roses didn't appear until long after the last stegosaur went extinct – see Chapter 14.

Box 10.1   Dino doggerel

OK. So we're not talking about Keats, Blake, Dickinson, or Yeats here; dinosaur poems never aspired to that. Most have contrasted dinosaurian enormity with their putative lack of brain power (and, no doubt, social graces), with few recent efforts to balance such dismal views.

The most famous dinosaurian poem celebrates the mental achievements of *Stegosaurus*, in particular the cerebral gymnastics powered by its double brains. The piece, by Bert L. Taylor, a columnist in the 1930s and 1940s for the *Chicago Tribune*, goes like this:

> Behold the mighty dinosaur,
> Famous in prehistoric lore,
> Not only for his power and strength
> But for his intellectual length.
> You will observe by these remains
> The creature had two sets of brains —
> One in his head (the usual place),
> The other at his spinal base.
> Thus he could reason a priori
> As well as a posteriori
> No problem bothered him a bit
> He made both head and tail of it.
> So wise was he, so wise and solemn,
> Each thought filled just a spinal column.

> If one brain found the pressure strong
> It passed a few ideas along.
> If something slipped his forward mind
> 'Twas rescued by the one behind.
> And if in error he was caught
> He had a saving afterthought.
> As he thought twice before he spoke
> He had no judgement to revoke.
> Thus he could think without congestion
> Upon both sides of every question.
> Oh, gaze upon this model beast,
> Defunct ten million years at least.

As a counterpoint to the power of Mesozoic intellects, we also provide some thoughts, entitled *The Danger of Being too Clever*, by John Maynard Smith, the famous English evolutionary biologist.

> The Dinosaurs, or so we're told
> Were far too imbecile to hold
> Their own against mammalian brains;
> Today not one of them remains.
> There is another school of thought,
> Which says they suffered from a sort
> Of constipation from the loss

provides us with a hint that *Kentrosaurus* was **gregarious** (exhibited herding and other social behaviors), as does some *Stegosaurus* material from Montana. The combination of gregarious behavior and sexual dimorphism suggests that at least some stegosaurs may have interacted through display, something like we see in modern birds. Of course, feathers – or anything approaching them – are not known in any stegosaur. And the signals likely came from the spines and plates that grace the back of stegosaurs.

Spines and plates

In keeping with their thyeophoran heritage, the majority of stegosaurs had at least two rows of osteoderms down the back; one edge embedded deeply in the thick tough hide (Figure 10.11). These osteoderms generally take the form of spines, spikes (Figure 10.12), blunt cones, or plates. We've seen that they are dimorphic. Originally, the idea was that they were all about protection and defense. But defense, if it is any part of the story, isn't the whole story.

The shapes and patterns of plates and spines in stegosaurs are nearly always **species specific**; that is, diagnostic for a particular species (in this case stegosaurs). Moreover, they have their maximum visual effect when viewed from the side. These observations, in conjunction with those above, further suggest a display function – both for predators *and* for other stegosaurs. Predators surely would have seen a larger, more intimidating animal with the plates sticking up. If **intraspecific** display (display among members of the same species) was involved, it is likely that individual stegosaurs would have

Of adequate supplies of moss.
But Science now can put before us
The reason true why Brontosaurus
Became extinct. In the Cretaceous
A beast incredibly sagacious
Lived & loved & ate his fill;
Long were his legs, & sharp his bill,
Cunning its hands, to steal the eggs
Of beasts as clumsy in the legs
As Proto- & Triceratops
And run, like gangster from the cops,
To some safe vantage-point from which
It could enjoy its plunder rich.
Cleverer far than any fox
Or Stanley in the witness box
It was a VERY GREAT SUCCESS.
No egg was safe from it unless
Retained within its mother's womb
And so the reptiles met their doom.
The Dinosaurs were most put out
And bitterly complained about
The way their eggs, of giant size,
Were eaten up before their eyes,
Before they had a chance to hatch,

By a beast they couldn't catch.
This awful carnage could not last;
The age of archosaurs was past.
They went as broody as a hen
When all their eggs were pinched by men.
Older they grew, and sadder yet,
But still no offspring could they get.
Until at last the fearful time, as
Yet unguessed by Struthiomimus
Arrived, when no more eggs were laid,
And then at last he was afraid.
He could not learn to climb with ease
To reach the birds' nests in the trees,
And though he followed round and round
Some funny furry things he found,
They never laid an egg – not once.
It made him feel an awful dunce.
So, thin beyond all recognition,
He died at last of inanition.
MORAL
This story has a simple moral
With which the wise will hardly quarrel;
Remember that it scarcely ever
Pays to be too bloody clever.

used these structures not only to tell each other apart, but also to gain dominance in territorial disputes and/or as libido-enhancers during the breeding season.[3]

The evidence that exists suggests that the idea is not completely out to lunch. The few juvenile stegosaurs known appear not to have had large spines or plates on their backs. The absence of these features in small, sexually immature individuals suggests that only when maturity was reached did looking big and sexy acquire importance. Here, then, is a strong argument for the importance of plates in display: if the plates develop only when you reach maturity, then they likely had little role to play in immature activities, and must have been associated with adult behaviors.

Thyreophorans as a group tend to be defensively armored, and stegosaurs were no exception. Along with the plates and spines, a flexible pavement of small osteoderms covered the neck region.

Hot plates?

The surfaces of the plates of *Stegosaurus* are covered with an extensive pattern of grooves, while the insides are filled with a honeycomb of channels (Figure 10.13). These external grooves and internal channels most likely formed the bony walls for an elaborate network of blood vessels. With such a rich supply of blood from adjacent regions of the body, could the plates have been used to cool the

---

[3] Which gender had the big plates and which had the narrower, smaller plates is an open question. If living birds and reptiles are any reasonable model it is likely that it was the males with the larger parts; but many exceptions surely exist.

Figure 10.11. Perhaps the best specimen (NMNH 4349) of *Stegosaurus* as it was found in the field. Note that the plates do not articulate directly with the vertebrae; they were embedded in the skin; so exactly how they and the tail spikes were oriented on the living animal has been somewhat hard to determine. Paleontological consensus, however, now has them as two parallel, vertically oriented rows down the back of the animal. The Prehistoric Museum in Price, UT, USA paleontologist K. Carpenter reconstructs the tail spines, however, oriented pointing outward; he noted, reasonably, that their evident functionality would have been impaired if they pointed directly upward like the plates. Image is the original drawing from 1887 publication by O. C. Marsh (see Chapter 15). O. C. Marsh. 1887. *American Journal of Science*, **34**, 413–417.

Figure 10.12. The skeleton of *Kentrosaurus*, a spiny stegosaur from the Late Jurassic of Tanzania.

body by dissipating heat as air passed over them, or to warm the body by absorbing solar energy? In short, could the plates have been used for **thermoregulation** (temperature control)? As a test of this idea, paleontologist J. O. Farlow and colleagues tested the ability of the plates to radiate (or absorb) heat. Arranged in symmetrical pairs, as they are in life, the plates provided significant heat dissipation, suggesting that thermoregulation may also have been a role of the plates.

Figure 10.13. Lateral view of one of the dermal plates of *Stegosaurus*. Note the great number of parallel grooves, used to convey nutrients through blood vessels across the outer surface of the plate to the keratin covering the bony plate.

Creative and attractive as this idea is, it was strongly critiqued by Harvard biologist R. P. Main and colleagues.[4] Their 2005 analysis of plate microstructure argues that no system of conduits exists, that would have circulated blood in and out of the plates radiator-style; rather, the grooves preserved on the surface of the plates were likely used to bring nutrients to the keratinized sheath that covered the bony plate in life, much as one sees in the bony underpinnings of nails, claws, and beaks.

So we come back to where we began: as an appearance-enlarging device, the plates may have served an important role in defense. Yet, there has never been any doubt about the role of the long, pointed tail spikes! These were likely slashed from side to side on the powerful tail. Older reconstructions show these spikes pointing primarily upward; recent work by dinosaur specialist Ken Carpenter (see caption, Figure 10.11) suggests that the spikes actually splayed out to the sides, producing a much more effective defensive weapon.

So stegosaurs appear as a mass of contradictions: chewers that may not have chewed all that well; animals in which supposed sexual display functions were prominent but in which there is little evidence for social behavior. Until very recently, we didn't even know the positions of the plates in *Stegosaurus*, the best-known genus. So much of what is apparently contradictory about stegosaurs is likely due to how little we know about them – a situation that we'd like to see changed!

Eurypoda: Ankylosauria – mass and gas

Ankylosaurs were masters of the art of defense-by-hunkering. As their name implies, ankylosaurs were encased in a pavement of bony plates and spines – each embedded in skin and interlocked with adjacent plates – that formed a continuous shield across the neck, throat, back, and tail (Figures 10.14, 10.15, and 10.16). In many cases, it covered the top of the head, cheeks, and even eyelids.

---

[4] We will see several examples of creative and attractive, but potentially discredited ideas as this book unfolds. If science is, as we asserted in Part I, fundamentally a creative process, then these ideas are important not only for what they teach us about dinosaurs, but for what they teach us about the creative process of hypothesis-development and testing.

Figure 10.14. *Euoplocephalus*, the armored, club-tailed ankylosaur.

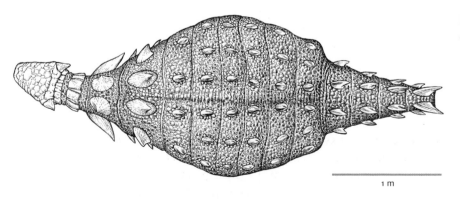

Figure 10.15. Dorsal view of the body armor of *Euoplocephalus*; reconstruction.

1 m

Under all that armor, ankylosaurs were round and very broad (see Figure 10.14); clearly their girth accommodated a large gut. The head was low and broad (Figure 10.17; 10.18a), and equipped with simple, leaf-shaped teeth for pulverizing whichever plants an ankylosaur chose to eat (Figure 10.18b).

Figure 10.16. Dorsal view of the body armor of *Sauropelta*.

Figure 10.17. Left lateral view of the skulls of (a) *Shamosaurus*, (b) *Ankylosaurus*, (c) *Pinacosaurus*, (d) *Tarchia*, (e) *Silvisaurus*, and (f) *Panoplosaurus*.

Among the most striking features of the skulls of ankylosaurs are the complex internal nasal passageways, which have been revealed by CT scans (see Box 11.1) of the skulls. These passageways were elaborate, and in some cases, convolute (see below, Figure 10.19). Their precise function is uncertain, although they suggest that olfaction, the sense of smell, was an important sense in this group of dinosaurs.

Ankylosaurs were mid-sized dinosaurs, rarely exceeding 5 m in length, although some (such as *Ankylosaurus*) ranged upward of 9 m. As in stegosaurs, the limbs were of different lengths, with the hindlimb exceeding the length of the forelimb by as much as 150 per cent.

Like stegosaurs, ankylosaurs also had a global distribution, coming predominantly from North America and Asia, but also from Europe, Australia, South America, and Antarctica (Figure 10.20). As a group, they reached their peak diversity during the Cretaceous.

The best-preserved fossils come from Mongolia and China, where stunning specimens have been found inland, far from the ocean (either then or now), nearly complete and articulated, and in most cases preserved upright or on their sides. In contrast, in North America, only partial skeletons have been found, and these are often upside-down, sometimes in rocks deposited along the seashore or even in rocks representing open marine environments. The North American forms evidently lived sufficiently close to the sea that their bloated carcasses might have been carried out with the tide, flipping upside-down because of their heavily armored backs.

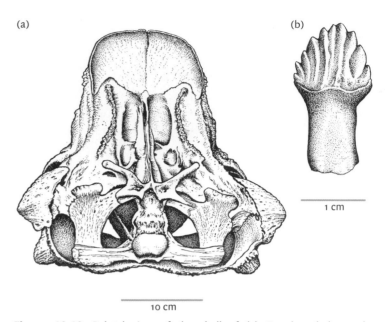

Figure 10.18. Palatal view of the skull of (a) *Euoplocephalus* and (b) a tooth of *Edmontonia*.

Ankylosaur finds most commonly consist of individual skeletons or isolated partial remains; there is only one ankylosaur bonebed known. Perhaps this indicates that these animals had solitary habits or lived in very small groups. Even from our incomplete window on the past, it appears reasonably certain that ankylosaurs did not enjoy the company of huge herds.

Ankylosauria consists of two great clades: **Nodosauridae** (*nodo* – knot; referring to the rounded osteoderms) and **Ankylosauridae** (Figures 10.21 and 10.22).

Nodosauridae

Nodosaurids had relatively long snouts and well-muscled shoulders, reflected by the presence of a large knob of bone on the shoulder blade, the **acromial process** (Figure 10.22), an attachment site for the heavy shoulder musculature that characterizes nodosaurs. Nodosaurs also had flaring hips, and pillar-like limbs. Many had tall spines at the shoulder (parascapular spines). Nodosaurids are known principally from the Northern Hemisphere (North America and Europe), although new discoveries in Australia and Antarctica have extended the geographical range of these animals deep into the Southern Hemisphere.

Ankylosauridae

Members of Ankylosauridae give the impression of impregnable, mobile fortresses. All are well armored (see Figures 10.15 and 10.21), but there are fewer tall spines along the body than in nodosaurids. The tail ends in a massive bony club, in some instances with several paired knobs or triangular spikes along its length. The head is shorter and broader in ankylosaurids than in nodosaurids and there are large triangular plates attached to the rear corners of the skull (termed "squamosal horns;" see Figure 10.18a).

Ankylosaur lives and lifestyles

Mouths to feed

Ankylosaurs doubtless had a very low browsing range, foraging for plants no more than a meter or so above the ground. The different beak shapes suggest different feeding preferences. The narrow beak of nodosaurids may suggest selective feeding, plucking or biting at particular kinds of foliage

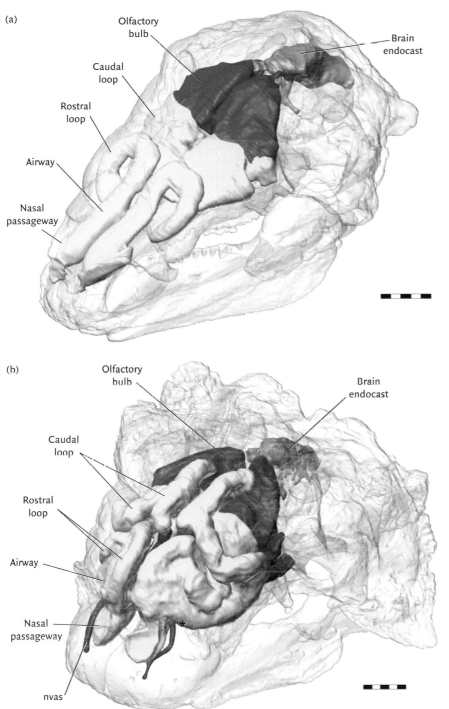

(a)

Olfactory
bulb

Caudal
loop

Rostral
loop

Airway

Nasal
passageway

Brain
endocast

(b)

Olfactory
bulb

Caudal
loop

Rostral
loop

Airway

Nasal
passageway

nvas

Brain
endocast

Figure 10.19. The skulls of nodosaurids (above) and ankylosaurids (below) compared. The skull of the ankylosaurid is broader, and was held at a more downward angle than that of the nodosaurid. Three-quarter view CT-scan reconstructions of (a) the nodosaur *Panoplosaurus* and (b) the ankylosaurid *Euoplocephalus*. In both cases, convolute nasal sinuses are revealed, running from the external nasal opening towards the braincase. These sinuses doubtless were involved in a heightened sense of smell; we don't know if ankylosaurs smelled good, but the evidence suggests that they smelled well! The small, darkened region to the rear of the skull is the brain. Scale bars = 5 cm.

and fruits with the sharp edge of the rhamphotheca. In contrast, the very broad beak of ankylosaurids may imply less selective feeding, in which plant parts were indiscriminately bitten off from the bush or pulled from the ground.

How the food was then prepared for swallowing is a bit of a mystery. Like stegosaurs (and pachycephalosaurs; see Chapter 11), the triangular teeth of both nodosaurids and ankylosaurids

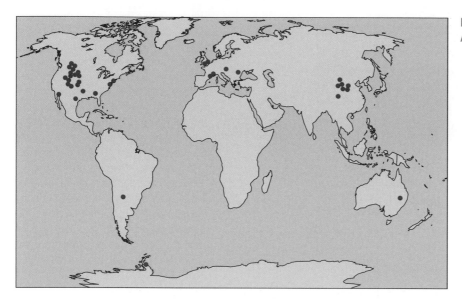

Figure 10.20. Global distribution of Ankylosauria.

are small, not particularly elaborate, and less tightly packed than animals with well-developed chewing behaviors (see Figure 10.18b). However, the wear marks on the teeth indicate that grinding took place. In addition, it is likely that ankylosaurs had a long, flexible tongue (in their throats, they have large **hyoid bones** that support the base of the tongue) and an extensive **secondary palate**, which allowed them to chew and breathe at the same time. Moreover, deeply inset tooth rows suggest well-developed cheeks to keep whatever food was being chewed from falling out of the mouth.

(a)

(b)

Figure 10.21. (a) An ankylosaurid (*Euoplocephalus*) and (b) a nodosaurid (*Sauropelta*) compared. The ankylosaurid is stockier, and has a tail-club. The nodosaurid is more lightly built, and bears distinctive dermal armor.

Figure 10.22. Left lateral view of the skeleton of *Sauropelta* without its armor shield. Note the projection on the scapula (indicated by arrow), known as the acromial process.

The jaw bones themselves were relatively large and strong (although lacking enlarged areas for muscle attachment). Indeed, most ankylosaur jaw features – except for tooth design and placement – suggest that ankylosaurs were reasonably adept chewers.

Perhaps the paradox of simple teeth in strong, cheek-bound jaws can be understood by looking not at how much chewing was done prior to swallowing (there obviously was some), but at the very substantial rear end of the animal, where the bones show a very deep rib cage circumscribing an enormously expanded abdominal region (see Figure 10.14). Commodious abdomens mean huge guts, and huge guts suggest that that digestion took place in a very large, perhaps highly differentiated, fermentation compartment(s) in these armored dinosaurs. Bacteria likely lived symbiotically within the stomach (or stomachs, because ankylosaurs may well have had a series of them). The stomach(s) must have served as a great fermentation vat(s), bacterially decomposing even the toughest woody plant material. Among modern mammals, this method of breaking down tough plant material is well known in ruminants such as cows.

The combination of browsing at low levels and having anatomy indicative of chewing and fermenting places limits on what kinds of plants ankylosaurs may have fed on. At the levels where ankylosaurs could forage, the undergrowth consisted of low-stature ferns, gymnosperms such as cycads, and shrubby **angiosperms** (flowering plants), a rich array of plants to choose from within the first few meters above the ground (see Chapter 14).

Brains and senses

Ankylosaurs may not have been particularly adept at making foraging choices, however, because their brain power was close to the bottom of the dinosaur range (only sauropods had smaller brains for their size; see Chapter 9 and Box 13.4). Still, had there been roses (see footnote 1, this chapter), ankylosaurs, too, might have stopped to smell them; the skulls were equipped with enlarged, convolute nasal passages (Figure 10.19), suggesting that olfaction may have been an important sense for these dinosaurs. Hand in hand with slow thinking, ankylosaurs were among the slowest moving of all dinosaurs for their body weight. Estimates suggest that they were able to run no faster than 10 kmh and walked at a considerably more leisurely pace (about 3 kmh). Ankylosaurs were built for digestion, not for speed.

Defense

Although ankylosaurs were slow on their feet, other aspects of the limb skeleton suggest that they could actively defend themselves against predators. In all ankylosaurs the entire upper surface of the

Figure 10.23. (a) The bony tail-club of *Euoplocephalus*. (b) Tail spikes in the tail of *Pinacosaurus*. Visible along the length of the tail are the ossified tendons.

body – the head, neck, torso, and tail – was covered by a pavement of bony plates (see Figures 10.15 and 10.16).

In the case of nodosaurids, the shoulder region, as we have seen, was exceedingly powerful, with tall spines. This pumped-up, well-defended front end, in conjunction with relatively long hindlimbs and wide stance, may have had the effect of dropping the center of gravity of the animal forward, providing nodosaurids with an aggressive frontal, spine-based defense.

Ankylosaurids went even further, augmenting their armor by a powerful tail-club (Figure 10.23), constructed of paired masses of bone set at the end of a tail (itself about half the length of the body). In some cases the tail was also equipped with spikes along its length (Figure 10.23b).

Thanks to CT-scans and careful muscle reconstructions, we now know a lot about the design of the ankylosaurid tail-club. The paired masses of bone varied in density, with the densest bone located outermost (Figure 10.24). While the tail was flexible at its base, the rear half was stiffened by modified, interlocking vertebrae, as well as by a series of tendons that had become **ossified** – turned to bone, and thus inflexible – running down its length. This part of the tail was not meant to bend, and it provided firm attachment for the powerful muscles. Using the flexibility of the tail vertebrae at the base and the stiffened tail farther out, the club could have been forcefully swung side to side like a bat.

Figure 10.24. (a) CT-scan image through an ankylosaur tail-club. (b) Interpretation of (a), showing densest bone (white), bone of medium density (light gray), and bone of lowest density (dark gray). Scale bar equals 5 cm.

Just how forcefully? Paleontologist Victoria Arbour theoretically calculated the velocity and thus the power of the swing; her estimates of the velocity, depending upon the type of muscle reconstruction used, are as high as 18.9 m/s, producing bone-crunching impulses of 376 kgm/s. The idea, clearly, would be to avoid any such contact with an ankylosaur tail-club in the first place!

For nodosaurids, a proactive defense must have been a head-first (or shoulder-first) proposition, keeping the parascapular spines pointed toward the predator. Ankylosaurids, on the other hand, likely first planted their hindlimbs and then rotated their forequarters with their strong forelimb muscles, ever keeping watch on the threatening opponent. Then they would have wielded that massive club at their opponent's legs and feet.

In both ankylosaurids and nodosaurids, however, the last resort must have been to hunker down defensively. With its legs folded under its body, a 3500 kg ankylosaur would have been formidably immobile. Safe under protective armor, both nodosaurids and ankylosaurids were the best-defended fortresses of the Mesozoic.

The evolution of Thyreophora
Thyreophora

We can be reasonably confident that the evolution of thyreophorans embodied increasing size and a return to quadrupedality, because the cladograms show that the quadrupedal eurypodan clades were all derived relative to primitive thyreophorans. Primitive thyreophorans suggest an evolutionary sequence from gracile, small, bipedal creatures like *Scutellosaurus* to larger, quadrupedal dinosaurs like *Scelidosaurus* (see Figures 10.3 and III.3).

Eurypoda

It is not difficult to imagine the evolution of an ankylosaur from an armored, primitive quadruped like *Scelidosaurus*. The cladogram suggests that the basal eurypodan must have looked something like *Scelidosaurus*, and the step to a larger, more powerful, more heavily armored primitive ankylosaur or stegosaur is easy to conceive.

Stegosauria

Stegosauria is a monophyletic clade of ornithischian dinosaurs, diagnosed on the basis of a number of important features shown in Figure 10.25. The ancestral stegosaur must have been an animal with spine-shaped osteoderms and fore- and hindlimbs of not too dissimilar lengths. Within Stegosauria, the basal split is between *Huayangosaurus* on the one hand and remaining species on the other (Figure 10.26). This divergence took place sometime before the latter half of the Middle Jurassic. *Huayangosaurus* itself has a number of uniquely derived features shared by a more inclusive group of stegosaurs, Stegosauridae (Figure 10.26).

Within Stegosauridae, *Dacentrurus* represents the most basal form. The remainder of Stegosauridae includes *Stegosaurus*, *Wuerhosaurus*, *Kentrosaurus*, and *Tuojiangosaurus*. The evolution of this group was evidently characterized by an increase in the difference in length between the fore- and hindlimbs (Figure 10.26).

Finally, there is *Stegosaurus* itself, the best-known, most common stegosaur (see Figure 10.5). *Stegosaurus* must have evolved its distinctive plates from the spiny, conical osteoderms present in its ancestry. Plates, however, are only known in *Stegosaurus*, and their evolution occurred sometime during the Middle or early Late Jurassic.

(a)

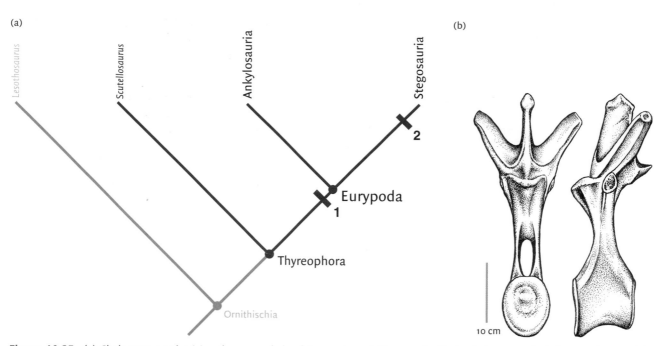

(b)

10 cm

Figure 10.25. (a) Cladogram emphasizing the monophyly of Eurypoda and Stegosauria. Derived characters include: at **1** (Eurypoda), bones that fuse to the margins of the eye sockets, loss of a notch between the quadrate (see Figure 4.6) and the back of the skull, and enlargement of the anterior part of the ilium; at **2** (Stegosauria), back vertebrae with very tall neural arches and highly angled transverse processes, loss of ossified tendons down the back and tail, a broad and plate-like acromial process, large and block-like wrist bones, elongation of the pre pubic process, loss of the first pedal digit, and loss of one of the phalanges of the second pedal digit, and a great number of features relating to the development of osteoderms, and formation of long spines on plates from the shoulder toward the tip of the tail. (b) The front and left lateral view of one of the back vertebrae of *Stegosaurus*. Note the great height of the neural arch; a diagnostic stegosaurian character (at **2**).

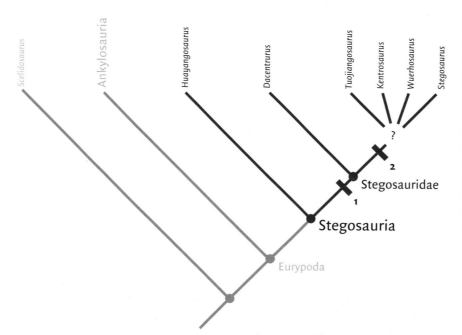

Figure 10.26. Cladogram of Stego-sauria, with Ankylosauria and *Scelido-saurus* as successively more distant relatives. Derived characters include: at **1**, large antitrochanter, long prepubic process, long femur, absence of lateral rows of osteoderms on the trunk; at **2**, widening of the lower end of the humerus, an increase in femoral length, and an increase in the height of the neural arch of the back and tail verte-brae. Relationships of genera of the ultimate node on the cladogram remain uncertain.

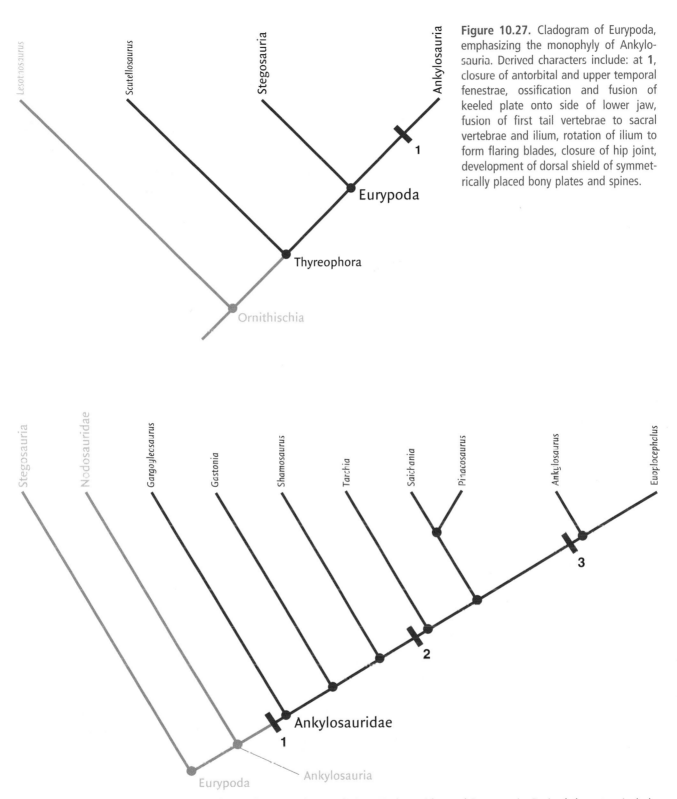

Figure 10.27. Cladogram of Eurypoda, emphasizing the monophyly of Ankylosauria. Derived characters include: at **1**, closure of antorbital and upper temporal fenestrae, ossification and fusion of keeled plate onto side of lower jaw, fusion of first tail vertebrae to sacral vertebrae and ilium, rotation of ilium to form flaring blades, closure of hip joint, development of dorsal shield of symmetrically placed bony plates and spines.

Figure 10.28. Cladogram of Ankylosauridae, with its two closest relatives, Nodosauridae and Stegosauria. Derived characters include: at **1**, pyramidal squamosal boss, shortening of premaxillary palate, presence of premaxillary notch, rostrolaterally directed mandibular ramus of pterygoid, tail-club; at **2**, rostrodorsal- and caudoventral-arching of palate, vertical nasal septum, rugose and crested basal tubera; at **3**, caudal end of post-axial cervical vertebrae dorsal to cranial end, fusion of sternals.

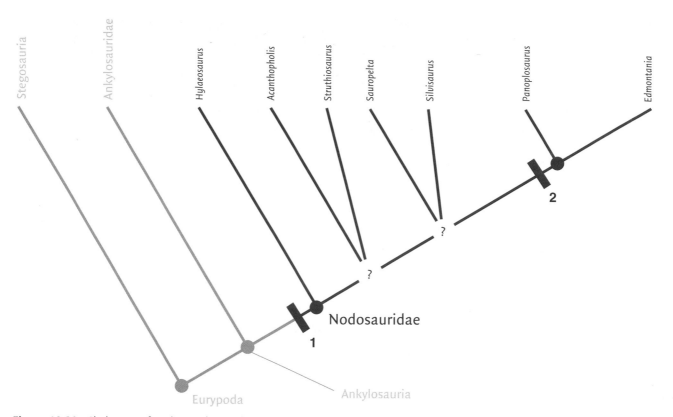

Figure 10.29. Cladogram of Nodosauridae, with its two closest relatives, Ankylosauridae and Stegosauria. Derived characters include: at **1**, knob-like acromion on scapula, occipital condyle derived from but one bone (the basioccipital), large knob (supraorbital boss) above the eye; at **2**, premaxillary teeth lacking, distinct rostrodorsal feature of secondary palate.

Ankylosauria

Reflecting the importance of heavy armor to ankylosaurs and the ease of its preservation, it is not surprising that armor and/or its support comprise the majority of derived features uniting the clade Ankylosauria (Figure 10.27). Ankylosaur evolution followed two principal pathways since the origin of the group sometime in the Jurassic: Ankylosauridae and Nodosauridae.

Primitively in all ankylosaurs, the beak was scoop-shaped but relatively narrow (although slightly broader than in stegosaurs) and remained so in the nodosaurids and in the ankylosaurid *Shamosaurus*. In all other ankylosaurids, by contrast, the beak became very broad, which matched the general broadening of the animal and the development of tail-clubs.

Ankylosauridae

Ankylosaurids share a suite of derived features. Figure 10.28 highlights the relationships among ankylosaur genera, as well as some of the key characters supporting these relationships.

Nodosauridae

Turning to the other great clade of ankylosaurs, nodosaurids share a number of derived features, as shown in Figure 10.29, particularly the well-developed acromial process. This musculature, and the parascapular spines that accompanied it, may have played a role in their defensive behavior. Nodosaurids changed little during their tenure on Earth; however, various diagnostic characters allow us to learn something of their relationships (Figure 10.29).

SUMMARY

Thyreophorans were small- to medium-sized quadrupedal ornithischians with rows of scutes down the back. Aside from some primitive forms, the two great groups of thyreophorans were Stegosauria and Ankylosauria, linked together within a monophyletic Eurypoda. Eurypodans, as a group, were not renowned for their intellects; some of the lowest brain : weight ratios known for dinosaurs come from Eurypoda.

Stegosaurs are relatively poorly known eurypodans with hindlimbs significantly longer than the forelimbs, and paired rows of plates or spines down the back, terminating in a tail with elongate spines: likely a defensive weapon. Their fossil record extends from the Middle Jurassic to the Early Cretaceous.

Stegosaurs are rare finds, and they likely lived in isolation rather than gregariously. The plates in *Stegosaurus* may have been involved in thermoregulation. Stegosaurs – like all eurypodans – had cheeks, which suggest chewing. The teeth, however, occluded relatively poorly, suggesting inefficient grinding. Due in part to their rarity, the reproductive strategies and behavior of stegosaurs are still largely unknown.

Ankylosaurs were squat, armored quadrupeds, coated with a pavement of osteoderms, and bony protection everywhere. They lived from the mid-Jurassic to latest Cretaceous. Two groups are known: Nodosauridae and Ankylosauridae. Nodosaurids were slightly more lightly built, with tall parascapular spines, while ankylosaurids were more squat, and equipped with a large tail-club.

Ankylosaurs were low-browsing animals. Their teeth and cheeks were rather like those of stegosaurs, but ankylosaurs had secondary palates, which may have aided in the efficiency of mastication. Unquestionably, though, much of their energy was obtained through gut fermentation, as suggested by the striking breadth of their girths. Ankylosaur bonebeds suggests that, unlike stegosaurs, these dinosaurs may have been gregarious animals. Little is known of their reproductive strategies and, by morphology, they were evidently animals that relied heavily upon defense, either by simply hunkering down or, in the case of ankylosaurids, by wielding tail-clubs.

SELECTED READINGS

Buffrenil, V. de, Farlow, J. O., and de Ricqlès, A. 1986. Growth and function of *Stegosaurus* plates: evidence from bone histology. *Paleobiology*, 12, 459–473.

Carpenter, K. (ed.) 2001. *The Armored Dinosaurs*. Indiana University Press, Bloomington, IN, 512 pp.

Carpenter, K. 2012. Ankylosaurs. In Brett-Surman, M. K., Holtz, T. R., and Farlow, J. O. (eds.) *The Complete Dinosaur*, 2nd edn. Indiana University Press, Bloomington, IN, pp. 505–525.

Coombs, W. P., Jr. and Maryańska, T. 1990. Ankylosauria. In Weishampel, D. B., Dodson, P., and Osmólska, H. (eds.) *The Dinosauria*. University of California Press, Berkeley, pp. 456–483.

Coombs, W. P., Weishampel, D. B., and Witmer, L. M. 1990. Basal Thyreophora. In Weishampel, D. B., Dodson, P., and Osmólska, H. (eds.) *The Dinosauria*. University of California Press, Berkeley, pp. 427–434.

Galton, P. M. and Upchurch, P. A. 2004. Stegosauria. In Weishampel, D. B., Dodson, P., and Osmólska, H. (eds.) *The Dinosauria*, 2nd edn. University of California Press, Berkeley, pp. 343–362.

Giffin, E. B. 1991. Endosacral enlargements in dinosaurs. *Modern Geology*, 16, 101–112.

Hopson, J. A. 1980. Relative brain size in dinosaurs – implications for dinosaurian endothermy. In Thomas, R. D. K. and Olson, E. C. (eds.) *A Cold Look at the Warm-Blooded Dinosaurs*. AAAS Selected Symposium no. 28, pp. 278–310.

Jerison, H. J. 1973. *Evolution of the Brain and Intelligence*. Academic Press, New York, 482 pp.

Maryańska, T. 1977. Ankylosauridae (Dinosauria) from Mongolia. *Palaeontologia Polonica*, **37**, 85–151.

Norman, D. B., Witmer, L. M,. and Weishampel, D. B. 2004. Basal Thyreophora. In Weishampel, D. B., Dodson, P., and Osmólska, H. (eds.) *The Dinosauria*, 2nd edn. University of California Press, Berkeley, pp. 335–342.

Sereno, P. C. and Dong, Z.-M. 1992. The skull of the basal stegosaur *Huayangosaurus taibaii* and a cladistic analysis of Stegosauria. *Journal of Vertebrate Paleontology*, **12**, 318–343.

Vicaryous, M., Maryańska, T., and Weishampel, D. B. 2004. Ankylosauria. In Weishampel, D. B., Dodson, P., and Osmólska, H. (eds.) *The Dinosauria*, 2nd edn. University of California Press, Berkeley, pp. 363–392.

TOPIC QUESTIONS

1. What are the diagnostic characters of Thyreophora? How are thyreophorans related to other ornithischians?

2, What are the diagnostic characters of Stegosauria?

3. What are the diagnostic characters of Ankylosauria?

4. How do ankylosaurids differ from nodosaurids?

5. Describe the evidence for there being two brains in *Stegosaurus*. What is the prevailing view on this issue now?

6. Did ankylosaurs defend themselves passively, or could they actively defend themselves?

7. What are the apparent contradictions in the mouths of stegosaurs? That is, what features appear to be well designed for chewing and what features appear not to be so well designed for chewing?

8. Do ankylosaurs have the same apparent contradictions in their chewing mechanisms as stegosaurs? Elaborate.

9. What might the plates of *Stegosaurus* have been used for? Would this use apply to all stegosaurs? Why?

10. Do the small brains in stegosaurs appear to have hindered their evolutionary success? Why?

11. What is the evidence for intraspecific competition in stegosaurs?

12. What are some of the differences between ankylosaurid defense and nodosaurid defense?

13. Why do we think ankylosaurs potentially ruminated?

14. Summarize what is known about herding behaviors in thyreophorans.

15. How does the interpretation of thermoregulation in stegosaurs correlate with their inferred activity levels?

16. What is known about sexual dimorphism in thyreophorans?

CHAPTER OBJECTIVES

- Introduce Marginocephalia and its two large constituent groups, Pachycephalosauria and Ceratopsia
- Develop familiarity with current thinking about lifestyles and behaviors of marginocephalians
- Develop an understanding of marginocephalian evolution using cladograms, and an understanding of the place of Marginocephalia within Dinosauria

Figure 11.1. Representatives of the two great groups of Marginocephalia. Left: *Prenocephale*, a pachycephalosaur. Right: *Protoceratops*, a ceratopsian.

Marginocephalia
Bumps, bosses, and beaks

11

Marginocephalia
Who were marginocephalians?

Marginocephalia (margin – edge; *cephal* – head): not a name you'll hear from the local 5-year-old dino-it-all. Yet, Marginocephalia reflects an important phylogenetic link between two major, somewhat different-looking, groups of dinosaurs: **Pachycephalosauria** (*pachy* – thick;) and **Ceratopsia** (*kera* – horn; *tops* – face). Together with Euornithopoda (Chapter 12), marginocephalians make up the taxon known as Cerapoda (Figure 11.2).

Marginocephalians all bear a ridge, or shelf, of bone running across the back of the skull. The size of this feature varies greatly, but in all cases, when viewed from above, it blocks from sight the bones at the back of the skull.

Although marginocephalians come in many shapes and sizes, they were restricted to the Northern Hemisphere during the Cretaceous Period.

Marginocephalia: Pachycephalosauria – in domes we trust

Pachycephalosaurs were bipedal ornithischians with thickened skull roofs (Figure 11.3). In the North American forms, this took the form of high domes, but in Asia, along with the high domes, certain varieties had flattened, thickened skulls (Figure 11.4); some, however, are considered to be juvenile forms of fully adult dome-headed pachycephalosaurs. Figure 11.5 shows the distinctively Northern Hemisphere distribution of pachycephalosaur sites.

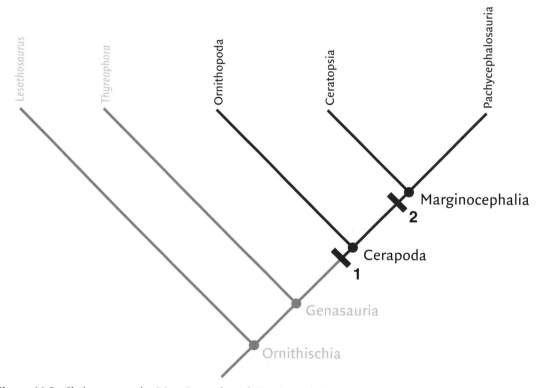

Figure 11.2. Cladogram emphasizing Cerapoda and Marginocephalia in Ornithischia. Derived characters include: at **1**, significant diastema between premaxillary and maxillary teeth, five or fewer maxillary teeth, finger-like anterior trochanter; at **2**, narrow parietal shelf obscuring occipital elements in dorsal view, lateral portions of shelf formed by squamosal.

Figure 11.3. The flat-headed, thick-headed *Homalocephale*, best known of all pachycephalosaurs.

Pachycephalosaur lives and lifestyles

Where did a pachycephalosaur call home?

In Asia, pachycephalosaurs apparently lived in a Sahara-like desert, punctuated by ephemeral streams in small drainage basins. Their remains are commonly found as nearly complete skulls and beautifully articulated skeletons. These fossils show little evidence of transport, and were apparently fossilized close to where the living animal died.

In North America, by contrast, the environments were very different. The rocks where marginocephalian remains are found represent a broad, Cretaceous coastal plain in a then-temperate

(a)

10 cm

(b)

10 cm

(c)

10 cm

(d)

10 cm

(e)

10 cm

Figure 11.4. Left lateral view of (a) *Homalocephale*, (b) *Prenocephale*, (c) *Stegoceras*, (d) *Pachycephalosaurus*, and (e) *Tylocephale*.

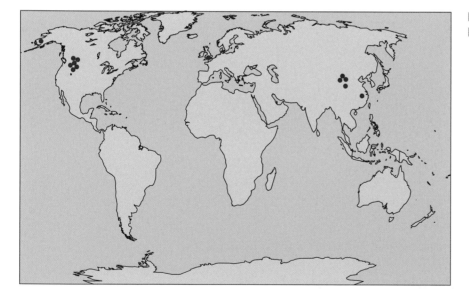

Figure 11.5. Global distribution of Pachycephalosauria.

climate – built from sediment eroded as the Rockies mountain range rose to the west (Figure 11.6). The most common finds are isolated, thickened skull caps, whose water-worn appearance suggests that they were transported long distances in rivers before burial and fossilization (Figure 11.7). Indeed, a recently (2010) discovered pachycephalosaur, *Texacephale langstoni*, is known only from its dome! The high proportion of skull cap finds suggest that only the most robust of bones – in this

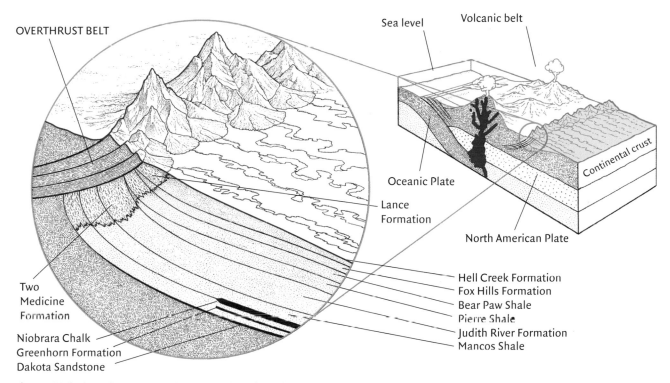

Figure 11.6. Late Cretaceous paleogeographic of North America. As the ancestral Rocky Mountains were uplifted and then drained and eroded by rivers, a thick sedimentary sequence was deposited into geological basins directly to the east of the rising Rockies. Fossil material was carried from the highlands by those rivers, and deposited onto the flat coastal plain to the east.

Figure 11.7. A museum tray filled with the isolated skull caps of pachycephalosaurs. Camera is 10 cm long.

Figure 11.8. An upper cheek tooth of *Pachycephalosaurus*. This is an example of a typical primtive ornithischian "leaf-shaped" tooth; similar teeth are found in other ornithischians, including heterodontosaurids (Figure III.8a) and thyreophorans (Figure 10.18b).

5 mm

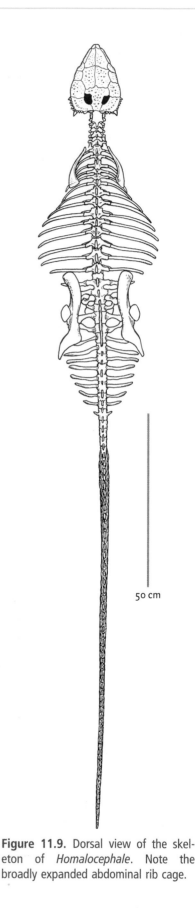

Figure 11.9. Dorsal view of the skeleton of *Homalocephale*. Note the broadly expanded abdominal rib cage.

50 cm

case, the skull caps – survived long journeys. The relative absence of articulated specimens in coastal plain sediments implies that, in life, North American pachycephalosaurs likely lived toward, or even in, mountains, where sediments were more likely to be eroded rather than deposited.

Feeding

The herbivorous habits of pachycephalosaurs are evident not only from their teeth, but also from the impressive volume of their abdominal regions. At the front of the mouth, the jaws contained simple, peg-like gripping teeth, the last of which were sometimes enlarged in a canine-like fashion (see Figure 11.4). A small rhamphotheca is thought to have been present. Further back, the cheek teeth of pachycephalosaurs were uniformly shaped with small, triangular crowns; the typical primitive ornithischian "leaf-shaped" tooth (Figure 11.8). The front and back margins of the crowns of such teeth bear coarse serrations, for cutting or puncturing plant leaves or fruits.

Turning to the opposite end of the gastrointestinal system, the rib cage of pachycephalosaurs was very broad, extending backward to the base of the tail. Its size suggests that it accommodated a large stomach (or stomachs) that broke down tough vegetation via bacterial fermentation (Figure 11.9).

Pachycephalosaur brains

Pachycephalosaurs had typical ornithischian brains, and doubtless aspired to typical ornithischian thoughts. Atypical, though, were the enlarged olfactory lobes of the brain, suggesting a better-than-average sense of smell. What they smelled, however, is a secret that died with them.

For all its unremarkable-ness, the pachycephalosaur brain was uniquely oriented in its skull. The back half of the brain is angled downward, which is reflected in the rotation of the back of the skull (the **occiput**) to face not only backward (as it does conventionally), but also slightly downward. It's been shown that, the higher the dome, the more downward the orientation of the

Figure 11.10. Head-on pushing and butting in (a) *Homalocephale* and (b) *Stegoceras*.

occiput. And that might be related to pachycephalosaurs using their heads for something other than profound thought.

Using your head … for a battering ram?

The morphology of the domes has suggested to many scientists that, incredibly, pachycephalosaurs used their thickened skull roofs as battering rams (Figure 11.10). Internally, the structure of the dome is very dense, with the bone fibers oriented in columns approximately perpendicular to the external surface of the dome. Such an arrangement may be ideal for resisting forces that come from strong and regular thumps to the top of the head and for transmitting such forces around the brain, much as the helmet worn by sparring boxers is supposed to channel forces around the head.

Using special clear plastic cut to resemble a cross-section of the high-domed pachycephalosaur *Stegoceras* (Figure 11.11), paleontologist H.-D. Sues stressed the model in a way that simulated head-butting. The stress lines, seen under ultraviolet light, mimicked the orientation of the columnar

Figure 11.11. Stress patterns in pachycephalosaur domes during head butting. (a) Vertical section through the dome of *Stegoceras*. Note the radiating organization of internal bone. (b) Plastic model of the dome of *Stegoceras* in which forces were applied to several points along its outer edge and seen through polarized light. Note the close correspondence of the stress patterns produced in this model and the organization of bone indicated in (c) the left side of the skull of *Stegoceras*.

bone, reinforcing the suggestion that the fibrous columns evolved to resist stresses induced by head-butting.

Building a better head-butter

If head-butting was the preferred means of pachycephalosaur expression, the body had to be set up to allow for it. Recall that the back of the pachycephalosaur skull is angled down and back when the head is held up. With the dome of the head pointing forward – the only position that makes sense for head-butting – rotation of the back of the skull minimizes the chance of violent twisting or even dislocation of the head on the neck.

We might hope to see some protective measures in the neck as well; unfortunately, the neck of pachycephalosaurs is not well understood. Still, it is clear from the large muscle-attachment sites on the occiput (Figure 11.12) that the neck musculature was unusually well developed and very strong. We surmise that it was used to position and hold the head correctly for head-butting.

Further down the spinal column, and utterly unique to pachycephalosaurs, the vertebrae have distinctive tongue-and-groove articulations, which must have provided rigidity to the back. These

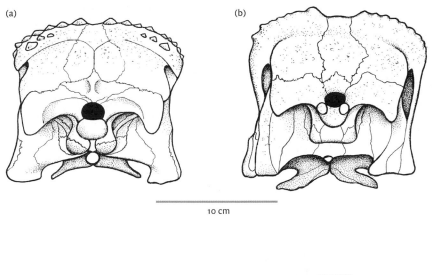

Figure 11.12. Rear (occipital) view of the skull of (a) *Homalocephale* and (b) *Stegoceras*.

Figure 11.13. Two forms of the dome of *Stegoceras*. The shallower dome (a) is thought to pertain to a female, while the other dome (b) may pertain to a male.

articulations would have prevented the kinds of violent lateral rotations of the body that would otherwise have been suffered at the time of impact.

Conscientious objectors?

But not so fast – or at least not so hard. The thickened skull cap of one North American pachycephalosaur – "*Stygimoloch*" – has been shown to contain abundant microscopic openings for blood vessels. With so much vascularization, the skulls of pachycephalosaurs may not have done well with either front or side impacts; this leaves the domes in this genus, and probably others, principally as display structures rather than WMDs.

Socializing pachycephalosaur style

Either way (or both!) – as display or as battering ram – social behavior is strongly implied, and some degree of sexual dimorphism might be anticipated. And so this idea was tested using a large sample of a single pachycephalosaur species, *Stegoceras validum*. It turns out that the domes of *Stegoceras* can be segregated into two groups on the basis of relative size and dome shape (Figure 11.13). One group had larger, thicker domes than the other. Strikingly, the ratio of larger-domed to less-large-domed individuals was one-to-one; exactly what you might expect if one population was male and the other, female. Which was which, however, is anybody's guess.

Sexual selection

Pachycephalosaurs all had – along with, and perhaps including the dome – a suite of features related to visual display. Firstly, there are the canine-like teeth. These could have been used in threat display

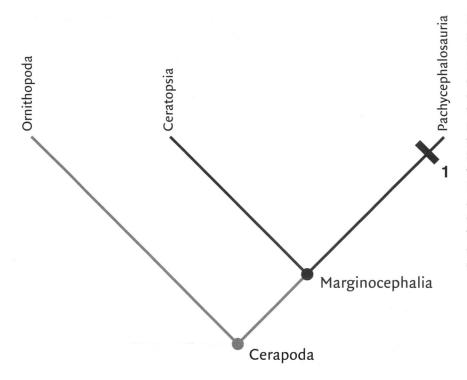

Figure 11.14. Cladogram of Marginocephalia emphasizing the monophyly of Pachycephalosauria. Derived characters include: at **1**, thickened skull roof, frontal excluded from orbital margin, tubercles on caudolateral margin of squamosal, thin, plate-like basal tubera, double ridge-and-groove articulations on dorsal vertebrae, elongate sacral ribs, caudal basket of fusiform ossified tendons, ilium with sigmoidal border, medial process on ilium, pubis nearly excluded from acetabulum, tubercles on squamosal, broad expansion of squamosal onto occiput, free ventral margin of the quadratojugal eliminated by contact between jugal and quadrate.

or biting combat between rival individuals, much like pigs and some deer do today. Equally suggestive are the knobby and spiny osteoderms that covered the snout, the side of the face, and most extensively on the back of the marginocephalian shelf (see Figure 11.2, node **2**; see also Figure 11.4). These distinctive features were likely all about showing off and establishing dominance.

The establishment of dominance gives one gender – males, if living reptiles are any guide – preferred reproductive access to the other gender (females, in that case), which then select with whom they'll mate. The males must also fend off competitors and establish dominance. In general, this practice of establishing dominance hierarchies constitutes **sexual selection**, selection by one gender (today generally, females), of the opposite gender (today, generally males). This is quite distinct from natural selection, in which particular morphologies are favored for reproductive success.

In pachycephalosaurs, domes, knobs, and spikes all acted in display: ritual, and/or potentially violent, clashes. The winner, likely the male with the best-fashioned cranial hardware, got the opportunity to be selected by a female. But he always had to be vigilant for other males that wanted to literally knock his block off – or at least knock him off the block.

The evolution of Pachycephalosauria

Pachycephalosaurs share a host of derived features, most of them cranial (Figure 11.14). Figure 11.15 maps general trends in pachycephalosaur evolution. The most primitive pachycephalosaurs on the cladogram are Asian, suggesting an Asian origin for the group. Because all but two pachycephalosaurs are from the Late Cretaceous, we infer that they underwent considerable evolution during that time.

The cladogram also shows that there was an early tendency to thicken the skull roof as well as develop nerves for the sense of smell. The increased dome heights in some later forms may reflect the evolution of head-butting from simple pushing to complicated head-to-flank or head-to-head butting, to ritualized display.

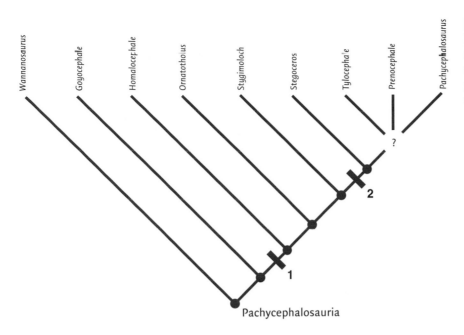

Figure 11.15. Cladogram showing evolutionary changes within selected pachycephalosaurs. Derived characters include: at **1**, broad parietal bone, broad medial process on ilium; at **2**, nasal and postorbitals incorporated into dome.

The broad girth of pachycephalosaurs is clearly derived relative to the narrower more primitive girths seen in most other ornithischians. It suggests a backward migration and enlargement of the digestive tract to occupy a position between the legs and under the tail. As was the case for thyreophorans (see introduction to Part III: Ornithischia; and Chapter 10), simple styles of chewing must have combined with fermentation-based digestion to increase the nutrition available to pachycephalosaurs from the plants they ate.

Marginocephalia: Ceratopsia – horns and all the frills

From the time of their discovery in the second half of the 1800s to the present day, there has hardly been a group of dinosaurs that has evoked more fascination than ceratopsians. Some of these quadrupedal, horned, frilled dinosaurs roamed the Great Plains of North America in the Late Cretaceous. They were rhino-like, ranging upward of 6 or 7 metric tons (Figure 11.16). Equally famous, but for other reasons, is a host of smaller, lighter (25 and 200 kg) non-horned Asian ceratopsians from slightly earlier in the Cretaceous (Figure 11.17).

We know a lot about ceratopsians: the fossil record from Asia and North America is one of the most outstanding of any dinosaur group (Figure 11.18). Primitively small bipeds, these animals evolved into powerful quadrupedals early in their history, developing thick hooves on all toes and reaching sizes to rival that of small tanks.

With or without horns, it is easy to recognize the ceratopsian clan stamp: ceratopsians all had skulls that were narrow, with a hooked beak in front and a skull that flared deeply in the cheek region (Figure 11.19). And at the tip of the snout in the upper jaw was the uniquely evolved **rostral bone** (Figure 11.20).

As befits their name, many ceratopsians had horns; however, some did not. Those ceratopsians that had them, though, grew some of the most impressive horns ever seen on any vertebrate (Figure 11.21). Like the horns of many mammals, the skulls only preserve the bony **horn cores**, covered by keratin sheaths that actually comprised the working end of the horn. The horn visible

Figure 11.16. *Chasmosaurus*, a ceratopsian from the Late Cretaceous of the Western Interior of North America.

on the head of the living animal was therefore significantly larger than the horn core alone (Figure 11.22).

Equally memorable is the ceratopsian **frill**, the eponymous marginocephalian shelf run amok. Extending from the back of the skull, frills vary considerably in almost every respect: size, ornamentation, and shape (see Figure 11.21). The largest reach 2 m in length.

Ceratopsian lives and lifestyles

Dressed and ready to chew

Ceratopsians chewed. A hooked rhamphotheca, blocks of cheek teeth in both upper and lower jaws, a sturdy coronoid process, and inset tooth rows, evidence for the existence of fleshy cheeks, all scream CHEWING.

(a)

(b)

Figure 11.17. Two small, Asian, horn-less ceratopsians. (a) *Psittacosaurus;* (b) *Protoceratops.*

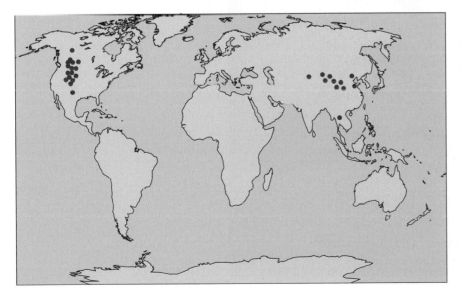

Figure 11.18. Global distribution of Ceratopsia. Ceratopsians were restricted to the Cretaceous of North America and Asia.

(a)

5 cm

(b)

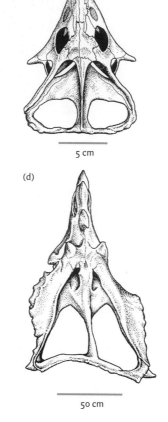

5 cm

Figure 11.19. Dorsal view of the skull of (a) *Psittacosaurus*, (b) *Protoceratops*, (c) *Styracosaurus*, and (d) *Chasmosaurus*.

(c)

50 cm

(d)

50 cm

Figure 11.20. Ceratopsian snout; rostral bone highlighted. Photo from Museum of Natural History in Vienna.

Figure 11.21. Left lateral view of the skull of (a) *Psittacosaurus*, (b) *Leptoceratops*, (c) *Bagaceratops*, (d) *Centrosaurus*, (e) *Styracosaurus*, (f) *Pachyrhinosaurus*, (g) *Pentaceratops*, (h) *Arrhinoceratops*, (i) *Torosaurus*, (j) *Achelousaurus*, (k) *Einiosaurus*, (l) *Regaliceratops*, (m) *Wendiceratops*, and (n) *Titanoceratops*.

Figure 11.22. The skull of *Triceratops* showing the horn core covered by the keratinized (finger-nail and claw material) horn as it would have been in life.

The business end of the ceratopsian mouth was the narrow, hooked, beak-tipped snout, suggesting the potential for careful selection of the plants for food. Individually the cheek teeth were relatively small, but they grew stacked and overlapping together into a large, single functional slicing block in each jaw, the **dental battery** (Figure 11.23). Worn teeth were constantly replaced, so that the active chewing surface of each of the four dental batteries was continually refurbished. Inexplicably, and unique in the animal kingdom, the orientation of the grinding surfaces migrated, becoming more and more vertical, until, in the large, highly derived North American forms, they occurred nearly vertically along the *sides* of the teeth comprising the dental battery (Figure 11.23b).

The force behind this high-angle mastication derived from a great mass of jaw-closing musculature, which in the frilled forms crept through the upper temporal opening and onto the base of the frill. The other end of this muscle attached to a massive, hulking coronoid process on the mandible (Figure 11.24). All in all, the chewing apparatus in ceratopsians was among the most highly evolved in any vertebrate.[1]

Beyond the mouth

Based upon their narrow girth of the body (by comparison with thyreophorans and pachycephalosaurs), the digestive tract does not appear to have been disproportionately large in ceratopsians, and did not likely rely upon wholesale bacterial fermentation for extracting nutrients from plants. Nevertheless, it must have been big enough to accommodate what must have been an endless parade of foliage that formed the diet of these animals.

---

[1] The primitive ceratopsians *Yinlong* and *Psittacosaurus* lacked the highly refined chewing specializations of the North American ceratopsians, but *Psittacosaurus*, at least, is known to have harbored a packet of gastroliths lodged in its gizzard, which would have doubly pulverized its meal. Gastroliths are known from no other ceratopsian.

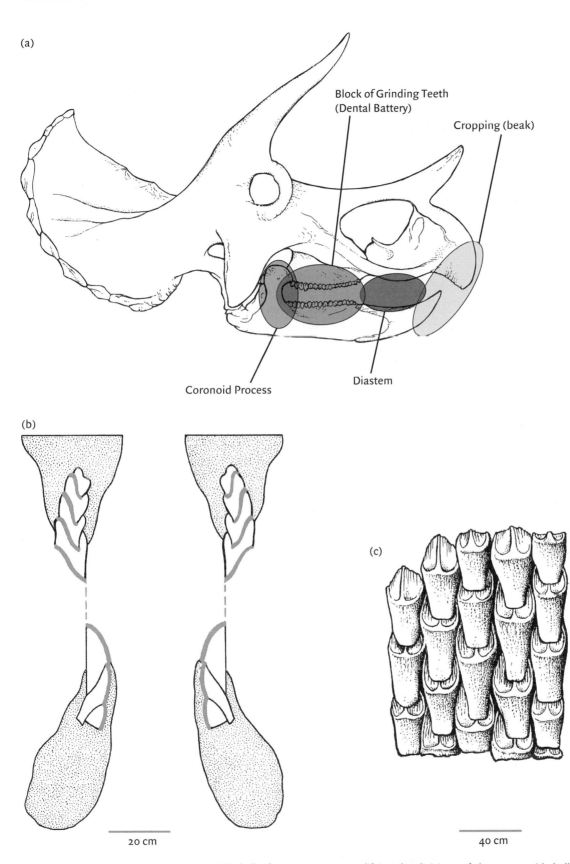

(a)

Block of Grinding Teeth
(Dental Battery)

Cropping (beak)

Coronoid Process

Diastem

(b)

20 cm

(c)

40 cm

Figure 11.23. Chewing in ceratopsids. (a) Skull of *Triceratops*, exemplifying the divisions of the ceratopsid skull: anterior cropping section (beige); diastema (maroon); block of grinding cheek teeth (blue); coronoid process (brown). See also Figure III.5. (b) Cross-section through the upper and lower jaws of *Triceratops* showing high-angle grinding motion of the dental batteries; (c) internal view of the dental battery in the lower jaw of *Triceratops*.

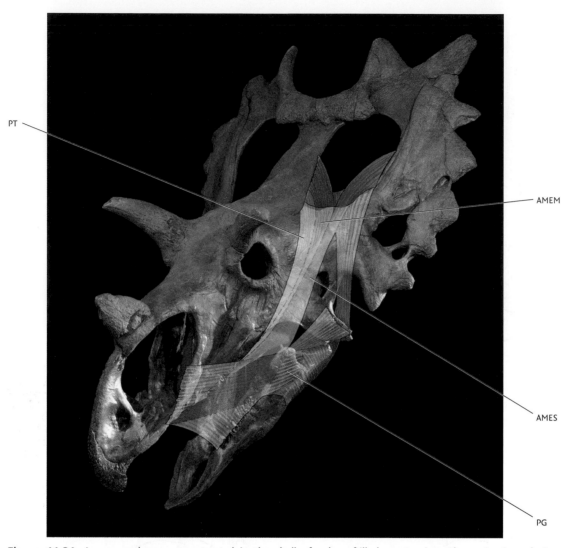

Figure 11.24. Jaw musculature reconstructed in the skull of a long-frilled ceratopsian. The major jaw closing (adductor) muscles are (1) the adductor mandibularis externus superficialis (AMES); (2) the adductor mandibularis externus medialis (AMEM); (3) the pseudotemporalis (PT); (4) the pterygoideus (PG).

Even the largest quadrupedal ceratopsians never browsed particularly high above the ground. The browse height of the largest was probably less than 2 m. Nevertheless, they may have been able to knock over trees of modest size in order to gain access to choice leaves and fruits.

Which plants were preferred by ceratopsians remains a mystery. The principal plants whose statures match browsing heights of ceratopsians were a variety of shrubby angiosperms, ferns, and small conifers (see Chapter 14).

Locomotion

The primitive ceratopsian *Psittacosaurus* appears to have been fully bipedal and must have walked in typical bipedal ornithischian fashion. But the rest were quadrupeds all, and while the back legs were fully erect, the orientations of the front limbs are somewhat controversial. Some have argued that the front limbs were directly under the body, as they are in mammalian quadrupeds. Others have argued, based on the shape of the bones of the front legs, that a more sprawling posture in the front legs is indicated (Figure 11.25).

Figure 11.25. Two end-points in the spectrum of reconstructions of the front limbs in ceratopsians: (a) fully erect stance; (b) semi-erect stance.

These considerations naturally affect how we understand the way ceratopsians ran. The sprawling front legs would likely entail slower speeds, and perhaps a more unhurried lifestyle. The mammal-like reconstructions, with fully erect stances for both front and rear legs, suggest faster speeds, reminiscent of a large, Cretaceous rhinoceros. With such uncertainty about stance, tentative estimates for walking speeds range somewhere between 2 and 4 kmh (see Box 13.3), while maximum running speeds range from 30 to 35 kmh.

Bringing up baby

The first dinosaur egg nests ever found, in 1922, were thought to belong to the small Asian ceratopsian *Protoceratops*, and it was with this dinosaur that, for the next 70 years, the eggs were posed in museum displays all over the world. So it was, uh, *informative* to learn – after some 70 years – that the embryos inside those "*Protoceratops* eggs" actually belonged to the theropod *Oviraptor* (see Chapter 6; Figure 6.32).

Despite the confusion with *Oviraptor*, however, juvenile ceratopsians from Asia are now relatively well known. Hatchlings of *Psittacosaurus*, no more than 23 cm long (with tail!) have been found. And recently, complete skeletons of *Protoceratops* hatchlings – grown somewhat past the newly hatched stage – were found in nests (Figure 11.26). This means that some parental care at the nest after birth is implied for *Protoceratops*. Moreover, all of the growth stages from hatchling to adult have been documented in *Protoceratops*, making the **ontogeny** – or growth and development – of this dinosaur perhaps the best understood of all dinosaurs. Similarly, the ontogeny of a number of other ceratopsian dinosaurs is known, including that of the latest Cretaceous *Triceratops* (Figure 11.27). And the ontogeny of ceratopsians turns out to hold key clues about their behavior.

Horns, frills, and ceratopsian behavior

Ceratopsian horns were once thought to have functioned to ward off predators at close quarters. More recent interpretations have not completely ruled this out (see below), but have instead focused on intraspecific behaviors such as display, ritualized combat, defense of territories, maturity and species identification, and establishment of social ordering.

(a)

Figure 11.26. Two nests of juvenile ceratopsians from Asia. (a) A nest of 34 *Psittacosaurus* juveniles from the Lower Cretaceous Yixian Formation of Liaoning Province, China. An adult *Psittacosaurus* skull, presumably a parent, is visible on the left center of the photograph. **Following page:** (b) A nest with 15 juvenile *Protoceratops*, from the Upper Cretaceous Djadokhta Formation of the Gobi Desert, Mongolia. An adult *Protoceratops* is about 2 m long; each of these babies is ~10 cm, suggesting that they were perhaps within their first year of life. Ruler is marked in cm.

(b)

Figure 11.26. (*cont.*)

Figure 11.27. A growth series for the Late Cretaceous ceratopsian *Triceratops*. The authors identified four growth stages: baby, juvenile, sub-adult, and adult.

The link between dominance, defense, and horns comes from studies of mammals in their natural habitats. In the case of almost all horned mammals, larger males tend to have a reproductive advantage over smaller males. Dominance in these mammals (and in other tetrapods) is accentuated by the development of structures that "advertise" the size of the animal; these obviously include horns and antlers, as well as the bony horn-like knobs (ossicones) of giraffes and the nasal horns of rhinoceroses. In short, the variety of horn and antler shapes in mammals are known to reflect (1) species-recognition mechanisms that aid in preventing **interspecific** matings (that is, matings between different species), and (2) intraspecific differences so that displays and ritualized fighting behavior occur between appropriate individuals (e.g., equally eligible males, in the case of modern mammals).

Turning to ceratopsids, few have doubted that the horns were used for combat; the question has been, at whom were they aimed? Using modern horned mammals as analogs, current thought suggests that the large nasal and brow horns of ceratopsids functioned primarily during territorial defense and in establishing dominance. Similarly, the appearance and shape of elaborate scallops and spikes along the frill margin in many of the more highly derived ceratopsians separates one species from another, as well as juveniles from adults (see Figure 11.27). Thought of this way, the remarkable variations in the horns and frills in ceratopsians could be used for *interspecific* identification as well as the establishment of *intraspecific* dominance (Figure 11.28).

Statistical studies of *Protoceratops* show two populations of adult frill and facial morphologies – strong evidence of sexual dimorphism (Figure 11.29). Moreover, the frills don't appear too large in juvenile specimens – they only develop when the animals reach 75 per cent of adult body sizes. This suggests that frill growth is coordinated with sexual maturity and therefore that there is a reproductive connection to frill size and shape. Sound like sexual selection?

Sexual selection is now thought to also occur in other ceratopsians, among them *Centrosaurus* and *Chasmosaurus*. In many of these forms, the development of scallops and spikes on the frill margin would enhance the dimorphic nature of the frill; equally plausibly, however, such features would also be useful for identification and indicators of maturity: who represents the competition for mating opportunities, and who does not.

All of this implies social interactions, and it thus comes as no surprise to learn that ceratopsians lived in large herds. The evidence for this comes from an overwhelming catalog of ceratopsian **bonebeds**, mass accumulations of single species of organisms. Bonebeds are known for more than nine different species, including several bonebeds exceeding 100 individuals. Such gregariousness makes sense when putting frills and horns into their behavioral context: identification, territoriality, ritualized combat and display, and the establishment of dominance are to be expected in animals that come together in highly social circumstances such as herds.

These ideas suggest that we ought to find puncture wounds inflicted on faces, frills, and bodies of competing ceratopsians. In fact, such wounds are preserved in at least five different ceratopsians. In *Triceratops* and *Centrosaurus*, healed wounds on the cheek region and frill were correlated with the types of horns each animal bears. The horns of *Triceratops* (two large brow horns; a smaller nasal horn) are very different from those of *Centrosaurus* (a single large nasal horn), and if *intraspecific* competition was the main use of the horns, the wounds left on the frills and faces of these animals should look very different from one another. As it turned out, *Triceratops* had a significantly higher number of healed wounds on the frill than did *Centrosaurus*, suggesting that, given the mechanics of fighting, the damage was caused by other *Triceratops* with their massive, highly placed brow horns.

Figure 11.28. "Back off": frill display in *Chasmosaurus*. The very long frill could have provided a very prominent frontal threat display, not only by inclining the head forward but also by nodding or shaking the head from side to side.

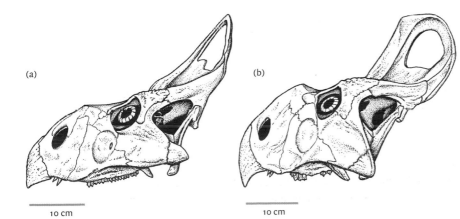

(a)

(b)

10 cm

10 cm

Figure 11.29. Sexual dimorphism proposed in *Protoceratops*. Note In (a), a presumed female, the frill is less pronounced and the nasal ridge is less prominent; quite the opposite of (b), a presumed male.

Figure 11.30. The Wonderful (and unexpected) World of *Psittacosaurus*. The tail region of the primitive ceratopsian *Psittacosaurus*, showing elongate (presumably) monofilamentous feathers (see arrow) along the back of the tail, interpreted as display features. Inset: complete specimen. Who knew!? Photographs taken from a cast of the specimen.

These results provide strong evidence of the blood-letting that comes from head-on engagements between competing members of the same species.

The argument for the importance of display in ceratopsians got an unexpected bump from the primitive ceratopsian *Psittacosaurus*. Feathers, or at least monofilamentous coats (as we learned in Part III), may have been primitive not just for theropods, but perhaps for all dinosaurs, because the basal ornithischian *Tianylong* has them (see Figure III.4). *Psittacosaurus*, too, had some kind of monofilamentous coating, evidently restricted to the back of the tail (Figure 11.30). These structures certainly have the appearance of a display feature; reinforcing the importance of display in both the origin of feathers and sociality in Ceratopsia.

Thoughts of a ceratopsian

We have a surprisingly good idea about the brains of dinosaurs, including ceratopsians, even though those brains are soft tissue, and were last seen on Earth millions of years ago (Box 11.1). Given the complex repertoire of inferred ceratopsian behaviors, it comes as a bit of a surprise that their brains were not particularly large (see also Box 13.4). Despite being near opposites in terms of body size and display-related anatomy, both *Protoceratops* and *Triceratops* had brains less than the size expected of a similarly sized crocodilian or lizard. Cerebrally, they were above sauropods, ankylosaurs, and stegosaurs, but commanded proportionally less gray matter than either euornithopods or theropods. Regardless, the variety of exotic morphology in and around the head suggests that ceratopsians, large-brained or no, may have had a relatively complicated behavioral repertoire.

Box 11.1   Dino brains

Brains, of course, are soft anatomy, and so finding one preserved is next to impossible. Yet, we have good evidence for how brains looked in many dinosaurs and, considering areas of enlargement, we can suggest that a particular dinosaur had a well-developed sense of smell, or perhaps good vision. How can this be?

Historically, if the bones of the braincase were well preserved, paleontologists initially obtained brain **endocasts**, three-dimensional models of the brain. This was not so easy: all of the interior rock needed to be prepared out of the braincase, a process that generally involved breaking the braincase, and then the thing in effect reassembled, so that its inside could be cast. Reassembly took a *very* skilled preparator. Casting was accomplished by painting several layers of latex on the interior of the braincase, and then, once it had dried, pulling the flexible material through the foramen magnum (the opening through which the spinal cord contacts the brain) if the braincase was intact; peeling it off of the exposed interior if it was not. Latex cast in hand, a mold could then be prepared that faithfully recorded the contours of the inside of the braincase. From this mold were cast the brain endocasts; in plaster, or perhaps resin.

Very much more recently, paleontologists have begun to use computer-aided tomography (CT) scanning techniques to successfully image the brain; in so doing acquiring revolutionary insights about dinosaur brains. As in humans, the technique is non-invasive, and responds to density differences between the fossil bone and the rock that encases it. Preparation is therefore not necessary. The instrument obtains a sequence of two-dimensional "slices" – images, in fact – which can then be assembled into a computer-generated composite image, and be rotated and visualized from any angle. The advantages of this are multiple: aside from obviating the need for preparation – which, as we have seen, is costly, time-consuming, and potentially destructive – the CT-scan provides as much detail as is preserved – the equal of, or better than, the very finest preparation jobs. For example, the CT-scanned image presented here (Figure B11.1.1) shows details of the cranial nerves (identified by Roman numerals) unavailable by conventional means. The CT dinosaur brain images we've shown in this book, as well as many others, are available on paleontologist and anatomist L. M. Witmer's laboratory ("Witmerlab") website: www.oucom.ohiou.edu/dbms-witmer/3D-Visualization.htm.

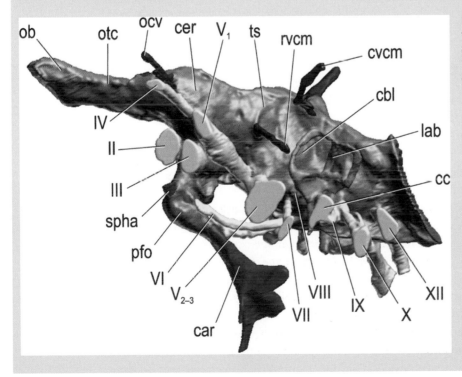

Figure B11.1.1. CT-scan reconstruction of the brain of the ceratopsid genus *Pachyrhinosaurus* (see Figure 11.21f). Scale bar = 4 cm. For abbreviations see Figure 6.22 on page 125.

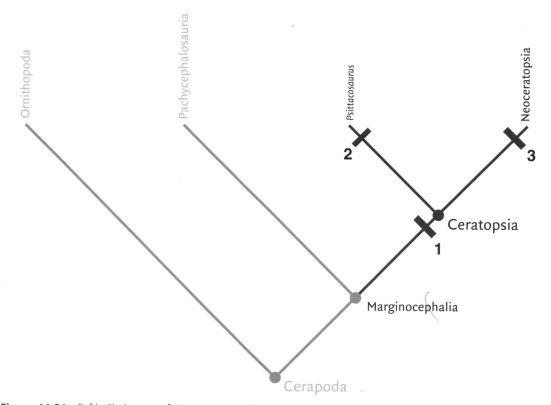

Figure 11.31. *(left).* Cladogram of Ceratopsia, emphasizing the monophyly of *Psittacosaurus* and Neoceratopsia. Derived characters include: at **1**, rostral bone, a high external naris separated from the ventral border of the premaxilla by a flat area, enlarged premaxilla, well-developed lateral flaring of the jugal; at **2**, short preorbital region of the skull, very elevated naris, loss of antorbital fossa and fenestra, unossified gap in the wall of the lacrimal canal, elongate jugal and squamosal processes of postorbital, dentary crown with bulbous primary ridge, manual digit IV with only one phalanx, manual digit V absent; at **3**, enlarged head, keeled front end of the rostral bone, much reduced quadratojugal, primary ridge on the maxillary teeth, development of humeral head, gently decurved ischium.

The evolution of Ceratopsia

Ceratopsia is a monophyletic taxon, indicated by a rich array of derived features (Figure 11.31). The two ceratopsians that are thought to represent the primitive condition for the group are *Yinlong* (Figure 11.32a) and *Psittacosaurus* (Figure 11.32b), both small, Asian bipeds. All more derived ceratopsians – Neoceratopsia – are quadrupeds. This underlines an important evolutionary event that we can read from the cladogram (Figure 11.31): relatively early in their history, ceratopsians, for whatever their reasons, adopted a quadrupedal stance.

Those early days also brought with them evidence of a major ceratopsian migration. Neoceratopsia (Figure 11.33) consists of a series of small, relatively primitive forms such as the Asian *Protoceratops* and *Bagaceratops*; the somewhat younger, though still primitive North American *Montanaceratops* and *Leptoceratops*, as well as the more derived, exclusively North American clade Ceratopsidae, that group of large, familiar ceratopsians such as *Triceratops* and *Centrosaurus* (Figure 11.34), as well as newly discovered ones like *Wendiceratops* (Figure 11.35).

When we compare the geographical locations of various neoceratopsians, that is, their **biogeography**, with primitive and advanced ceratopsians on the cladograms shown in Figures 11.33 and 11.34 it becomes clear that, early in neoceratopsian history, a primitive neoceratopsian – looking

Figure 11.32. Basal ceratopsians from Asia. (a)–(c) The skull of *Yinlong*, the most primitive ceratopsian known; (a) front view; (b) right side; (c) top view. Photograph courtesy of J. M. Clark. **Following page:** (d) The primitive ceratopsian *Psittacosaurus*, as displayed in the NHM, Vienna.

perhaps a bit like *Protoceratops* – migrated to the New World. The route of choice would likely have been briefly exposed land across the Bering Straits. Recent work not only confirms this pattern, but also suggests that several such migrations took place during ceratopsian history (Figure 11.36).

Once in North America, a few lineages retained the comparatively modest morphology of their more primitive forebears. However, the clade radiated into two spectacular and diverse groups of much larger, flashier ceratopsids: chasmosaurines, after *Chasmosaurus*; and centrosaurines, after *Centrosaurus* (see Figure 11.34). Chasmosaurines are generally called "long-frilled," after a tendency in the group to develop large, open frills, while centrosaurines are sometimes called "short-frilled," after a tendency in the group toward shorter frill lengths.

(b)

Figure 11.32. (*cont.*)

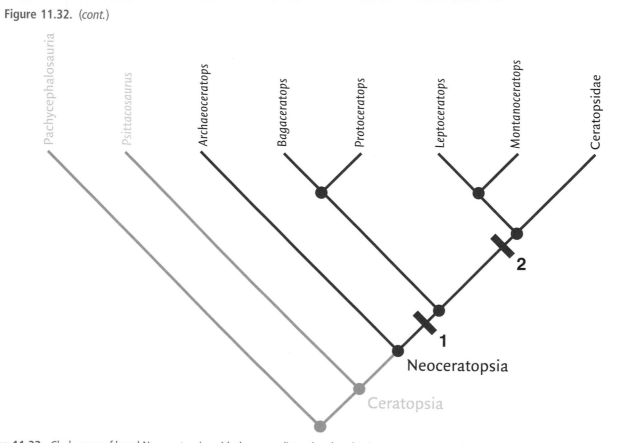

Figure 11.33. Cladogram of basal Neoceratopsia, with the more distantly related *Psittacosaurus* and Pachycephalosauria. Derived characters include: at **1**, elongated preorbital region of the skull, an oval antorbital fossa, triangular supratemporal fenestra, development of the syncervical (fusion of cervical vertebrae); at **2**, greatly enlarged external nares, reduced antorbital fenestra, nasal horn core, frontal eliminated from the orbital margin, supraoccipital excluded from foramen magnum, marginal undulations on frill augmented by epoccipitals, more than two replacement teeth, loss of subsidiary ridges on teeth, teeth with two roots, ten or more sacral vertebrae, laterally everted shelf on dorsal rim of ilium, femur longer than tibia, hoof-like pedal unguals.

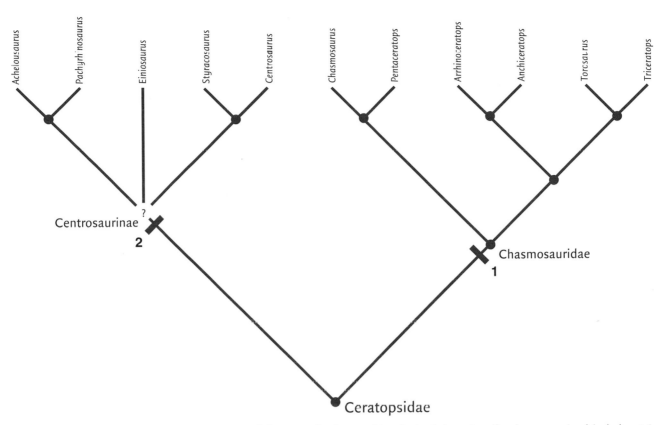

Figure 11.34. Cladogram of Ceratopsidae. Derived characters for Ceratopsidae. Derived characters (for chasmosaurines) include: at **1**, enlarged rostral, presence of an interpremaxillary fossa, triangular squamosal epoccipitals, rounded ventral sacrum, ischial shaft broadly and continuously decurved. Derived characters (for centrosaurines) include: at **2**, premaxillary oral margin that extends below alveolar margin, postorbital horns less than 15 per cent of skull length, jugal infratemporal flange, squamosal much shorter than parietal, six to eight parietal epoccipitals, predentary biting surface inclined steeply laterally.

Figure 11.35. The recently (2015) discovered *Wendiceratops*, a centrosaurine ceratopsid from Alberta, Canada. This ceratopsian is named after its discoverer, Wendy Sloboda, who celebrated her find by tattooing its image on her arm (see also Figure 11.21m).

Figure 11.36. Cladogram of ceratopsians superimposed on a map of North America and Asia, showing migration of more derived forms to North America. Arrow shows the general route of migration.

The evolution of behavior

If there is some correspondence between morphology and behavior, then the morphological trends identified by all the ceratopsian cladograms should give us insights into the evolution of neoceratopsian behavior. In those ceratopsians with relatively modest frills and horns – forms such as the Asian *Protoceratops*, and the North American *Leptoceratops* and *Montanoceratops* – display perhaps involved swinging the head from side to side. Should this have failed to impress, these animals may have rammed full tilt into the flanks of their opponent.

The more derived ceratopsids share more elaborate frills and either nasal or brow horns. Among the long-frilled ceratopsians (for example, *Chasmosaurus*, *Pentaceratops*, and *Torosaurus*), the display

Figure 11.37. "Crossing of the horns": combat between (likely male) *Centrosaurus*.

function of the frill may have been emphasized (see Figure 11.28). In contrast, most of the short-frilled ceratopsians (such as *Centrosaurus*, *Avaceratops*, and possibly *Pachyrhinosaurus*) were somewhat rhinoceros-like in their appearance, and perhaps tried to catch each other on their nasal horns, thus reducing to a degree the amount of damage inflicted on the eyes, ears, and snout (Figure 11.37).

Reality check

The elaborate frill ornamentation; the variety of horns all seem to scream "Sexual Selection!!" and, as we've noted in the past, horns, frills, and chewing seem almost to drive the evolution of ceratopsian dinosaurs. In this diverse group, we think we're witnessing a world where display, recognition, and competition were all-important. So what then to do with a well-supported, data-rich, 2014 catchily titled study ("Males resemble Females") which concluded that there was no measurable difference between populations (presumably "males" and "females") of *Protoceratops*? Is Figure 11.29 pure wishful thinking? The answer might be "yes." Does this negate everything paleontologists have inferred about ceratopsian social behavior? We think not; however, it is well to remember that, as noted by the authors of the study of *Protoceratops*, it could be that they didn't measure the key sexually dimorphic features in this genus. For all that we think we know about non-avian dinosaurs, it is well to remember how vastly it is exceeded by what we don't know.

SUMMARY

Marginocephalia consists of the bipedal Pachycephalosauria, the dome-headed ornithischians, and the quadrupedal Ceratopsia, the horned, parrot-beaked, frilled ornithischians. The group was largely restricted to the Cretaceous of Asia and North America, and is diagnosed by the presence of a variably sized, bony shelf that formed along the back of the skull.

Marginocephalians were gregarious animals, and sexual selection was likely a driving force in much of their behavior, a fact that is reflected in their morphology. The domes of pachycephalosaurs have been interpreted as structures designed for intraspecific competition: head- and flank-butting have been suggested. The striking morphological variety of horns and frill shapes, and cranial ornamentation, in ceratopsians suggests a high level of intraspecific competition. In both groups, sexual dimorphism has been recognized. Ceratopsian gregariousness is also reflected by the presence of large monospecific bonebeds, suggesting that herds of ceratopsians roamed what is now the Great Plains of Canada and the USA.

All marginocephalians are genasaurs, which means that they chewed their food to a greater or lesser extent. While pachycephalosaur teeth don't reveal evidence of remarkable chewing adaptations and pachycephalosaurs likely gut-fermented in their capacious stomachs, ceratopsians developed sophisticated chewing mechanisms including a robust coronoid process, dental batteries, a skull partitioned into cropping, diastema, and grinding sections, and powerful jaw adductor muscles that may have attached high on the frill.

Care of the young is known in ceratopsian dinosaurs. Nests of partially grown ceratopsians have been found, suggesting parental care at the nest.

Ceratopsian evolution was characterized by increasing size, as well as by one or more migrations across the Bering Strait land bridge, from Asia to North America. While intraspecific competition was likely an important behavioral aspect of even the earliest ceratopsians, later forms evolved elaborate frill or horn displays.

SELECTED READINGS

Dodson, P. 1996. *The Horned Dinosaurs*. Princeton University Press, Princeton, NJ, 346 pp.

Dodson, P. and Currie, P. J. 1990. Neoceratopsia. In Weishampel, D. B., Dodson, P., and Osmólska, H. (eds.) *The Dinosauria*. University of California Press, Berkeley, pp. 593–618.

Dodson, P., Forster, C. A., and Sampson, S. D. 2004. Ceratopsidae. In Weishampel, D. B., Dodson, P., and Osmólska, H. (eds.) *The Dinosauria*, 2nd edn. University of California Press, Berkeley, pp. 494–513.

Farke, A. A, Wolf, E. D. S., and Tanke, D. H. 2009. Evidence of combat in *Triceratops*. *PlosONE*, 4, e4252. doi:10.1371/journal.pone.0004252.

Farlow, J. O. and Dodson, P. 1975. The behavioral significance of frill and horn morphology in ceratopsian dinosaurs. *Evolution*, 29, 353–361.

Maiorino, L., Farke, A.A., Kotsakis, T., and Piras, P., 2014, Males resemble females: Re-evaluating sexual dimorphism in *Protoceratops andrewsi* (Neoceratopsia, Protoceratopsidae). *PlosONE*, 10(5): e0126464. doi:10.1371/journal.pone.0126464.

Makovicky, P. 2012. Marginocephalia. In Brett-Surman, M. K., Holtz, T. R., and Farlow, J.O. (eds.) *The Complete Dinosaur*, 2nd edn. Indiana University Press, Bloomington, pp. 527–549.

Sereno, P. C. 1990. Psittacosauridae. In Weishampel, D. B., Dodson, P., and Osmólska, H. (eds.) *The Dinosauria*. University of California Press, Berkeley, pp. 579–592.

Witmer, L. M. and Ridgely, R. C. 2008. Structure of the brain cavity and inner ear of the centrosaurine ceratopsid dinosaur *Pachycephalosaurus* based on CT scanning and 3D visualization. In Currie, P. J., Langston, W., and Tanke, D. H. (eds.) *A New Horned Dinosaur from an Upper Cretaceous Bone Bed in Alberta: National Research Council of Canada Monographs no. 49729*, pp. 117–144.

You, H.-L. and Dodson, P. 2004. Basal Ceratopsia. In Weishampel, D. B., Dodson, P., and Osmólska, H. (eds.). *The Dinosauria*, 2nd edn. University of California Press, Berkeley, pp. 778–793.

TOPIC QUESTIONS

1. Who are marginocephalians?
2. What are the diagnostic characters of Marginocephalia? How are marginocephalians related to other ornithischians?
3. What are the diagnostic characters of Pachycephalosauria?
4. What are the diagnostic characters of Ceratopsia?
5. Describe chewing as practised by ceratopsian dinosaurs.
6. What do we know about ceratopsian egg-laying and nesting?
7. Give a brief history of ceratopsian biogeography.
8. What is sexual selection? Intraspecific competition?
9. What is in the inferred function of the horns and frill in ceratopsians?
10. What is the inferred function of the dome in pachycephalosaurs?
11. How do marginocephalian features relate to intraspecific competition and sexual selection?

Figure 12.1. *Edmontosaurus*, a Late Cretaceous hadrosaurid ornithopod from the Western Interior of North America.

Ornithopoda
Mighty Mesozoic masticators

12

CHAPTER OBJECTIVES

- Introduce Ornithopoda
- Develop familiarity with current thinking about lifestyles and behaviors of ornithopods
- Develop an understanding of ornithopod evolution using cladograms, and an understanding of the place of Ornithopoda within Dinosauria

Ornithopoda

Ornithopods (*ornith* – bird; *pod* – foot) were the cows, deer, bison, wild horses, antelope, and sheep of the Mesozoic (Figure 12.1). Herbivores all, they were one of the most numerous, most diverse, and longest-lived groups in all Dinosauria: from the Middle Jurassic, when they first appeared, until the end of the Cretaceous, when they all went extinct, ornithopods evolved over 100 species at present count.

Ornithopods lived on every continent. They ranged from near the then-equator to such high latitudes as the north slopes of Alaska, the Yukon, and Spitsbergen in the Northern Hemisphere, and Seymour Island, Antarctica, and the southeast coast of Australia in the Southern Hemisphere (Figures 12.2 and 12.3). Local conditions in these regions varied widely, so ornithopods lived in quite diverse habitats and in a wide range of climates.

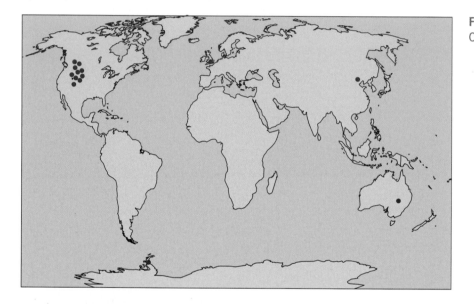

Figure 12.2. Global distribution of basal Ornithopoda.

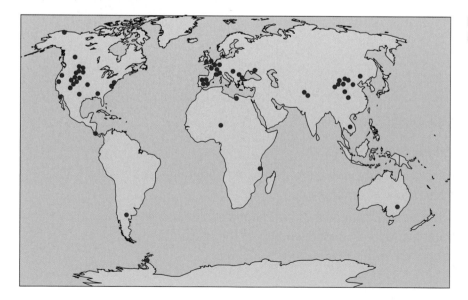

Figure 12.3. Global distribution of Iguanodontia.

They also evolved a range of sizes: early in their history, ornithopods were generally small (ranging from 1 to 2 m in length); however, later some members of the group got very big (upwards of 15 m; Figure 12.4).

We know as much about ornithopods as about almost any other group of dinosaurs: *Iguanodon* was a charter member of Sir Richard Owen's original 1842 Dinosauria (see Chapter 15). Hadrosaurids ("duckbills") are known from single bones to huge bonebeds. Their remains include skin impressions – patterns have even been used to distinguish different species – and ossified tendons, as well as delicate skull bones such as sclerotic rings (that support the eyeball), stapes (the thin rod of bone that transmits sound from the eardrum to the brain), and hyoid bones (delicate bones that support the tongue). We have hadrosaurid eggs with all growth stages represented, from hatchling, to "teenager," to adult. Ornithopod footprints and trackways abound in many parts of the world.

Who were the ornithopods?

As we have seen, ornithopods are genasaurian cerapodans (Figure 12.5). As ornithopod phylogeny is currently understood, there is a basic split between a few primitive ornithopods (generally small bipeds) such as *Orodromeus*, and the remaining ornithopods, **Iguanodontia**. Most of the dinosaurs we commonly think of as ornithopods, including *Iguanodon* and and the duck-billed dinosaurs (**hadrosaurids**; see Figures 12.4 and 12.1, respectively), are iguanodontians. The matter is somewhat complicated because a number of small bipedal herbivores, once considered euornithopods (such as *Agilisaurus* and *Hexinlusaurus*), are now thought to be outside of Ornithopoda, either more basal to it or immediately basal to Cerapoda as a clade called Neornithischia. These basic relationships are shown in Figure 12.6.

Ornithopod lives and lifestyles

Gettin' around

Ornithopods functioned as bipeds and quadrupeds, with both locomotor modes commonly occurring in the same beast. The smallest were predominantly bipedal, with gracile bones that suggest agility and speed. When eating or standing still, however, they also may have adopted a quadrupedal stance. Some of the larger ornithopods, however, such as *Iguanodon*, may have functioned more as full-time quadrupeds, only going bipedal

Figure 12.4. *Iguanodon bernissartensis*, the magnificent Early Cretaceous ornithopod, as seen in the Museum of Natural History of Vienna in its original early 1880s mount. A more modern take on ornithopod posture has the backbone of the animal largely parallel with the ground.

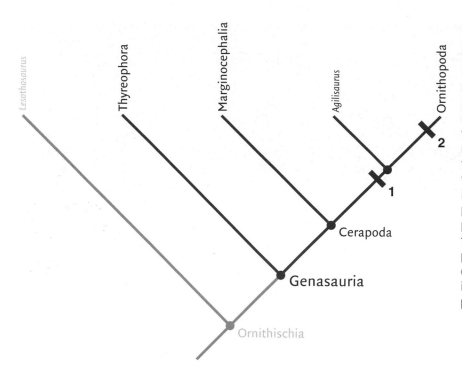

Figure 12.5. Cladogram of Genasauria (cheek-bearing ornithischians; see Part III), emphasizing the monophyly of Ornithopoda. Derived characters include: at **1**, pronounced ventral offset of the premaxillary tooth row relative to the maxillary tooth row, crescentic paroccipital processes, strong depression of the mandibular condyle beneath the level of the upper and lower tooth rows, elongation of the lateral process of the premaxilla to contact the lacrimal and/or prefrontal; at **2**, scarf-like suture between postorbital and jugal, inflated edge on the orbital margin of the postorbital, deep postacetabular blade on the ilium, well-developed brevis shelf, laterally swollen ischial peduncle, elongate and narrow prepubic process.

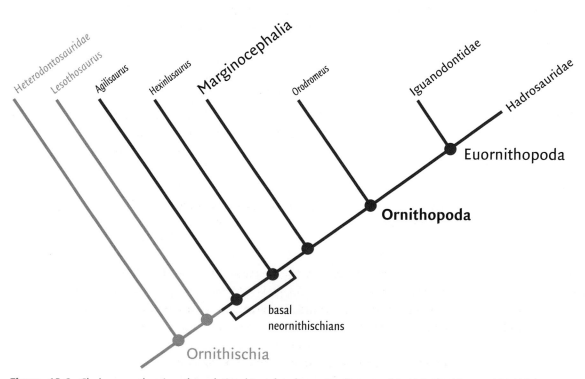

Figure 12.6. Cladogram showing the relationship of a few miscellaneous bipedal primitive ornithischian taxa, formally thought to be ornithopods. As we saw in Part III, primitive ornithischian relationships are in a state of flux; what is presented here is undoubtedly not the last word on them!

Figure 12.7. Ossified tendons (indicated by arrows) along the neural arches of *Iguanodon* at the Museum of Natural History in Vienna. This would have immobilized the backbone, providing strength in the pelvic region where the stresses were high during bipedal locomotion.

when in a hurry. A number of larger ornithopods have a sturdy wrist and a hand with thickened hoof-like nails on the central digits, a hand that clearly was capable of considerable weight support. Indeed, some hadrosaurs are now known to have been obligate quadrupeds; they were quadrupedal at all times. For all this, however, reconstructions of their hip musculature suggest more similarities with bipeds than with quadrupeds. Perhaps the strongly bipedal cast to their inferred hip musculature is a function of their ancestry: like ceratopsians, they undoubtedly evolved from bipedal ancestors, but unlike ceratopsians, their hip morphology appears not to have evolved as distinctly quadrupedal. Juveniles may have been more bipedal than their fully-grown, adult counterparts.

In all cases, the tail was long, muscular, strengthened by **ossified tendons**, that is, tendons that had mineralized and turned to bone, and held at or near horizontal, making an excellent counterbalance for the front of the animal. In the largest ornithopods, ossified tendons extended into the hip region, providing strength and rigidity at the pelvis, where considerable stress was placed, particularly during bipedal locomotion (Figure 12.7). In general, the powerful hindlimbs tend to be at least as long as, and in some cases more than twice the length of, the forelimbs.

How fast could these dinosaurs have traveled? Larger iguanodontians such as hadrosaurids may have been able to reach 15 to 20 kmh during a sustained bipedal run, but perhaps as high as 50 kmh on short sprints. Quadrupedal galloping appears unlikely given the rigidity of the vertebral column and the limited movement of the shoulder against the ribcage and sternum. For smaller ornithopods, running speeds could have been even higher (see Box 13.3).

Arms and hands

Many ornithopods appear to have had gracile forelimbs, and likely used their hands to grasp at leaves and branches, bringing foliage closer to the mouth so that it could be nipped off by the toothed beak.

Most iguanodontians had very specialized fingers and hands, indicating multiple functions (Figure 12.8). In *Iguanodon*, for example, the first digit (thumb) was conical and sharply pointed. It has been suggested that it was used as a stiletto-like, close-range defensive weapon, or for breaking into seeds and fruits (or both, or . . . ?). The middle digits (II, III, and IV) were tipped with hooves, and

Figure 12.8. The hands of *Iguanodon*. Digits labeled on right hand. Note the spiked thumb (digit I). Photo from Museum of Natural History in Vienna.

were evidently weight-bearing; these would have been key players when the animal was in a quadrupedal stance. Finally, the outer finger (V) was highly flexible, and could bend across the palm, very much as the thumb does in humans: a grasping, opposable pinkie.

Unlike other iguanodontians, in hadrosaurids digit I of the hand was lost, and digit V was relatively small. This left three main fingers, all tipped with hooves, with hardly any way to function other than to support the animal while standing. Hadrosaurids likely spent a lot of time on all fours.

Dietary fiber

Fine dining, ornithopod style, is relatively well understood. For hadrosaurids at least, "mummies," complete with stomach contents, have been found (Figure 12.9). These spectacular specimens, though not true mummies in the sense that original soft tissue and bone are preserved, apparently first dried to dinosaur jerky before burial. The dried, toughened muscle and sinew didn't decompose, so the whole package – tissue *and* bones – was replaced during burial (see Chapter 1). The startling result is preserved skin impressions, stretched tendons and muscles, and the last supper in the stomach and intestines. Hadrosaurids, we now know, ate twigs, berries, and coarse plant matter.

This selection of food correlates nicely with ornithopod height: they are thought to have been active foragers on ground cover and low-level foliage from conifers and in some cases from deciduous shrubs and trees of the newly evolved angiosperms; that is, the clade of all plants that bear flowers (see Chapter 14). Browsing on such vegetation appears to have been concentrated within the first meter or two above the ground, but the taller animals may have reached vegetation as high as 4 m.

Eating coarse, fibrous food requires some no-nonsense equipment in the jaw to extract enough nutrition for survival, and ornithopods had the necessary goods. Like all genasaurs, ornithopods had a beak in the front for cropping vegetation, a diastem, a group of cheek teeth for shearing, and a large, robust coronoid process for serious mastication (Figure 12.10). A deeply inset tooth row indicates large fleshy cheeks, and powerful, closely pressed cheek teeth developed in hadrosaurids

Figure 12.9. A hadrosaurid "mummy," complete with skin, ligaments, muscle tissue, stomach contents, and skeleton. The animal died on a floodplain, after which its carcass dessicated to dinosaur jerky. With burial, the dried dinosaur tissues were replaced with minerals, ultimately producing an exact replica of the original dried carcass. Courtesy of the American Museum of Natural History.

into true dental batteries (Figure 12.11), even more tightly packed than those we saw in ceratopsids. But beyond these basics, different ornithopods had different modifications of the jaw, and different kinds of jaw motions are believed to have been used for the processing of food.

Modern treatments of ornithopod jaw mechanics suggest some differences among ornithopod feeding behaviors. In basal ornithopods, the beak was relatively narrow, implying a somewhat selective cropping ability. Iguandontians, particularly the more highly evolved hadrosaurs, had broad snouts (Figure 12.12), and in some cases even developed a strongly serrated edge on the rhamphotheca. They were likely not too selective; instead, they hacked and severed leaves and branches without much regard for what they were taking in, confident they had the equipment to handle whatever went in. Basal ornithopods were more likely careful nibblers, while iguanodontians were wood chippers.

Beyond the diastema, the chewing began. Here it was aided by something that is utterly foreign to humans (but not so, for example, to snakes). Our skulls and lower jaws are **akinetic**, meaning that, except for the vertical motion of the lower jaws, the bones in our skulls are solidly fused and locked together. Not so with ornithopods. Above and beyond the familiar up and down compression of chewing, slight movements of particular, individual bones within the skull and lower jaws allowed the cheek teeth to grind past each other from *side to side*. A skull in which individual bones move is called **kinetic**.

Ornithopods likely evolved a unique, kinetic skull, in which they mobilized their upper jaws. This kind of mechanism, called **pleurokinesis**, involved a slight outward rotation of portions of the upper jaw, especially the maxilla (the bone that contains the upper teeth), with each bite (Figure 12.13). When the upper and lower teeth were brought into contact on both right and left sides, the opposing surfaces of the dental batteries sheared past one another. Pleurokinesis was an important advance for ornithopods, especially iganodontians, giving them the ability to chew the toughest, most fibrous plants.

Chewing reached its most refined state in hadrosaurids, in which the cheek teeth were fitted tightly together into a *dental battery*, which effectively acted as a single shearing or grinding tool in each jaw (see Figure 12.11c). With constantly replacing teeth, tooth wear was never an issue (as it is in mammals, which only replace teeth once, so their adult teeth have to last their entire lives). The toughest, most fibrous plants undoubtedly succumbed to the hadrosaurid combination of powerful jaw muscles operating on a pleurokinetic skull equipped with constantly replaced grinding surfaces.

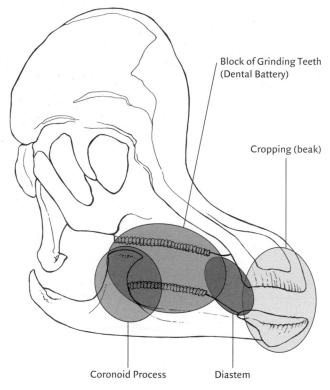

Figure 12.10. The skull of the Late Cretaceous hadrosaurid *Corythosaurus*, exemplifying chewing in hadrosaurids. The divisions of the skull are: anterior cropping section (beige); diastema (maroon); block of grinding cheek teeth (blue); coronoid process (brown). See also Figures III.5; 11.23.

As in all of the other ornithischians that have been discussed, once ornithopods chewed their food, it was swallowed and digested in a capacious bacteria-filled gut. Unsurprisingly, given their sophisticated chewing abilities, the proportionately largest guts were found in the largest iguanodontians (including hadrosaurs); these ornithopods were all about extracting the most nutrition out of a low-quality, high-fiber, high-volume diet.

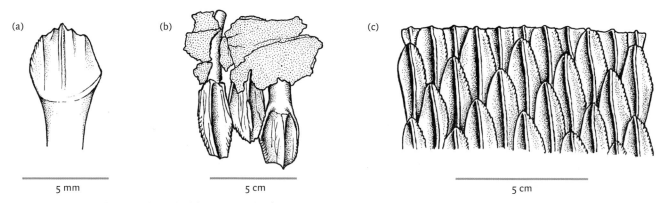

Figure 12.11. Teeth in ornithopods. (a) Upper tooth of *Hypsilophodon*, (b) three upper teeth of *Iguanodon*, and (c) lower dental battery of *Lambeosaurus*. The teeth throughout ornithopod evolution become progressively more tightly packed, culminating in the lambeosaur dental battery.

Figure 12.12. *Edmontosaurus*, a flat-headed hadrosaurid from the western USA, exemplifying the broad snouts of hadrosaurid ornithopods.

Thoughts of an ornithopod

By dinosaur standards, ornithopods were smart – as smart or smarter than might be expected of living archosaurs if they were scaled up to dinosaur size (see Box 13.4). For example, *Leaellynasaura*, a basal ornithopod from Victoria, Australia, was apparently quite brainy and had acute vision, as suggested by prominent optic lobes in the brain.[1] In general, euornithopod smarts may be related to greater reliance on sight, smell, and hearing for protection that, in the absence of other means, may

---

[1] The animal had an estimated encephalization quotient (EQ; see Box 13.4) of 1.8; J. A. Hopson estimated that the average EQ of other ornithopods is about 1.5.

have been their only defense. Moreover, brain size in these dinosaurs may have also been an integral part of a complex behavioral repertoire.

Socializing *à la* Ornithopoda

From the time of their discovery, ornithopods have attracted a good deal of attention, particularly for the extraordinary crests on the heads of hadrosaurids and the lumps on the forehead of *Ouranosaurus*. All of these features hint at sophisticated social behavior.

Song of the saurian

Hadrosaurids have attracted the most attention, in large part because some of them bore striking hollow crests (Figure 12.14). The hollow crest morphology was once thought to relate to putative aquatic habits of the group (nobody thinks they were aquatic any more!) or to increasing a sense of smell (still a possibility), but studies also suggest that the internal chambers of the crests would have made good resonating chambers, producing loud, low-frequency sounds[2] – a kind of Mesozoic *alphorn*. There is no point in making noise if there is nobody who can hear it; however, studies of the morphology of the hadrosaur hearing mechanism (inner ear) suggest that hadrosaurs could hear the low frequencies that may have been produced in the hollow crests.

Figure 12.13. Jaw mechanics in Ornithopoda, showing lateral mobility of the upper jaws (pleurokinesis).

With that insight, much of the discussion now centers on intraspecific competition (see Chapter 5) and sexual selection (see Chapter 6). To convey information about species, sex, and even rank, crests had to have been *visually* and, if part of their function was as a resonating chamber, *vocally* distinctive. Only then can they have promoted successful matings between consenting adults. How strongly is the role of sexual selection implied by hadrosaurid crests?

Paleontologist J. A. Hopson made five predictions that test the hypothesis that hadrosaurid crest morphology was all about sexual selection.

(1) If communication and display were important, hadrosaurids must have had both good hearing and good vision.
(2) If the crest served the dual role of visual display *and* as a vocal resonator, then its external shape might not necessarily mimic the internal shape of the resonating cavities inside.
(3) If crests acted as visual signals, then they should be species-specific in size and shape, and they should also be sexually dimorphic.
(4) If the crests were a visual cue, they ought to be increasingly distinctive as the number of hadrosaurids living together increases.
(5) The crests should become more distinctive through time as a consequence of sexual selection.

How did these hypotheses fare? Hypothesis (1) is relatively well supported, in that hadrosaurids, to judge from their **sclerotic rings** (see Figure 4.6), had relatively large eyes, implying acute vision.

---

[2] For all you musicians: from a G two octaves below middle C to the F# above middle C.

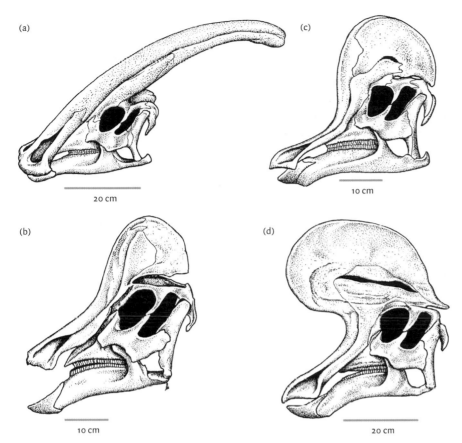

(a)

(c)

20 cm

10 cm

(b)

(d)

10 cm

20 cm

Figure 12.14. Left lateral view of the skull of some hollow-crested hadrosaurids. (a) *Parasaurolophus*, (b) *Hypacrosaurus*, (c) *Corythosaurus*, and (d) *Lambeosaurus*.

Similarly, preserved middle and inner ear structures suggest that a wide range of frequencies was audible to these animals. Hypothesis (2) is upheld in virtually all cases, in that the profile of the crest is more elaborate or extensive than the walls of the internal plumbing (Figure 12.15). Hypothesis (3) is amply upheld in large part thanks to studies on the growth and development in lambeosaurine hadrosaurids, which show that crests become most prominent when an animal approaches sexual maturity. Moreover, adult lambeosaurines are known to be dimorphic, particularly in terms of crest size and shape (Figure 12.16). Could these "morphs" have been male and female?

Hypothesis (4) is based on the idea that distinctiveness would be an advantage during the breeding season. It was tested at Dinosaur Provincial Park in Alberta, Canada, where five distinctively equipped species of hollow-crested hadrosaurid and one species of solid-crested hadrosaurid all lived together in multi-specific bliss. In support of the hypothesis, elsewhere where hadrosaurid diversity is lower, the distinctiveness of the crests is decreased. Interestingly, however, hypothesis (5) is not well supported, for lambeosaurines' crests, at least, arguably become less distinctive over time. Perhaps, then, as we saw in ceratopsians, some of this ornamentation was related to the identification of an individual's maturity, rather than his (her?) species.

If the crests were used for species recognition, ritualized display, courtship, parent–offspring communication, and social ranking, the accentuated nasal arch seen in genera such as *Gryposaurus*, *Maiasaura*, and *Brachylophosaurus* may have been used for broadside or head pushing during male–male combat (Figure 12.17a). Inflatable flaps of skin possibly covered the nostrils and surrounding regions (Figure 12.17b); if present, these could have been inflated and used for visual display, as well as noise-making – more a Mesozoic *bagpipe*. In *Prosaurolophus* and *Saurolophus*, in

Figure 12.15. Highly modified nasal cavities housed within the hollow crests on the heads of lambeosaurine hadrosaurids. (a) *Lambeosaurus*; (b) *Parasaurolophus*; (c) *Corythosaurus*.

Figure 12.16. Growth and sexual dimorphism in lambeosaurine hadrosaurids. (a) Juvenile and (b) adult *Corythosaurus*. (c) Male (?) and (d) female (?) *Lambeosaurus*.

which a solid, spiky crest points posteriorly, a sac might have extended onto the solid crest that extended above the eyes (Figure 12.17d), while in *Edmontosaurus*, with neither hollow crest nor solid spike (see Figures 12.1 and 12.12), a broadly excavated nostril region may have housed an inflatable sac. Unfortunately, these soft-tissue-based hypotheses are all speculative.

Figure 12.17. (a) The circumnarial depression in *Gryposaurus* (indicated by cross-hatched region) which may have supported an inflatable flap of skin in hadrosaurines; (b) speculative reconstruction of an inflatable sac in *Gryposaurus*; (c) the skull of *Saurolophus*, including the distinctive posterior-pointing solid spike; (d) speculative reconstruction of an inflatable sac in *Saurolophus*.

In lambeosaurine (hollow-crested) hadrosaurids, the crests perched atop the head must have provided for instant recognition (Figures 12.14). This would have been by visual cues as well as by low honking tones produced in the large resonating chamber of the crest (see Figure 12.15).

Other ornithopods

Other ornithopods show features potentially interpretable in terms of sexual selection and intraspecific competition. Low, broad bumps on top of the head of *Ouranosaurus* and the arched snout of *Muttaburrasaurus* and *Altirhinus* may well have behavioral significance relating to intraspecific competition and sexual selection. *Ouranosaurus* was equipped with extremely tall neural spines, which formed a high, almost sail-like ridge down its back (Figure 12.18). What these did – or for that matter what the spines did in *Spinosaurus* (see Figure 6.26) – is anybody's guess. It is possible that these long spines supported a membrane used by the animal to warm up and cool down, and/or the spines may have had a display function, providing the animal with a greater side profile than it would otherwise have had. We just don't know.

Display behavior in many ornithopods begins to make even more sense when considered in the context of the discoveries of bonebeds containing just one type of dinosaur. **Monotypic** bonebeds, that is, bonebeds containing only one type of animal, are known for *Dryosaurus*, *Iguanodon*, *Maiasaura*, and *Hypacrosaurus*, among ornithopods. In the case of hadrosaurids, at least, the evidence suggests that a single herd could have exceeded 10 000 individuals, rivaling the multiple mile-sized bison herds that roamed the Great Plains of North America before the unfortunate pairing of the transcontinental railroad with the .55 caliber Sharps rifle.

Figure 12.18. The Early Cretaceous iguanodontian *Ouranosaurus* from Niger.

1 m

Bringing up baby II

The secrets of ornithopod reproductive behavior are just beginning to be told. For the small, basal ornithopod *Orodromeus*, hatchlings had well-developed limb bones, with fully formed joints, indicating that the young could walk, run, jump, and forage for themselves as well as any adult. We thus infer minimal parental care.

As first discovered by paleontologist J. R. Horner, hadrosaurids did not take so *laissez-faire* an attitude toward child-rearing. *Maiasaura*, *Hypacrosaurus*, and probably most others nested in colonies, digging a shallow hole in soft sediments and laying, in the case of *Maiasaura*, up to 17 eggs in each nest. These nests were separated by about a mother's body length, strongly suggesting that they were regularly tended by a parent. *Maiasaura* hatchlings (Figure 12.19) look like typical babies (think puppies): big limbs, feet, and hands; large heads and eyes; shortened snouts! The babies were found amid an abundance of eggshell fragments, implying an extended stay at the nest that wreaked havoc on the eggs that once housed them.

With poorly developed joints and limbs, the offspring were helpless during the nest-bound time. They could hardly have foraged far from the nest and must have depended on their parents to provision and protect them. But to go from a 1 m hatchling to a 9 m adult, growth must have come hot and heavy: at approximately 12 cm per month, as fast as or faster than fast-growing mammals and birds (see Chapter 13). This means that hatchlings must have channeled into growth virtually all of the food that their parents brought them.

Recently the development of *Hypacrosaurus*, at least, has been taken to new levels with a detailed analysis of changes in the face as the animals grew up. A complete growth series of skull specimens representing embryos, nestlings, juveniles, subadults (teenagers!), and adults has been reconstructed (*à la Protoceratops*) and from this it is clear that these dinosaurs, like all juveniles, begin life

Figure 12.19. Left lateral view of the skull and skeleton of a hatchling hadrosaurid *Maiasaura*. Keeping in mind that the adult was ~4.5 m long, check out the scale bar on this figure (see also Figure 13.13).

25 cm

relentlessly cute: short snouts, large eyes, and rudimentary crest development! Just as in all other vertebrates, these features change, becoming less cute, as adulthood is reached.

A new take on ornithopod child-rearing was provided by the hypsilophodont *Oryctodromeus*, a dinosaur that evidently raised altricial young in a burrow (see Figure 1.6). Only one specimen of the animal is known, but this reveals a burrow with an end chamber, in which were found the remains of an adult and two juveniles. *Oryctodromeus* appears to have some digging specializations in its skull and thoracic region, suggesting a **fossorial**, or burrowing, lifestyle.

Ornithopod life, therefore, certainly involved much opportunity for interaction: within herds, as breeding pairs, and as families. All of this is part and parcel of the visual and vocal communication we postulated earlier, and affirms complex social behavior, particularly in hadrosaurids.

How do those ornithopods for which we have information conform to either of these two contrasting reproductive strategies outlined in Chapter 9? *Orodromeus* seems to have been an *r*-strategist, an inference that is based on the precocial nature of the young. In contrast, *Maiasaura*, *Hypacrosaurus*, and perhaps other hadrosaurids had nest-bound, altricial hatchlings, and were likely *K*-strategists.

The evolution of Ornithopoda

The basal split of Ornithopoda from the generalized cerapodan condition likely occurred in the latest Triassic or earliest Jurassic. The clade is diagnosed on the basis of a number of derived features (see Figure 12.5). As we've seen, an early split in Ornithopoda occurred between primitive ornithopods such as *Orodromeus* and *Agilisaurus*, and Iguanodontia, with iguanodontians containing much of the future diversity of the clade.

Ornithopoda, once comfortably thought of as the home of three-toed, bipedal ornithischians, has proven to be a rascally taxon under the rigorous scrutiny of cladistic methods (Box 12.1). It is best understood by its definition: all genasaurs more closely related to *Parasaurolophus* (a hadrosaur) than to *Triceratops*. On the cladogram, this means, in effect, all non-marginocephalian ceropodans (Figure 12.20). It consists of a host of relatively small, agile ornithopods such as *Hypsilophodon* and *Gasparinisaura*, as well as a few somewhat larger, more robust forms (*Parksosaurus* and *Thescelosaurus*), and the diverse clade Iguanodontia (Figure 12.21), residence of such dinosaurian luminaries as *Camptosaurus* (Figure 12.22), *Iguanodon*, and all the hadrosaurids.

Box 12.1.   Those pesky basal neornithischians!

It used to be really easy: if it was bipedal, ornithischian, herbivorous, and three-toed, it was likely an ornithopod. But when paleontologists got to thinking about relationships a bit more carefully, and looked through the lens of cladistic analysis, who was a diagnosable ornithopod and who was not, turned out to be a bit stickier.

The lid blew off the pot with heterodontosaurids: these three-toed, herbivorous bipedal ornthischians had appeared for years to be ornithopods. Careful cladistic analysis by R. J. Butler, P. M. Barrett, and colleagues (see Chapter 15), however, suggested that they properly belong way back at the base of Ornithischia.

Other supposed "ornithopods" also were problematic: *Othnielosaurus*, *Hexinlusaurus*, *Agilisaurus*, *Zephryrosaurus*, *Lesothosaurus*, *Pisanosaurus*, *Jeholosaurus*, *Yandusaurus*, *Stormbergia*, to name just a few! Worse yet, venerable clades like Hypsilophodontidae were shown to be paraphyletic, so that perplexed researchers, as noted by North Dakota paleontologist C. Boyd (2015, p. 2) "chose conservatively to refer to all non-marginocephalian, non-iguandontian. . .taxa as 'basal neornithischians.'"

Part of the problem stems from the strengths and weaknesses of phylogenetic systematics. As we've seen, the method depends upon synapomorphies for its power; thus, when dealing with primitive characters, it is at its weakest, because there are few synapomorphies to help distinguish the groups. The result is that it is hard to determine which of the many differing hypotheses of relationships – the cladograms – is the most parsimonious, and should thus be selected.

It turns out that most humans can live utterly comfortable lives without ever knowing the exact phylogenetic position of, for example, *Gasparinisaurus*. That being the case, we offer two simplified examples of the many cladograms that have been proposed for these basal neornithischian dinosaurs. Figure 12.20, in the text, is a version of the 2012 cladogram developed by paleontologists R. J. Butler and P. M. Barrett. Figure B12.1.1, here, is a version of the cladogram suggested by Boyd in 2015. The take-home message in all this is that ornithopod – or perhaps we might better say neornithischian – evolution, was not a simple straight line, but rather a complicated bushy tree, with many dead ends and false starts on the way to the most highly evolved group of ornithopods: hadrosaurs.

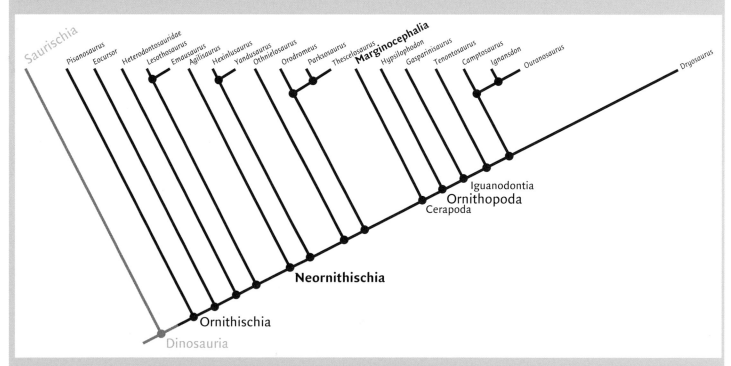

Figure B12.1.1. C. Boyd's 2015 proposal for the relationships among ornithischian dinosaurs. It is broadly comparable to the cladogram we show in Figure 12.20; however, it differs in some key aspects as well. For example, *Orodromeus*, considered by some to be a basal ornithischian, is here not even within the group. Likewise, *Hypsilophodon*, *Parksosaurus*, and *Thescelosaurus*, long regarded as fine, upstanding ornithopods, are here regarded as too primitive to be within Ornithopoda. *Gasparinisaurus*, considered by Butler and Barrett as rather close to *Hypsilophodon*, is here a member of Iguanodontia. It's bewildering, and the goal is not to give non-specialists hives; it is just to convey a sense of the complexity and instability of our understanding of basal neornithischian relationships as of 2016! Figure modified from Boyd, C. 2105. The systematic relationships and biogeographic history of ornithischian dinosaurs. *PeerJ*, **3**:e1523; DOI 10.7717/peerj.1523.

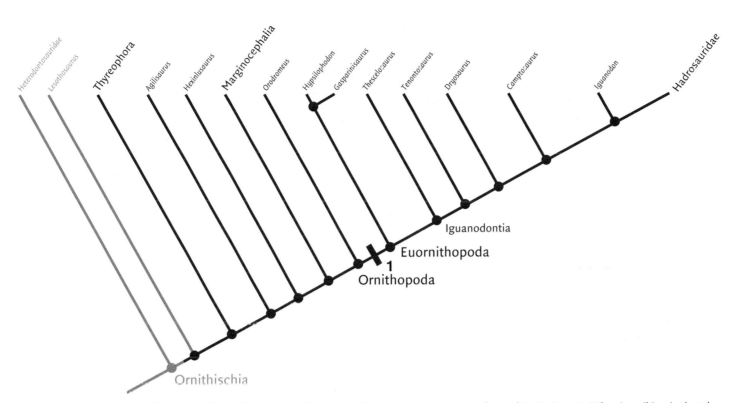

Figure 12.20. Cladogram of basal Ornithopoda, after the 2012 study of paleontologists R. J. Butler and P. M. Barrett. What is striking is that the majority of the most famous ornithopods are most precisely designated "euornithopods." We use the more general term "ornithopods" throughout this chapter, however, as a nod to convention. Derived characters include: at **1**, subcircular external antorbital fenestra, distal offset to apex of maxillary crowns, strongly constricted neck to the scapular blade, ossification of sternal ribs, hypaxial ossified tendons in the tail; at **2**, rectangular lower margin of the orbit, widening of the frontals, broadly rounded predentary, dentary with parallel dorsal and ventral margin, absence of premaxillary teeth, ten or more cervical vertebrae, six or more sacral vertebrae, presence of an anterior intercondylar groove, inflation of the medial condyle of the femur.

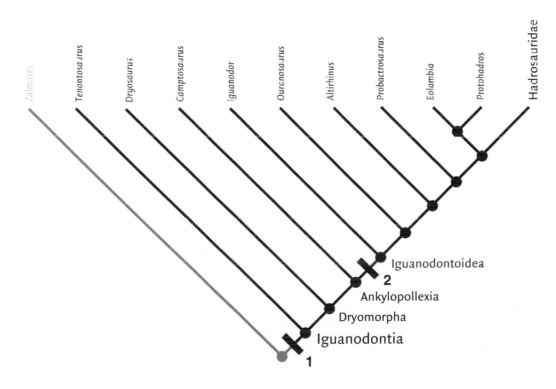

Figure 12.21. Cladogram of basal Iguanodontia. Derived characters include: at **1**, premaxilla with a transversely expanded and edentulous margin, reduction of the antorbital fenestra, denticulate margin of the predentary, deep dentary ramus, loss of sternal rib ossification, loss of a phalanx in digit III of the hand, compressed and blade-shaped prepubic process; at **2**, strong offset of premaxilla margin relative to the maxilla, peg-in-socket articulation between maxilla and jugal, development of a pronounced diastema between the beak and mesial dentition, mammillations on marginal denticles of teeth, maxillary crowns narrower and more lanceolate than dentary crowns, closely appressed metacarpals II–IV, deep triangular fourth trochanter, deep extensor groove on femur.

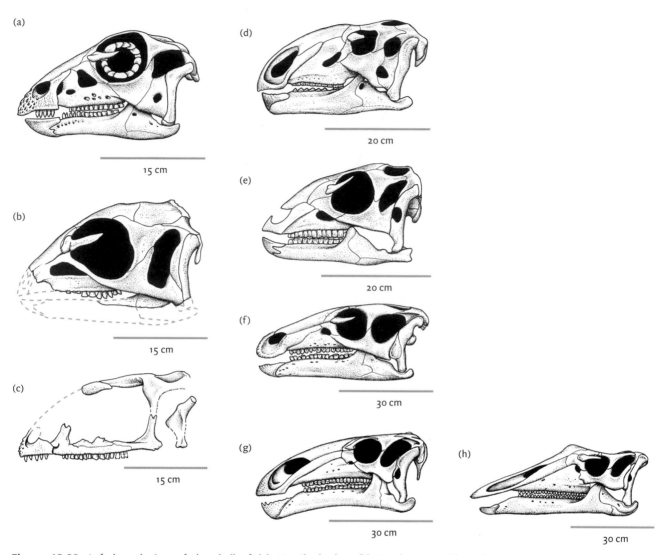

Figure 12.22. Left lateral view of the skull of (a) *Hypsilophodon*, (b) *Yandusaurus*, (c) *Zephyrosaurus*, (d) *Tenontosaurus*, (e) *Dryosaurus*, (f) *Camptosaurus*, (g) *Iguanodon*, and (h) *Ouranosaurus*.

In general, non-hadrosaur iguanodontians tended to reach their apogee in the Late Jurassic–Early Cretaceous interval.

Hadrosaurids are among the best known of all dinosaurs, with a fabulous fossil record that allows, as we have seen, insights into their behavior. Two major clades within hadrosaurids – Lambeosaurinae and Hadrosaurinae – constitute most of Hadrosauridae, with a few forms left whose relationships within Hadrosauridae are uncertain (Figure 12.23). A variety of hadrosaurids is shown in Figure 12.24.

Several interesting evolutionary trends are present within Ornithopoda. It is perhaps no coincidence that ornithopod diversity seems to parallel gymnosperm and angiosperm diversity (see Figure 14.11), suggesting that these dinosaurs and plants may have been involved in a kind of reciprocal *pas de deux*: as gymnosperms developed ways to discourage predation, ornithopods developed more and more efficient ways of extracting nutrients. The reciprocal evolution

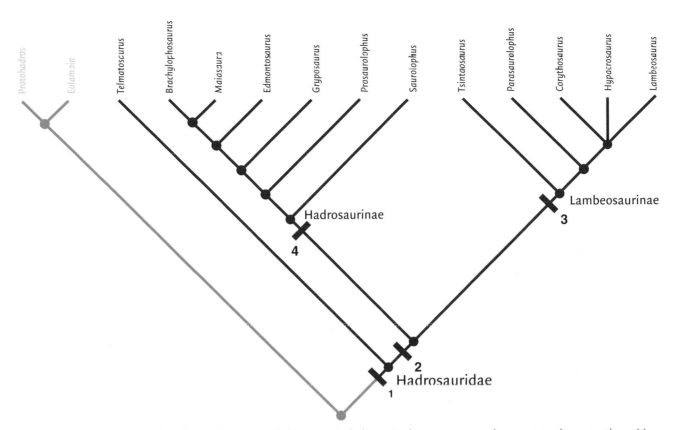

Figure 12.23. Cladogram of Hadrosauridae. Derived characters include: at **1**, three or more replacement teeth per tooth position, posterior extension of the dentary tooth row to behind the apex of the coronoid process, absence of the surangular foramen, absence or fusion of the supraorbital to the orbit rim, long coracoid process, dorsoventrally narrow proximal scapula, very deep, often tunnel-like intercondylar extensor groove; at **2**, absence of the coronoid bone, reduction in surangular contribution to coronoid process, double-layered premaxillary oral margin, triangular occiput, eight or more sacral vertebrae, reduced carpus, fully open pubic obturator foramen, absence of distal tarsals II and III; at **3**, maxilla lacking an anterior process but developing a sloping dorsal shelf, groove on the posterolateral process of the premaxilla, low maxillary apex, a parietal crest less than half the length of the supratemporal fenestrae; at **4**, presence of a caudal margin on the circumnarial fossa.

culminated in the highly efficient pleurokinetic hadrosaurid jaw, with its well-developed, integrated packages of teeth in continuously replacing dental batteries. Overall patterns within the Late Cretaceous of North America and Asia, at least, suggest that hadrosaurids ecologically replaced large non-hadrosaurid iguanodontians. Hadrosaurids arguably evolved the art of chewing to levels of sophistication unparalleled in the history of life; perhaps the requirement of a tough, fibrous gymnosperm diet.

Along with the specializations associated with chewing, ornithopod evolution may have also been characterized by a greater and greater investment by parents in their young. We have seen that *Orodromeus* produced relatively precocial offspring; its basal position within Ornithopoda (see Figures 12.6 and 12.20) suggests that precocity may be primitive for at least Ornithopoda. As the diversity of Ornithopoda increased, altricial behavior likely evolved in more derived ornithopods some time prior to the origin of Hadrosauridae, which all are thought to have given birth to altricial young.

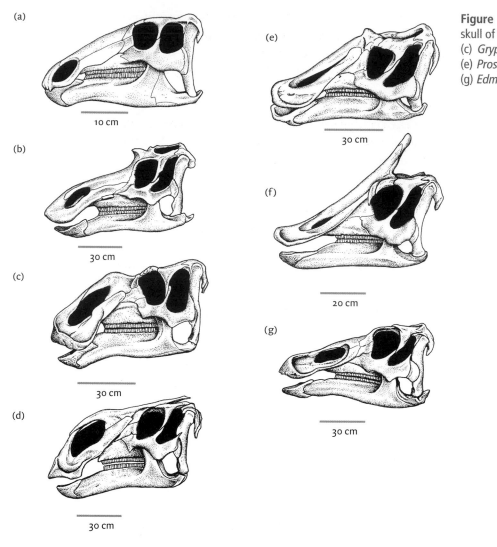

(a)

10 cm

(b)

30 cm

(c)

30 cm

(d)

30 cm

(e)

30 cm

(f)

20 cm

(g)

30 cm

Figure 12.24. Left lateral view of the skull of (a) *Telmatosaurus*, (b) *Maiasaura*, (c) *Gryposaurus*, (d) *Brachylophosaurus*, (e) *Prosaurolophus*, (f) *Saurolophus*, and (g) *Edmontosaurus*.

SUMMARY

Ornithopods were the most numerous and diverse herbivores of Dinosauria, consisting of iguanodontians, hadrosaurids (duck-bills), and a few distinctive, more primitive forms. Although equipped with robust back legs and long tails, and undoubtedly capable of sustained bipedal locomotion, hooves on certain fingers of hadrosaurids and iguanodontians suggest that these animals spent time in a quadrupedal stance as well.

Ornithopods ranged in size from very small (<2 m) to rather large (>15 m), and evidently colonized virtually every inhabitable region of the globe.

Ornithopods had very advanced chewing capabilities. The skull has an inset tooth row indicating cheeks, as well as the tripartite division of chewing herbivores (including a cropping beak, a diastem, and closely appressed cheek teeth for grinding), and in virtually all cases, a large coronoid process suggests strong jaw adductor muscles. Hadrosaurs took chewing to unprecedented heights with the evolution of a pleurokinetic skull combined with dental batteries. Coprolites and stomach contents suggest hadrosaurs needed all the teeth they had: their diet appears to have been coarse and fibrous.

Ornithopods were very social animals, none more so than hadrosaurs. The discovery of many bonebeds attests to this, as do a remarkable and complex variety of sexually dimorphic head and skull features found in many ornithopods, suggesting that life as an ornithopod involved intensive sexual selection. In hadrosaurs at least, communication may well have been enhanced by a variety of vocalizations.

The genus *Maiasaura* has given us a view of child-rearing, duckbill style. Complete growth series, from hatchlings at nests to adults, are known, and considerable evidence exists that hadrosaurs grew remarkably quickly. Nesting apparently was communal, and parental care was expended on altricial young: duck-bills were likely *K*-strategists. Interestingly, other euornithopods, such as the genus *Orodromeus*, may have raised precocial young and favored an *r*-strategy of child-rearing.

Ornithopod evolution was characterized by an increase in the sophistication of chewing specializations. Among the large ornithopods, it could be argued that, in North America and Asia at least, the highly advanced hadrosaurs ecologically replaced non-hadrosaurid iguanodontians.

SELECTED READINGS

Boyd, C. 2015. The systematic relationships and biogeographic history of ornithischian dinosaurs. *PeerJ*, 3:e1523; DOI 10.7717/peerj.1523.

Butler, R. J. and Barrett, P. M. 2012. Ornithopods. In Brett-Surman, M. K., Holtz, T. R., and Farlow, J. O. (eds.) *The Complete Dinosaur*. Indiana University Press, Bloomington, IN, pp. 551–556.

Eberth, D. A. and Evans, D .C. (eds.) 2014. *Hadrosaurs*. Indiana University Press, Bloomington, IN, 619 pp.

Hopson, J. A. 1975. The evolution of cranial display structures in hadrosaurian dinosaurs. *Paleobiology*, 1, 21–43.

Horner, J. R. 1984. The nesting behavior of dinosaurs. *Scientific American*, 250, 130–137.

Horner, J. R., Weishampel, D. B., and Forster, C. A. 2004. Hadrosauridae. In Weishampel, D. B., Dodson, P., and Osmólska, H. (eds.) *The Dinosauria*, 2nd edn. University of California Press, Berkeley, pp. 438–463.

Norman, D. B. 2004. Basal Iguanodontia. In Weishampel, D. B., Dodson, P., and Osmólska, H. (eds.) *The Dinosauria*, 2nd edn. University of California Press, Berkeley, pp. 413–437.

Norman, D. B., Sues, H.-D., Coria, R. A., and Witmer, L. M. 2004. Basal Ornithopoda. In Weishampel, D. B., Dodson, P., and Osmólska, H. (eds.) *The Dinosauria*, 2nd edn. University of California Press, Berkeley, pp. 393–412.

Ostrom, J. H. 1961. Cranial morphology of the hadrosaurian dinosaurs of North America. *Bulletin of the American Museum of Natural History*, 122, 33–186.

Varricchio, D. J., Martin, J. A., and Katsura, Y. 2006. First trace and body fossil evidence of a burrowing, denning dinosaur. *Proceedings of the Royal Society Series B*, 274, 1361–1368.

Weishampel, D. B. 1984. The evolution of jaw mechanisms in ornithopod dinosaurs. *Advances in Anatomy, Embryology and Cell Biology*, 87, 1–110.

Weishampel, D. B. 1997. Dinosaurian cacophony. *BioScience*, 47, 150–159.

TOPIC QUESTIONS

1. Who are the ornithopods? What are the diagnostic characters of Ornithopoda? How are ornithopods related to other ornithischians?
2. What are the major divisions of Ornithopoda, and what are their diagnostic characters?
3. How could a single ornithopod be both bipedal and quadrupedal?

4. Describe the hands of non-hadrosaurid iguanodontians. How did those differ from the hands of hadrosaurids?
5. Highlight what is known of nests and nesting in ornithopods.
6. What is pleurokinesis? How did it function in the jaws of hadrosaurids?
7. Name another vertebrate with a kinetic skull.
8. Why is it that most paleontologists now think that hadrosaurid head structures were related to intraspecific competition and/or sexual selection?
9. Give a non-dinosaur example of a *K*-strategist, and an *r*-strategist.
10. How does ornithopod diversity parallel gymnosperm and angiosperm diversity?
11. Use a cladogram to make the argument that a *K*-strategy evolved *at least* two times in the history of vertebrates.

PART IV
ENDOTHERMY, ENDEMISM, AND EXTINCTION

Non-professionals often imagine paleontology much as we've so far described it in this book (minus the cladograms, of course!): as the discovery of extinct, wondrous beasts. The excitement in the field, it is generally assumed, is in exotic travel; collecting the fossils; in studying and naming new and (to us) bizarre or unexpected life forms. And no active paleontologist would deny that these are all exciting and rewarding parts of the job.

But for most of us, good as these are, they really don't touch the full intellectual richness of the work. It's the *big* picture – the synthesis – where paleontology really makes its mark. Paleontology allows us to ask highly pertinent, Earth-sized questions, such as how entire ecosystems have evolved over time; whether evolution has any kind of intrinsic direction (like whether organisms tend to evolve towards greater intelligence or whether life through time is moving towards increasing diversity). The historical perspective allows us to investigate the great rhythms of life on time scales that can barely be grasped by a human mind.

Particularly relevant in a world in which dramatic climate change is causing rampant global extinction, paleontologists are uniquely positioned to ask whether we are in the middle of a mass extinction right now. What happens when the Earth undergoes a major mass extinction? How do ecosystems respond? Are some organisms more resilient to mass extinctions than others? Are some *ecosystems* more resilient to mass extinctions than others? How long does it take for things to return to normal (called the "**recovery**")? We like to think of ourselves as unique, and living in unique times, but it turns out that the Earth has experimented with almost everything in the past (although, perhaps, not an industrial age); for example, global climate change and rising sea level have occurred many times in the past and we have some ideas about how the biosphere responds.

In fact, humans developed and have flourished in an unusually cold period in Earth history. In the past, things were generally warmer; we might ask, therefore, whether global warming is actually the new (old) normal. Paleontology, with its vast sweep of Earth history, is uniquely positioned to provide insights to these very timely questions.

In this section of the book, therefore, we take a small step towards thinking about larger synthetic questions that naturally arise from studies of the past.

Dinosaur thermoregulation: some like it hot

13

CHAPTER OBJECTIVES

- Vertebrate metabolic strategies
- Reconstruction of past metabolic strategies
- Potential dinosaur metabolism(s)
- Scientific inference in a historical science

The way they were

Since the beginning, paleontologists have been of two minds about dinosaurs. Owen originally described them as mammal-like (see Chapter 15). Yet within 50 years, the prevailing idea that the closest living relatives of dinosaurs might be crocodiles suggested to many scientists that crocodylians were a better model for dinosaur behavior. And so, for much of the twentieth century, dinosaurs were thought of as, in effect, dolled-up crocodiles. That perpsective got revisited with the revolutions of the 1960s and early 1970s and things have been a bit more uncertain – and a lot more interesting – ever since.

If you'll think back to what you've read and seen in the pages preceding this chapter, we've seen a lot of morphology and inferred behaviors that just don't say "crocodile." If not that, with all those bird-like qualities, should we think of Mesozoic dinosaurs as, perhaps, birds of a different feather?

So here is a multiple-choice question that forms the core of this chapter: Mesozoic dinosaurs were

(a) cold-blooded, slow, and "reptilian;"
(b) hot-blooded beasts that functioned like modern mammals and birds; or
(c) something else entirely.

Physiology: temperature talk

We begin with a bit of vocabulary and perhaps a concept or three. **Physiology** is the study of the various integrated functions of an animal; here we discuss the aspect of physiology known as **metabolism**, in this case, the chemical reactions and pathways an organism uses to obtain energy and put it to work (Box 13.1). We often hear expressions like "warm-blooded" and "cold-blooded"; but it turns out that these terms don't mean much. For example, most "cold-blooded" vertebrates have warmed blood when they are active. A more meaningful distinction is between **endotherms** (*endo* – inside; *therm* – heat), organisms that regulate their temperature internally, and **ectotherms** (*ecto* – outside), organisms that use external sources of heat to regulate their temperatures.

Considered from another perspective, in some organisms, called **poikilotherms** (*poikilo* – changing), temperature fluctuates, but in others, called **homeotherms** (*homeo* – same), the temperature remains constant. Humans are *endothermic homeotherms*: when we are unable to maintain our body temperature, we get sick. Ectotherms, such as lizards, can *tolerate* decreases in core temperature, while endotherms must internally *regulate* their core temperatures.

Ectothermy and endothermy are two biochemically and biophysically different methods of obtaining heat. The terms poikilotherm and homeotherm, however, are endpoints in a spectrum that runs from maintaining a constant temperature to having a fluctuating temperature. While many organisms cluster at the familiar metabolic endpoints (endothermic homeotherm; ectothermic poikilotherm), many do not (Box 13.2).

In this chapter, we'll take a tour of some of the data that give us clues about dinosaur metabolism. We'll start with some of the older evidence, generally marshaled in the 1970s and 1980s, and move to more current studies. It turns out that the metabolisms of dinosaurs, as we shall see, likely did not closely match those found in living vertebrates: "(c)" is thus the best answer to the multiple-choice question posed above. But the *Jurassic Park*s and *Jurassic World*s notwithstanding, we can't stick a thermometer into the vent of a Mesozoic dinosaur, so how can we tell?

Box 13.1   Chain of fuels

We all know that somehow we get energy from the family of carbon-based molecules called carbohydrates (think "Candy bar!"). Not quite so familiar, perhaps, is how this works. Simply put, the energy stored in the bonds of carbohydrates is transferred to energy stored in the bonds of a molecule called **ATP** (adenosine triphosphate). Then we – and all living organisms – access that energy by breaking the ATP molecule to produce a closely related molecule called **ADP** (adenosine diphosphate) and thereby releasing some of that energy.

Cellular respiration

In respiration, chemical bonds in carbohydrates (such as the sugar glucose) are broken via a type of reaction called **oxidation**. These reactions occur as complex, linked, series involving a number of intermediate steps. The breakdown of a single molecule of glucose (a simple carbohydrate) through this suite of reactions can produce 36 new molecules of ATP through a series of reactions called the "citric acid cycle," so named because citric acid is produced as an intermediate step in the carbohydrate breakdown (Figure B13.1.1).

This type of metabolism, called *aerobic* (involving oxygen), however, is not 100 per cent efficient: ATP production captures about 40 per cent to 60 per cent of the energy of the bonds of the carbohydrates. The remainder is released as heat.

Organisms respire oxygen because energy storage as ATP involves, as we have seen, oxidation reactions. As the energy output of the organism is increased, the amount of ATP needed is increased, and hence more oxygen is consumed and more heat produced. This is why breathing, heart rate, and temperature increase when we exercise: we are using more energy, requiring more ATP to be generated, and thus we need more oxygen. And give off more heat!

There is a point, however, at which the volume of oxygen supplied by breathing is insufficient. Under such conditions, a different reaction path called *glycolysis* is followed. The process of generating energy through glycolysis is a type of metabolism called anaerobic (without oxygen). Glycolysis bypasses the citric acid cycle, and instead directly produces

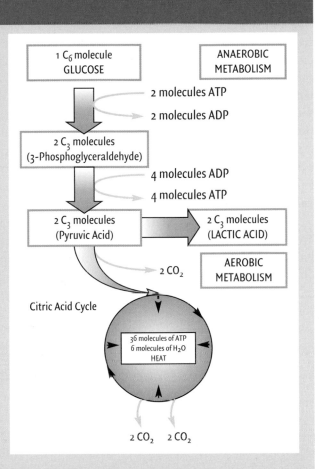

Figure B13.1.1. Cellular respiration consists of the breakdown of carbohydrates to produce energy that is stored in ATP. In this example, the 6-carbon molecule glucose is broken down. Two pathways are shown: the aerobic path, in which ATP is produced via the citric acid cycle, and the anaerobic path, in which lactic acid is ultimately produced via glycolysis.

two 3-carbon molecules called pyruvic acid. The pyruvic acid in turn generates lactic acid, which accumulates in the muscles and causes the familiar ache after extreme exercise (see Figure B13.1.1). After hard exercise, we breathe heavily to replenish our depleted oxygen supply, and eventually the lactic acid is removed from the muscles.

What about dinosaurs?

Given what we know now, we begin with the most obvious piece of evidence: who dinosaurs are. As we have seen, birds are dinosaurs and modern birds are surely endothermic. So we ask the simple question: at what point during dinosaur evolution did "avian" endothermy evolve?

An important clue comes with insulation. All small- to medium-sized modern endotherms are insulated with fur or feathers. There is a certain sense to this; if an ectotherm depends upon external sources for heat, why develop a layer of protection (insulation) from that external source? And can a small endotherm, constantly eating to maintain its metabolism, afford to lose heat? *Archaeopteryx*, with its plumage, is therefore usually considered to have been endothermic.[1] The discovery of non-avian, feathered (insulated) theropods from China (see Chapter 10) gives us a clue that endothermy must have occurred well before Coelurosauria, and perhaps at an even more basal level within Theropoda (Figure 13.1).

Of course, using just this criterion, it's looking more and more like a lot of animals could have been endotherms. We saw that the primitive ornithischian, *Tianyulong*, was likely feathered (see Part III: Ornithischia and "Feathers without flight" in Chapter 5); the evidence suggests that pterosaurs were covered with a filamentous coat. On the other hand, crocodiles aren't. If feathers only evolved once, that is, the monofilamentous feathers found in ornithischians are *homologous* with those found in theropods (saurischians), then the chances are that feathered insulation (and the

Box 13.2   Warm-bloodedness: to have and to have hot

Although endothermy is *characteristic* of birds and mammals, it is by no means restricted to these groups. For some time, physiologists have known of plants (!) that can regulate heat in a variety of ways, the most common being to decouple metabolism (described in Box 13.1) from respiration, so that energy from the breakdown of ATP is simply released as heat. Several snakes are known to generate heat while brooding eggs, although this is accomplished by muscle exertion. Certain sharks and tunas can retain heat from their core muscles by counter-current circulation, and a variety of insects, including moths, beetles, dragonflies, and bees, are known to regulate their body temperatures.

Endothermy is not characteristic of these groups of organisms. Simply, it is known that some of them maintain temperatures warmer than those of the medium (air or water) in which they are living. Indeed, it has been estimated that maintaining a temperature against an external gradient has evolved independently at least 13 different times.

This, of course, differs from the idea that endothermy is *diagnostic* of a particular group. Indeed, endothermy is characteristic of but two groups: birds and mammals. And even in these, the familiar human (or cat/dog) endothermic homeothermy is not always what one finds. For example, many endotherms maintain basal resting rates, which are significantly lower than their active rates. Familiar examples of such animals would include hummingbirds, bats, and hibernating bears. These animals are all endotherms, but they are not properly considered homeotherms. Many other strategies and varieties abound, and it is surely the case that metabolic strategies are in reality far more complex and varied than the simplified endpoints we've presented in this chapter.

---

[1] In 1992, J. A. Ruben suggested that *Archaeopteryx* could have been an ectotherm. His idea was based upon the amount of energy needed for flight, and upon the amount of energy available from an ectothermic metabolism. Since the bones of *Archaeopteryx* show limited adaptations for sustained, powered flight, Ruben argued that an ectothermic metabolism would have been more than sufficient for the kind of limited flight that apparently characterized *Archaeopteryx*. While powered flight may be possible in an ectothermic tetrapod, none (save perhaps *Archaeopteryx*) ever evolved it. Moreover, it seems to us that the presence of insulation (feathers) in *Archaeopteryx* is incompatible with an ectothermic metabolism.

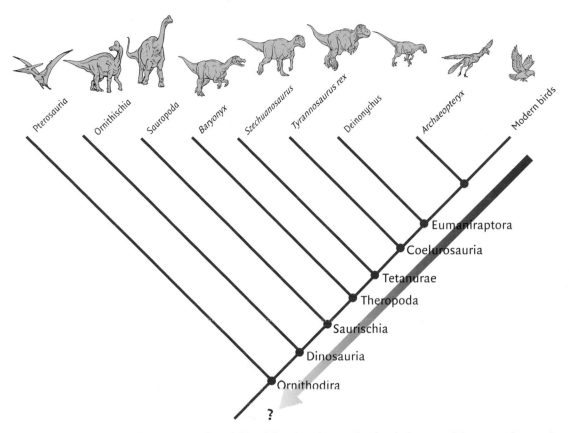

Figure 13.1. Cladogram showing the inferred "depth" within Theropoda of endothermy. While we can be certain the avialans were all endotherms, it is not clear how far back – or how deep within the cladogram – endothermy extends. With the discovery of the feathered *Yutyrannus* (Chapter 7), we now know that it goes back further than coelurosaurs. With monofilamentously feathered ornithischians, perhaps it goes back as far as basal Dinosauria.

metabolism type that it implies) could characterize all Dinosauria (Figure 13.1; see also Part III: Ornithsichia). Using this same logic, if the monofilamentous feathers of dinosaurs are homologous with those in pterosaurs, feather coats might have been invented as early as the first ornithodirans! Unless, of course, those first feathers were sparse; not true coats, but display or sensory features. Does the mere presence of feathers guarantee an endothermic metabolism?

Air sacs and unidirectional breathing in living organisms are certainly indicative of high metabolic rates and thus endothermy. As we've seen, pneumaticity – indicative of expanded air sacs – and thus unidirectional breathing, are well developed in saurischian dinosaurs as a group (Chapters 8–10). These features are indicative of highly evolved respiratory systems, and suggest higher metabolic rates than commonly associated with ectotherms, as well as incipient endothermy.

Seminal studies – early evidence for dinosaur endothermy that fueled the dinosaur renaissance

Along the way, some very clever attempts were made to gain insights into the metabolisms (metabolisms, because one size need not necessarily fit all) of dinosaurs; here we review some of the most important.

Some of the earliest evidence that dinosaurs might be endothermic was marshaled by paleontologist R. T. Bakker (see Chapter 15). These arguments are particularly interesting, because they are examples of how a creative individual, looking at the fossil record in unconventional ways, can infer things from it that far transcend the mere bones of the animals. Bakker appreciated the metabolic significance of the bird–dinosaur relationship we've just reviewed: in 1974, he and paleontologist P. M. Galton proposed that, on the basis of their presumably shared endothermic metabolism, a whole new *class* of vertebrates be recognized: Dinosauria (including birds)! Within 12 years, phylogenetic systematics demonstrated that their insight was correct.

What kinds of evidence led Bakker to conclude that dinosaurs had to be endothermic? A simple observation was that all non-avian dinosaurs had a fully erect stance. Among living vertebrates, a fully erect stance occurs only in birds and mammals, both of which are endothermic. The fully erect limb position in dinosaurs was therefore suggestive to him of endothermy.

Great coincidence; but is there a causal relationship between stance and endothermy? The original idea was that the fine neuromuscular control necessary to maintain a fully erect stance would only be possible within the temperature-controlled environment afforded by an endothermic metabolism. Since Bakker, further work (see also Box 4.3) suggests that, when an animal with sprawling stance moves, the trunk of the organism flexes from side to side as the animal walks. Such flexion reduces the amount of air that can fill the lungs on the scrunched side (Figure 13.2), so that when the animal needs the most air, it gets the least. The evolution of a fully erect stance may thus have been a means by which lung volume could be maximized during high-speed

Figure 13.2. A sprawling vertebrate running quickly. The trunk alternately compresses the lung capacity on each side as the animal runs.

locomotion. Considered in this way, a fully erect posture could be a prerequisite for an endothermic metabolism.

Two other simple correlations have since been noted between anatomy and endothermy. The first is that long-leggedness is characteristic of living endotherms while living ectotherms possess relatively stubby limbs. Certainly many dinosaurs possessed rather long limbs. The second correlation between anatomy and endothermy is the observation that, among living tetrapods, the only bipeds are endotherms. A recent study reconstructed the energy required for bipeds to walk and slow run. The amount of energy required for maintaining these gaits was theoretically calculated for both large and small dinosaur bipeds, and the somewhat surprising conclusion – surprising because it ran counter to prevailing wisdom – was that the amount of energy required by large bipeds, such as hadrosaurs and tyrannosaurs, to walk and slow run was significantly greater than that which would be available to ectotherms. The data were more equivocal for smaller bipeds, but even these pushed the maximum energy available with an ectothermic metabolism to its limit. Large bipedal dinosaurs, the authors concluded, must thus have been endotherms.

A study in 2009 suggested that this should be no surprise. The authors used models that related locomotor anatomy to metabolic rates as measured by inferred oxygen consumption, and then applied the locomotor parameters of thirteen extinct bipedal species to determine their oxygen consumption. In all cases, including a non-dinosaur dinosauromorph (*Marasuchus*; see Chapter 5), an oxygen consumption rate calling for an endothermic metabolism was indicated, particularly so in the five largest dinosaurs (*Tyrannosaurus*, *Allosaurus*, *Plateosaurus*, *Dilophosaurus*, and *Gorgosaurus*) and in small active bipeds such as *Heterodontosaurus*, *Compsognathus*, and *Velociraptor*. All of this, of course, suggests that in these bipeds, at least, an endothermic metabolism was present.

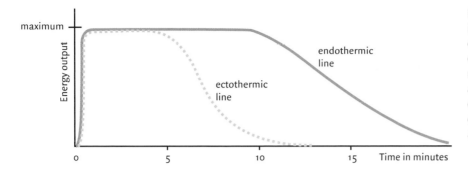

Figure 13.3. Energy output versus time in ectotherms and endotherms. The curves show that the muscles of both endotherms and ectotherms achieve their maximum energy output virtually instantaneously. In general, however, an endotherm *sustains* maximum energy output for more than twice as long as an ectotherm does.

Legs and lungs

Endotherms produce more energy than ectotherms; it takes longer, during heavy use, for the muscles of endotherms to enter an anaerobic physiological state (characterized by **lactic acid** production). This means that, generally, endothermic tetrapods are capable of higher levels of activity sustained over longer periods of time than are ectothermic tetrapods (Figure 13.3).

A variety of small- to medium-sized bipedal dinosaurs such as dromaeosaurids and ornithomimids are characterized by gracile bones, in which the thigh is short relative to the length of the calf. This, in turn, suggests high levels of sustained running – behavior generally not characteristic of modern ectotherms.

And what of the larger dinosaurs, especially those that were not bipedal? Here the issue becomes murkier. The walking speeds of all tetrapods can be calculated from a combination of footprint spacing (stride length) and the length of the hindlimb (Box 13.3). But, of course, the walking that produced most trackways was generally not pedal-to-the-metal running.

Could quadrupedal dinosaurs have run like the fastest mammals today? Ancestry gives a hint. In mammals, a fully erect posture evolved in a quadrupedal ancestor; however, in dinosaurs the fully erect posture evolved in a biped. Quadrupedal dinosaurs are thought to have evolved their four-legged stance secondarily (see Chapters 9, 11, and 12) and thus the front limbs of dinosaur quadrupeds look, and likely functioned, differently from those of mammals (see Figure 11.25). Yet, as we have seen above, quadupedal gaits modeled in *Edmontosaurus*, at least, were very fast.

Of predators and their prey

Perhaps Bakker's most brilliant insight was his inference of metabolism from the numbers of predators and prey found in the fossil record. He began with a simple reality of energy budgets, based in modern animals: *endothermy is much more costly in terms of energy use than ectothermy.* He estimated that it costs 10 to 30 times as much energy to maintain an endothermic metabolism as to maintain an ectothermic metabolism, in part because so much energy is expended on maintaining a constant body temperature.

Given that fact, Bakker reasoned, if predators are endothermic, they should require more energy than if they were ectotherms, and this should be in turn reflected in the weight proportions of predators to prey, or **predator/prey biomass ratios**. He estimated that predator/prey biomass ratios for *ectothermic* organisms are around 40 per cent, while predator/prey biomass ratios for *endothermic* organisms are 1 per cent to 3 per cent. Here, then, was an order of magnitude difference in the biomass ratios, which ought to be recognizable in ancient populations. Now, by counting specimens of predators and presumed prey in major museums and by estimating the specimens' living weights,

Box 13.3   In the tracks of dinosaurs

Trackways, the most tangible record of locomotor behavior, provide evidence for one aspect of an animal's walking and running capabilities, and the only independent test of anatomical reconstructions. When footprints are arranged into alternating left–right–left–right patterns, they demonstrate that all dinosaurs walked with a fully erect posture. But how can trackways also give us an indication of locomotor speed?

We begin with stride length; that is, the distance from the planting of a foot on the ground to its being planted again. When animals walk slowly, they take short strides, and when animals are walking quickly or running, they take considerably longer strides. This much is intuitive for anyone trying to catch a bus about to pull away from the curb. Now, consider the situation when you are being chased by something smaller than you. The creature chasing you must take long strides for its size, and more of them too, just to keep up. So there is clearly a size effect during walking and running, and these will likely be different for different kinds of animals under consideration.

How, then, to relate stride, body size, and locomotor speed? British biomechanist R. M. Alexander provided an elegant solution to this problem by considering **dynamic similarity**. Dynamic similarity is a kind of conversion factor: it "pretends" that all animals are the same size and that they are moving their limbs at the same rate. With these adjustments for size and footfall, it doesn't matter if you're a small or large human, a dog, or a dinosaur. All will be traveling with "dynamic similarity"; only speed will vary. That variable Alexander terms "dimensionless speed."[a] It is dimensionless speed that has a direct relationship with relative stride length. Stride length, of course, can be measured from trackways, which in turns allows us, for the first time, to calculate locomotor speed in dinosaurs.

To see how all this works, let's use Alexander's example of the trackway of a large theropod from the Late Cretaceous of Queensland, Australia. The tracks are 64 cm long, which Alexander, from other equally sized theropods, estimated must have come from a theropod with a leg length of about 2.56 m. The stride length of these tracks is 3.31 m, so the relative stride length (stride length : leg length) is 1.3. The dimensionless speed for a relative stride length of 1.3 is 0.4. And from all these measures, this Australian theropod must have been traveling reasonably quickly, at about 2.0 m/s, or 7.2 kmh.

As complicated as this approach appears, it represents the best method for estimating the actual speeds implied by trackways. But what about the fastest speeds a dinosaur might have been capable of? In 1982, R. A. Thulborn of the University of Queensland developed a method by which absolute locomotor abilities could be calculated. Thulborn's work relied heavily upon Alexander's slightly earlier studies on speed estimates from footprints and the relationship between body size, stride length, and locomotor speed

[a] Dimensionless speed may appear oxymoronic, but in fact is equivalent to real speed divided by the square root of the product of leg length and gravitational acceleration.

Bakker was armed with a tool from modern ecosystems that he believed could reveal the energetic requirements of ancient ecosystems.

His results seemed unequivocal: among the dinosaurs, the predator/prey biomass ratios were very low, ranging from 2 per cent to 4 per cent. He interpreted this low number to indicate that predators and prey in dinosaur-based food chains were endothermic (Figure 13.4).

This study, for all its originality, had some problems. For example, the assumption that prey are approximately the same size as the predators is clearly not correct (consider a bear eating a salmon), and when they are not similarly sized, this has drastic effects on the resultant ratio. Most significantly, the predator/prey biomass calculation assumes that all deaths are the result of predation; that there can be no mortality due to other causes. This is simply not the case in modern ecosystems.

There are other problems associated with using fossils. Most obvious are difficulties in estimating dinosaur weights (Box 9.1). Moreover, the preservation of dinosaur material is subject to a variety of biases. How can one ever be sure that the proportions of the living community are represented?

among living animals. For both approaches, Thulborn determined that relative stride length has a direct relationship with locomotor speeds at different kinds of gaits (for example, walking, running, trotting, galloping).

Explicitly (for the quantitative among you):

$$\text{Locomotor velocity} = 0.25(\text{gravitational acceleration})^{0.5}$$
$$\times (\text{estimated stride length})^{1.67}$$
$$\times (\text{hindlimb height})^{-1.17}$$

Thulborn used this equation to estimate a variety of running speeds for more than 60 dinosaur species. The first group of estimates were for the walk/run transition, where stride length is approximately two to three times the length of the hindlimb. A potentially more important estimate – especially for dinosaurs fleeing certain death or pursuing that all-important meal – is maximum speed, which Thulborn calculated using maximum relative stride lengths (which range from 3.0 to 4.0) and the rate of striding, called limb cadence (estimated at $3.0 \times$ hindlimb length$^{-0.63}$).

Although we report some of these speeds elsewhere in this book, it is of some value to summarize the overall disposition of Thulborn's estimates. For small bipedal dinosaurs – which would include certain theropods and ornithopods – all appear capable of running at speeds of up to 40 kmh. Ornithomimids, the fastest of the fast, may have sprinted up to 60 kmh. The large ornithopods and theropods were most commonly walkers or slow trotters, probably averaging no more than 20 kmh. Thus the galloping,

sprinting *Tyrannosaurus*, however attractive the image, did not impress Thulborn (or us) as likely.

Then there were the quadrupeds. Stegosaurs and ankylosaurs walked at a no more than a pokey 6 to 8 kmh. Sauropods likely moved at 12 to 17 kmh. And ceratopsians – galloping along full throttle like enraged rhinos? Thulborn estimated that they were capable of trotting at up to 25 kmh.

Are these estimates accepted uncritically by all? P. Dodson has argued that these calculations would suggest that humans can run as quickly as 23 kmh, which in life they cannot. So are other means of determining dinosaur speeds possible?

Enter the computer. Analogous with the computerized weight estimates (Box 9.1), vertebrate biologist W. I. Sellers and colleagues developed a series of complex computer models, using anatomy and what they called "mechanically and physiologically plausible" gaits to model locomotion in a duck-billed dinosaur (*Edmontosaurus*). An interesting aspect of their model was, given the parameters of the duck-bill and its gait, the generation of virtual trackways, which could then be compared with existing trackways. Their model suggested that hopping was the fastest gait (17 m/s), followed by quadrupedal galloping (16 m/s), and finally bipedal running (14 m/s), a result that is contrary to the widely held impression that when duck-bills wanted to go fast, they went bipedal. Clearly there's lots more to be done in this research direction.

Because we can't, paleontologists commonly talk about fossil **assemblages**, which are simply the groups of fossils that we have collected in a particular locality, or representing a time interval or geographic area, may have nothing to do with the proportions of the same animals in the living *community* in which those animals lived.

Finally, Bakker obtained his data by counting specimens in museum collections, specimens that were likely collected because they were rare or particularly well preserved. Museum collections thus tend to represent assemblages of well-preserved organisms with a higher percentage of rare organisms than was present in the original fauna.

Several other studies have since attempted to duplicate Bakker's study, with mixed and/or equivocal results. Ultimately, too much uncertainty for the results to be definitive was introduced through the brilliant, but likely flawed, idea of predator/prey biomass ratios. But Bakker's evidence and advocacy kindled the flame of endothermic dinosaurs, and they would never be thought of as tricked-out crocodiles again!

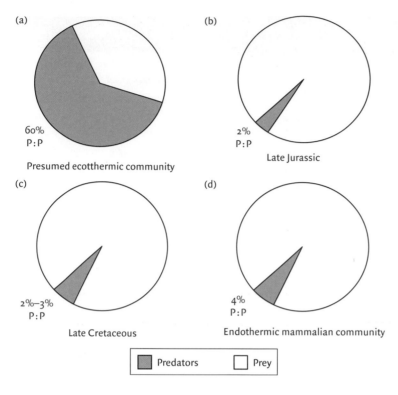

(a)

60%
P:P

Presumed ecotthermic community

(b)

2%
P:P

Late Jurassic

(c)

2%–3%
P:P

Late Cretaceous

(d)

4%
P:P

Endothermic mammalian community

■ Predators   □ Prey

Figure 13.4. The proportions of predators to prey (P : P) in selected faunas in the history of life, as reconstructed by R. T. Bakker. Predators are shaded; prey are unshaded. (a) Early Permian of New Mexico; (b) Late Jurassic of North America; (c) Late Cretaceous of North America; (d) Middle Cenozoic of North America. The Cenozoic fauna (d) is mammalian and obviously endothermic, providing clear guidelines on what predator/prey ratios are in an endothermic fauna. As reconstructed by Bakker, the predator/prey ratios of the dinosaur communities clearly are closer to those seen in the known endothermic mammalian community (Middle Cenozoic) than in the inferred ecothermic community (Early Permian).

Hearts and minds

All living endotherms possess four-chambered hearts. The four-chambered heart system, in which the oxygenated blood is completely separated from the deoxygenated blood, may be a prerequisite for endothermy. Endothermy requires relatively high blood pressures in order to perfuse complex, delicate organs such as the brain with a constant supply of oxygenated blood. Such high blood pressures, however, would "blow out" the alveoli in the lungs. For this reason, mammals and birds separate their blood into two distinct circulatory systems: the blood for the lungs (pulmonary circuit) and the blood for the body (systemic circuit). The two separate circuits require a four-chambered heart – a pump that can completely separate the circuits.

Would such a heart be possible in dinosaurs? The nearest living relatives of dinosaurs, birds and crocodiles, possess four-chambered hearts; thus it is likely that a heart with a double-pumping system was present in basal Dinosauria.

This idea was strongly reinforced by the discovery in 2000 of what was controversially inferred to be a four-chambered heart with an aorta. The "heart" was preserved as an ironstone mass within the thoracic cavity of the basal ornithopod *Thescelosaurus*, and identified using computed tomography (a CT-scan). Doubters doubted; advocates advocated; and the issue remains unresolved.

In the late 1970s, attempts were made to assess the intelligence of dinosaurs using the **encephalization quotient** or **EQ** (Box 13.4). The idea was that living endotherms (birds and mammals) have significantly higher EQs than do living ectotherms (reptiles and amphibians), presumably because their more refined levels of neuromuscular control require an endothermic metabolism. Encephalization quotient was reconstructed in dinosaurs using **brain endocasts**, internal casts of the braincases of dinosaurs (see Box 11.1). Based upon EQ, coelurosaurs were likely as active as many birds and mammals, while large theropods and ornithopods were somewhat less active than birds and

Box 13.4   Dinosaur smarts

How can we measure the intelligence of dinosaurs? The short answer is "Not easily." However, it is clear that, at a very crude level, there is a correlation between intelligence and brain : body weight ratios. Brain : body weight ratios are used because they allow the comparison of two differently sized animals (that is, brain : body weight ratios allow comparison of chihuahua and St. Bernard dogs). The correlation suggests that, in a general way, the larger the brain : body weight ratio, the smarter the organism. Indeed, mammals have higher brain : body weight ratios than fish and are generally considered to be more intelligent.

But how smart could a very large dinosaur with a minuscule brain be (for example, see Figure 10.10 and Box 11.1)?

It is well known that organisms change proportions as they increase in size; this is **allometry**. And it turns out that brain : body weight ratios follow allometric principles as well: brains do not increase in size proportionally to the rest of the animal. For example, the brain of a 0.5 m rattlesnake is proportionally larger than the brain of a 3 m anaconda. Does this mean that the anaconda is significantly stupider than the rattler? Obviously not. So, when considering how big or small a brain is in an animal, there has to be a way to compensate meaningfully for size. A quantitative method of doing this was first proposed by evolutionary psychologist H. J. Jerison, who, in the early 1970s, developed a measure called the "encephalization quotient" (EQ). Jerison constructed an "expected" brain : body weight ratio for various groups of living vertebrates (reptiles, mammals, birds) by measuring many brain : body weight ratios among these animals. Jerison noted that, on the basis of EQ, living vertebrates cluster into two groups, endotherms and ectotherms. The endotherms show greater encephalization (higher EQs) and the ectotherms showed lower encephalization (lower EQs). Thus, for Jerison, living endotherms and ectotherms could be distinguished by brain size. Having constructed a range of expected brain : body weight ratios, he could account for size in different organisms (and accommodate what might at first seem like an extraordinarily large or small brain). Noting that some organisms still didn't exactly fit in his ectotherm or endotherm group (by virtue of having brains either larger or smaller than expected), he measured the amount of deviation, and then termed this EQ.

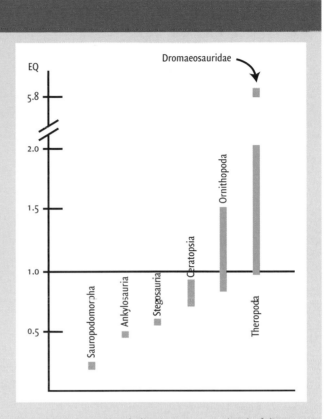

Figure B13.4.1. Encephalization quotients (EQs) of dinosaurs compared. The line at 1.0 represents the crocodilian "norm," and suggests that many groups of dinosaurs had larger brains than would be predicted from a conventional reptilian model (the crocodile). Note also the break between 2.0 and 5.8; if these measures mean anything, apparently coelurosaurs significantly outdistanced other dinosaurs in brain power. (Data from Hopson, 1980.)

Paleontologist J. A. Hopson, now knowing what he could expect for living vertebrates, measured how much the estimated brain : body weight ratio EQ of extinct vertebrates deviated from the expected brain : body weight ratios of their living counterparts. Figure B13.4.1 shows the EQs for several major groups of dinosaurs as reconstructed by Hopson. Because dinosaurs are "reptiles," he measured the deviation of various groups relative to a "reptilian" norm, in this case a living crocodile. Significantly, many ornithopods and theropods show a brain : body weight ratio that is significantly larger than would be expected if a modern reptilian level of intelligence is being considered.

(a)

5 cm

(b)

10 cm

Figure 13.5. Cross-sections (solid shading) through the nasal regions of (a) an extinct dinosaur (*Velociraptor*) and (b) a living bird (*Rhea*); skulls and positions of the cross-sections are shown to the left. The nasal cavity of the bird shows convoluted respiratory turbinates, while that of the dinosaur does not.

mammals, but more active than typical living reptiles. Using EQ as representative of activity levels, other dinosaurs appear to have been in the range of living reptiles.

The nose shows

Endothermy requires the lungs to replenish their air (ventilate) at a high rate. And high rates of ventilation lead to water loss, unless something is done to prevent it. What modern mammals and birds do is to grow convoluted sheets of delicate, tissue-covered bone, called **respiratory turbinates**, in the nasal cavaties. The mucus-covered surfaces of the turbinates pull moisture out of the air before it leaves the nose, thus conserving water (Figure 13.5).

Biologist John Ruben wondered if the fact that modern endotherms require a means to conserve moisture during respiration might provide a clue about dinosaur metabolism. Although a number appear to have had olfactory turbinates, indicative of a well-developed sense of smell, apparently none – as far Ruben knew – had turbinates to allow them to recoup respired moisture. Considered exclusively on this basis, Ruben concluded that dinosaurs could not have been endothermic in the way that most mammals and birds are today.

In the meantime, however, the full details about nasal turbinates appear to be more nuanced than Ruben's original conclusions. First, the delicacy of the turbinates being what it is, a number of workers have pointed out that their absence doesn't mean that they weren't there; they simply might not have been preserved. And – more significantly – respiratory turbinates are now known from two dinosaurs: the pachycephalosaurs *Stegoceras* and *Sphaerotholus*. Even their function is being revisited: new research suggests that nasal turbinates do more than conserve moisture; they may also direct air to heighten the sense of smell as well as – more importantly – cool the brain. It is likely that more respiratory turbinates will turn up on other dinosaur genera.

Where the wild things are

The distribution of dinosaurs around the globe far exceeds the current distribution of modern ectothermic vertebrates, which are generally not found above and below, respectively, latitudes

Figure 13.6. The latitudinal distribution of ectothermic tetrapods on Earth today. The larger terrestrial tetrapods do not get much beyond about latitude 20° North and South. These include large snakes and lizards, crocodilians, and tortoises.

45° north and 45° south. Large modern ectotherms rarely occur above latitude 20° north and below latitude 20° south (Figure 13.6).

Correcting for continental movements, Cretaceous dinosaur-bearing deposits have been found close to latitudes 80° north and 80° south of the equator. Both the northern and southern sites experienced extended periods of darkness, and we can be reasonably sure that, at least occasionally, temperatures in winter fell below freezing.

The Arctic assemblage, from North America, includes hadrosaurids, ceratopsids, tyrannosaurids, and troodontids. The Antarctic dinosaur assemblage, from Australia, includes a large theropod and some basal euornithopods (including many juveniles). Along with these dinosaurs are fish, turtles, pterosaurs, plesiosaurs, birds (known solely from feathers), and, incredibly, a improbable late-surviving **temnospondyl** (an amphibian group that apparently went extinct in the Early Jurassic everywhere else in the world; see Figure 14.6).

This mix of animals is not particularly easy to interpret in terms of endothermy or ectothermy. Several members of the southern dinosaur assemblage had large brains and well-developed vision, potentially useful during periods of extended darkness. Others, however, were not so well equipped. Burrowing may have been an option for some, but not for all.

In the case of the Northern Hemisphere faunas, only *Troodon* is of a size that could make burrowing feasible. Migration was potentially a solution to inclement winter weather, although the

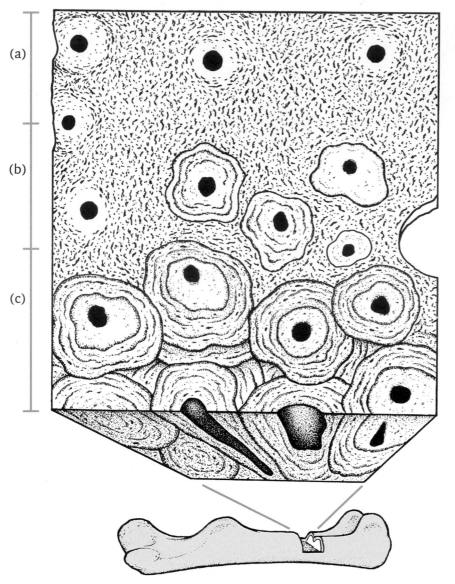

Figure 13.7. Primary bone in the process of being replaced by Haversian bone in the leg of a hadrosaurid. Longitudinal canals (at the top of the figure) in primary lamellar bone (a) are resorbed (b) and then reconstituted as Haversian bone (c).

dinosaurs would have had to migrate large distances before temperatures warmed sufficiently. There is evidence neither for nor against this possibility.

Haversian bone

Bones grow by **remodeling**, which involves the resorption (or dissolution) of bone first laid down – **primary bone** – and redeposition of a kind of bone called **secondary bone**. Secondary bone is deposited in the form of a series of vascular canals called **Haversian canals**, and resorption and redeposition of secondary bone can occur repeatedly during remodeling. When remodeling occurs, a type of Haversian bone known as **dense secondary Haversian bone** is formed. This bone has a distinctive look about it (Figure 13.7), visible in modern as well as fossil bone (Figure 13.8).[2]

---

[2] Fossil bone can preserve fine anatomical details. To see these, a thin slice of bone can be cut, mounted on a glass slide, and ground down to ~30 microns, so thin that light can be transmitted through it. The slide with the slice of bone can then be studied under a microscope.

(a)

(b)

Figure 13.8. (a) Dense Haversian bone in *Tyrannosaurus*. (b) Magnified view of dense Haversian bone in *Archaeornithomimus*.

Dense secondary Haversian bone is found in many mammals and birds – all, of course, endotherms. Among *extinct* vertebrates, dense secondary Haversian bone has been observed in dinosaurs, pterosaurs, and therapsids (including Mesozoic and Cenozoic mammals). Given the distribution of dense secondary Haversian bone in vertebrates, it is not too great a leap to suppose that dinosaurs, too, must have been endotherms.

Although this idea was initially promising, dense secondary Haversian bone forms due to a variety of factors, only one of which is endothermy. Secondary Haversian canals correlated with size and age, and possibly with the type of bone being replaced, the amount of mechanical stress undergone by the bone, and **nutrient turnover** (the metabolic interaction between soft tissue and developing bony tissue). The last two, however, are still related to metabolism, and as we shall see (below) bone **histology**, the study of bone as a type of living tissue, has continued to play a key role in understanding dinosaur metabolism.

Newer approaches

Since the 1980s, there have been some tremendous strides in understanding dinosaur physiology using advances in our understanding of bone histology, as well as some new geochemical techniques that appear to be very promising.

LAGs

Concentric growth rings have been observed in the bones and teeth of dinosaurs. Among modern tetrapods, such growth rings are typically found in ectotherms, where they are believed to represent seasonal cycles. During times of slowed metabolism (such as dry seasons in the tropics, or cold seasons in more temperate latitudes), growth is stymied – hence the term **"lines of arrested growth,"** or **LAGs** (Figure 13.9).

So bone records a pattern of ring-like deposits representing cycles of growth and stasis. Among living homeothermic endotherms, on the other hand, such patterns are rarer, because the relatively constant, elevated metabolic rates ensure growth at a constant rate.

Most dinosaurs – with the exceptions of some theropods, hypsilophodontids, iguanodontids, and sauropods – show well-developed LAGs, suggesting to researchers that growth in dinosaurs might have been more susceptible to external climatic influences than had been predicted by the simple homeothermic endothermic view of dinosaur metabolism. In two small flyers and one large flightless

Figure 13.9. Lines of arrested growth (LAGs) in a *Tyrannosaurus* fibula (forearm). Arrows indicate LAGs.

enantiornithine (*Patagopteryx*) bird, the presence of LAGs led to the conclusion that the early birds' metabolism(s) were subject to seasonal growth, even though the birds clearly bore feathers. The presence of LAGs in these early birds could mean that, ultimately, they had not quite attained the level of endothermy seen in living birds (see below).

Enantiornithine bone histology contrasts with that of ornithurine birds (for example, *Hesperornis* and *Ichthyornis*). In enantiornithines, the bone tissue looks very similar to that in modern birds. So too does the bone tissue of the primitive, Early Cretaceous *Confuciusornis* (see Figure 8.12).

LAGs as time markers

But are the LAGs truly seasonal? Our best evidence suggests that indeed they are. Of course, there is no numerical way to date them (see Chapter 2), so the evidence is circumstantial. Yet, it comes from several sources and those sources all produce a coherent story. For one thing, comparable lines known to form annually are found in a wide variety of groups closely and not-so-closely related to Dinosauria. Moreover, the amount of bone that forms between the lines is consistent with annual (seasonal) cycles. Finally, the spacing of the lines shows a steady decrease, consistent with a decrease in growth rate as the animals get older. Indeed, LAGs have never been found to come from anything other than annual growth.

And so if we accept that LAGs are annual signals, they suddenly give us insights into a heretofore elusive aspect of dinosaur paleobiology: how fast they grew, and how long they lived.

Growth

How, then, does one determine the age of a dinosaur? Paleohistologists A. de Ricqlès, J. Horner, and K. Padian describe their technique:

(1) Count LAGs, which yield a number presumably corresponding to the age of the dinosaur specimen;

(2) measure bone thickness between each LAG, and divide that thickness by the number of days/ year, which should produce the amount of bone deposited per day; and

(3) measure total bone thickness, and develop an estimate of how much time it would reasonably take to form.

From those steps, they infer the age of the particular specimen in question.

But how about entire growth trajectories, that is, the pattern of growth through an animal's life? To take a familiar example, human growth trajectories aren't linear: we start with rapid growth, decreasing gradually until the onset of sexual maturity, after which growth, if it occurs at all, slows dramatically. By the time most humans have completed the first quarter of their lives, growth has stopped. By contrast, many animals, such as crocodiles, have continuous growth throughout their lives (although growth rates are fastest when they are juveniles; Figure 13.10).

Two approaches can be used to learn about the growth of a dinosaur. The lengths of the long bones – femur or tibia, generally – can be measured as representative of growth; alternatively, the animal's weight can be estimated (see Box 9.1) and considered as representative of its growth. Figure 13.11 shows estimated growth curves for the sauropod *Janenschia*, and for some tyrannosaurs; a growth curve for *Archaeopteryx* is shown in Figure 8.9.

So in the end, what does all this say about the subject of this chapter, dinosaur endothermy? It has become well established that metabolic rates are well correlated with maximal growth rates. Fast growth rates – such as are found in modern birds – are highly suggestive of an endothermic metabolism. Figure 13.12 contrasts the collective growth rate of dinosaurs as a group (derived from six dinosaur genera) with those of some familiar groups of living animals.

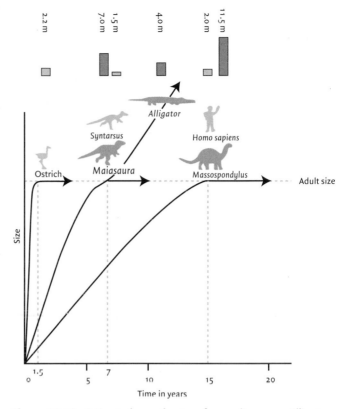

Figure 13.10. Estimated growth rates of some dinosaurs, *Alligator*, and a human. Unlike the other tetrapods presented, *Alligator* grows continuously throughout its life; hence, it has no fixed "adult size." Sexual maturity usually comes within six to eight years. (Estimates for *Syntarsus* and *Massospondylus* from the work of A. Chinsamy; estimates for *Maiasaura* from the work of J. Peterson and J. Horner.)

While some larger dinosaurs may have undergone very rapid growth to attain large size (Figure 13.13), the growth rates of dinosaurs as a group were generally similar to modern endothermic growth rates. This in turn suggests that their metabolism may have been more closely allied to those seen in birds and mammals, than that seen in crocodiles and lizards.

Stable isotopes

Conveniently, fossil vertebrates carry around their own thermometers. These come in the form of **stable istopes**; isotopes that, unlike their unstable brethren, do *not* spontaneously decay. Of particular interest to us are the istopes of the element oxygen. There are three: ^{16}O, by far the most common, ^{17}O, and ^{18}O.[3] The last, ^{18}O, is particularly interesting, because its proportion to ^{16}O varies as temperature varies. This process involves **fractionation**, which is the separation or division of the heavy and light isotopes during a phase transition.[4] Therefore, if a substance contains oxygen, one can learn

---

[3] The isotope ^{16}O comprises 99.763 per cent, ^{17}O 0.0375 per cent, and ^{18}O 0.1905 per cent of total atmospheric oxygen.

[4] In this case, the phase transition in which the fractionation occurs is from liquid (the water that the organism ingests) to the solid (incorporation of the oxygen molecules into the bones and teeth). We will also see fractionation occurring during the phase transition from a gas (the atmosphere) to a liquid (rainwater, condensed from the atmosphere).

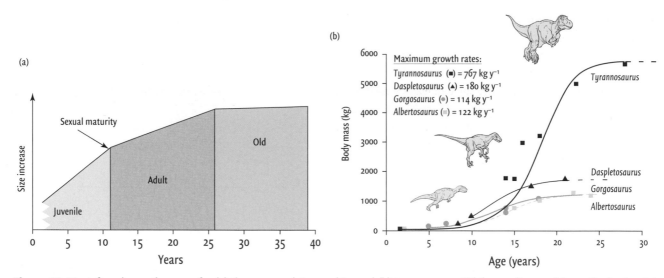

Figure 13.11. Inferred growth curves for (a) the sauropod *Janenschia*; and (b) tyrannosaurs. While age (in years) is on the horizontal axis in both curves, the vertical axis of the sauropod curve is based upon inferred size, while the vertical axis of the tyrannosaur curve is based upon inferred weight. *T. rex*, strikingly heavier than other tyrannosaurids, evidently underwent a considerable growth spurt between 10 and 25 years; other tyrannosaurids had proportionately smaller growth spurts at the same time interval or slightly earlier. From Erickson, G. M. 2005. Assessing dinosaur growth patterns: a microscopic revolution. *Trends in Ecology and Evolution*, **20**, 667–684.

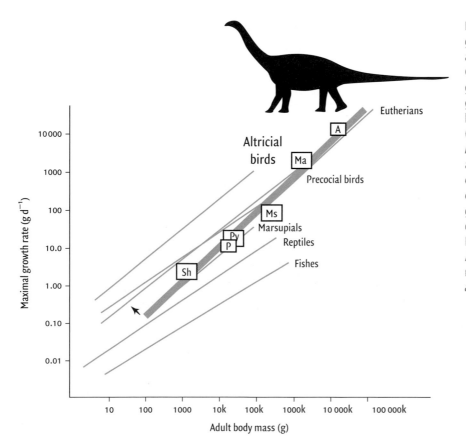

Figure 13.12. Maximum composite growth rates of various groups of animals, across a broad range of sizes. Composite rates constructed from growth rates of several animals in each group. In the case of the dinosaurs (**bold** line), it is constructed from *Shuvuuia* (Sh), *Psittacosaurus* (P), *Syntarsus* (Sy), *Massospondylus* (Ms), *Maiasaura* (Ma), and *Apatosaurus* (A). The difference in composite slope between dinosaurs and other vertebrate groups in the graph suggests that dinosaurs as a group grew at different rates from these other groups. Redrawn from Erickson, G. M. 2005. Assessing dinosaur growth patterns: a microscopic revolution. *Trends in Ecology and Evolution*, **20**, 677–684.

something about the temperature at which that substance formed by the ratio $^{18}O/^{16}O$. In the case of bone and teeth, the oxygen is contained in phosphate (PO$_4$) that forms part of the mineral matter in the bone. Thus, knowing the oxygen isotopic composition of the bone, one can learn something of the temperature at which the bone formed. A number of different approaches bearing on the question of dinosaur endothermy have been tried, including core-to-extremities measurements, direct temperature estimates using isotope ratios, measurements of temperatures using latitudinal gradients, and, most recently, "clumped isotope geothermometry."

Core-to-extremities

If dinosaurs were poikilothermic, reasoned paleobiologist R. Barrick and geochemist W. Showers, there should be a large temperature difference between parts of the skeleton located deep within the animal (that is, ribs and trunk vertebrae) and those located toward the exterior of the animal (that is, limbs and tails; Figure 13.14).

If, however, dinosaurs were homeothermic, there should be little temperature difference between bones deep within the animal and those more external, because the body would be maintaining its fluids at a constant temperature. The difference in temperatures – or lack thereof – should be reflected in the proportions of ^{18}O to ^{16}O (that is, $^{18}O/^{16}O$).

Barrick and Showers showed that bones from the cores of some of the dinosaurs tested (*Tyrannosaurus*, *Hypacrosaurus*, *Montanoceratops*, *Orodromeus*, and a juvenile *Achelousaurus*) showed a $\pm$ 2 °C temperature variation, suggesting that they were formed under near homeothermic conditions. A nodosaurid ankylosaur that they tested, on the other hand, had a large isotopic variation (an inferred temperature variation of ~11 °C) that is well beyond the limits of conventional homeothermy. Barrick and Showers concluded that all of the dinosaurs they tested, except nodosaurs, were homeotherms that experienced some "regional heterothermy" (Figure 13.15).

Direct temperature measurements

As we have seen, Barrick and Showers' work depended upon isotope ratios, but there was a catch. The calculation of temperature from isotopic ratios in bone requires knowing the isotopic composition of the environments in which the animal lived; usually, a function of the temperature of the water that it drank. This means that the isotopic

...osaurs grew very fast, an-
...er sign of their warm-
...odedness. Here, a newborn
...iasaura's shinbone (actual
...e) is compared to that of a
...-year-old adult.

Figure 13.13. Femur of the hadrosaur *Maiasaura* hatchling compared to that of an adult. Note the size of the human hand.

Figure 13.14. Core-to-extremity temperatures in a poikilothermic ectotherm. Because this tetrapod's temperature fluctuates with the ambient temperature, when it's cold outside, its extremities are much colder than its core.

Figure 13.15. Estimated maximum temperature variations between bones located in the core of the body and those located at the extremities in living and extinct vertebrates, reconstructed with the use of oxygen isotopes. Living vertebrates are represented by the Komodo dragon (an ectothermic lizard), a bird, and a selection of mammals (endotherms). Note that the greatest variation between core and extremities occurs in the opossum (conventionally considered to be an endotherm), reinforcing the idea that even endotherms can undergo significant fluctuations between core and extremities (in this case the long tail). The researchers concluded that all dinosaurs tested, except the ankylosaur, matched the definition of homeotherms.

composition of the bone is a function of the temperature of the animal, as well as the temperature of the water. But how could one ever know that? Surely that water was as long gone as the dinosaurs being tested! Barrick and Showers, however, had designed a very clever experiment: they didn't need to worry about this because they were comparing *relative* temperatures – variations between cores and extremities.

A number of later studies by a variety of paleontologists and geochemists attempted, however, to obtain the actual temperatures at which various dinosaurs functioned. This was done by calibrating the estimated temperatures of known ectotherms to obtain those of dinosaurs. So, for example, crocodiles were an important feature of many dinosaur ecosystems. Knowing the temperatures at which modern crocodiles function, one could calibrate the estimated isotopic composition of the water, until the calculated temperature of fossil crocodiles was within a couple of degrees of the known temperatures of modern crocodiles. Once the isotopic composition of the water was properly estimated, the real temperatures of dinosaurs could then be determined.

The results of these types of studies have been consistent. For Cretaceous theropods, sauropods, ornithopods, and ceratopsians, the very considerable body of data collected suggests that these animals maintained body temperatures of between 30 °C and 37 °C, well within the range of modern endotherms, at least.

Latitudinal gradients

There is yet another way to skin this dinosaur – isotopically speaking. Geochemist H. Fricke and paleontologist R. Rogers first – and other workers later – looked at the isotopic patterns of teeth across a latitudinal (e.g., north–south) gradient. They used teeth because it turns out that teeth are tough and they preserve original isotopic signatures better than bone. And they looked across latitudinal gradients because temperature differences between the equator and the poles cause the $^{18}O : ^{16}O$ ratio in rain to decrease from the equator to the poles. This occurs through the fractionation of the oxygen isotopes as they go from a gaseous state in the atmosphere, to a liquid state, as rain (via condensation). So, as one goes northward (or southward) from the equator, rain – and thus the water that fills the lakes and streams – becomes "lighter;" that is, relatively more enriched in ^{16}O than water in lakes and streams closer to the equator (which, being more enriched in ^{18}O, is said to be "heavier"). Lake and stream water, of course, gets drunk by terrestrial organisms – including dinosaurs, and contributes its share to the isotopic content of the organisms that drank it. Fricke and Rogers called such water "body water."

As we said, the body water is incorporated into the tissues of the living animal, including its teeth. Now, the step of incorporation into the bones and teeth involves a *second* fractionation step: from the liquid of the body water to the solid of the teeth. In the case of an endothermic homeotherm, the amount of the second fractionation will not change, because the temperature of the animal does not change regardless of the latitude in which it lives. In the case of the ectothermic poikilotherm, however, the temperature of the animal changes, and thus the amount of second fractionation changes: in lower latitudes, with higher temperatures, there is less fractionation; in higher latitudes, with lower temperatures, there is more fractionation.

All of that gives us a very powerful tool with which to recognize endothermic homeotherms and ectothermic poikilotherms (Figure 13.16). And, in fact, a very large study, involving about 100 dinosaurs, including sauropods, theropods, ornithopods, and ceratopsians, used the same technique to arrive at the same conclusion that had first been reached by Fricke and Rogers: dinosaurs gave a very different slope from that of ectothermic poikilotherms (crocodiles) when the isotopic compositions of the teeth are plotted against latitude. These results signal clearly that dinosaurs were fractionating oxygen isotopes in a manner closer to modern endothermic homeotherms, and quite differently from Cretaceous crocodiles, for example, which produced a signal that is comparable to their modern counterparts today.

Clumped isotopes

The Achilles heel, as we have seen, of direct isotope-based temperature measurements is that that isotopic composition of the water that the animals drank has to be estimated. In the past six years, however, a new technique – **clumped isotopes** – has been applied to the problem of the body temperatures at which ancient bone and teeth formed. Clumped isotopes refers to the attraction – the "clumping" – of ^{18}O and ^{13}C at different temperatures. The amount of clumping is temperature-dependent, and

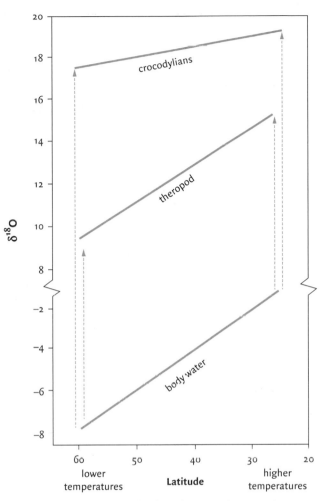

Figure 13.16. Using latitudinal gradients and oxygen isotopes to determine dinosaurian metabolic strategy. Latitude is plotted along the x-axis and oxygen stable isotopic content is plotted along the y-axis. The line labeled "body water" is the oxygen isotopic content of the water that organisms drink. Its oxygen isotopic content varies with latitude, as a result of fractionation due to latitudinal temperature differences. The lines labeled "theropod" and "crocodylians" are the oxygen isotopic content of the teeth for these animals along the same latitudinal gradient. The dotted arrows represent the isotopic fractionation that occurs during the formation of tooth enamel from body water. In the dinosaurs, regardless of latitude (outside temperature), the same amount of fractionation occurs, whereas in the crocodylians, the amount of fractionation that occurs as the oxygen is incorporated into the teeth is directly related to external temperature – and thus latitude. For this reason, the isotopic content relative to latitude of the crocodylians is a distinctly lower slope than that of the theropods – suggesting that the theropods' metabolism was that of an endothermic homeotherm, while that of the crocodylians is that of an ectothermic poikilotherm. Redrawn (and simplified) from Fricke, H. and Rogers, R. 2000. *GEOLOGY*, **28**, 799–802.

so with this technique, by measuring the degree of clumping, there is no need to estimate the isotopic composition of the water. Geochemist R. A. Eagle and colleagues applied this technique to a number of extinct and recent vertebrates, including the teeth of five Jurassic sauropod specimens. Having calibrated this geothermometer with the modern vertebrates, they determined that the dinosaurs operated at temperatures between 31 °C and 38 °C. This temperature is again suggestive of an endothermic metabolism.

Different strokes for different folks?

Where does all that leave us? What is unimpeachable at this point is that dinosaurs were not ectothermic poikilotherms *à la* crocodile *mode*. They were something else; the question is, "what?" Several different, potentially co-existing, general classes of ideas have been proposed, including:

(I) *Different endothermic strategies could have been adopted by different dinosaurs, including, of course, birds.* The disparities in size and in dinosaur behavior make it difficult to support a metabolic "one-size-fits-all" model. Some scientists have argued that large sauropods could have relied upon a strategy called **gigantothermy**: small surface : volume ratios (resulting from large size) retained core heat, allowing sauropods to maintain a homeothermic metabolism without the metabolic cost of being truly endothermic. Yet the results of the clumped isotopic studies suggest that sauropod body temperatures may have been too *cool* for a passive strategy like gigantothermy. To those researchers, sauropods must have actively maintained their temperatures: they must have been true endotherms. By contrast, considerable evidence suggests that small- to medium-sized theropods, and perhaps similarly sized ornithopods, were likely homeothermic endotherms throughout their active lives; not perhaps to the extent that a modern bird is, but perhaps somewhat like the kind of endothermy exhibited by animals like an opossum. The recent discovery of feathers in large, as well as small, theropods, supports this view.

(II) *Dinosaurs maintained an "intermediate" metabolic strategy, that is, a metabolism that was somehow in between modern ectothermic poikilotherms and endothermic homeotherms.* The late dinosaur paleophysiologist R. E. H. Reid argued for some years that dinosaurs had an "intermediate" metabolic strategy. In this view, dinosaurs would have been possessed of a four-chambered heart and a low metabolic rate, used gigantothermy and blood circulation to stabilize their temperature, and been tolerant of a range of temperatures. Reid saw the fast growth rates as *concordant with* endothermy, but not necessarily *requiring* endothermy, and pointed out in support of this idea that, unlike birds and mammals, non-avian dinosaurs never really produced truly tiny forms (which cannot rely at least in part upon their bulk for temperature stability). On the basis of this and other, generally histological, evidence, he argued that endothermy, as seen in modern birds and mammals, likely did not exist in non-avian dinosaurs. In this viewpoint, "intermediate" does not imply that the metabolism is on the way to evolving into modern-style endothermy; it is to be regarded as a fundamentally different and unique metabolic strategy that characterized non-avian dinosaurs. A similar take on this perspective was offered in 2014 by University of New Mexico physiologist J. M. Grady and colleagues, who concluded that reconstructed dinosaur growth rates matched neither modern endothermic homeotherms nor ectothermic poikilotherms, but rather were intermediate, a range of metabolic strategies that they characterized as "mesothermy."

(III) *Different metabolic strategies were in place during different ontogenetic stages of a dinosaur's development.* Some paleontologists suggest that some dinosaurs – particularly large ornithopods and theropods – maintained something close to endothermic homeo-thermy as fast-growing juveniles, but may have became closer to homeothermic ectotherms (via gigantothermy) as adults. While this may be true, the elevated tempera-tures consistently recorded by geochemists suggest that these animals maintained a temperature different from that of the air that surrounded them: the hallmark of an endothermic metabolism.

(IV) All – or some – of the above. This – the "I don't know/default option" – and thus our best call, is that dinosaurian physiology was likely a complex mix of various endothermic strategies, relating to size, behavior, and, potentially, environment. Because dinosaurs are a monophy-letic group, something of a basal dinosaurian metabolic strategy must have existed; this, however, could have evolved in different directions in different groups, according to behavior-based need.

Regardless, after this chapter, it should be clear that although the question of dinosaur metabolism has lost some of its immediacy and, uh, heat, the last word on this still simmering topic is very far from being said.

SUMMARY

Physiologists avoid terms like "warm-blooded" and "cold-blooded" and instead replace them by more meaningful terms like endothermy and ectothermy, which describe the heat source, and homeotherm and poikilotherm, which describe the degree to which temperatures fluctuate in organisms. All of these terms should be regarded as endpoints on spectra of metabolic strategies.

Anatomical indicators in dinosaurs suggestive of an endothermic metabolism include: the erect stance, which among living vertebrates is exclusively posessed by endotherms; the inferred presence of a four-chambered heart, necessary to sustain the higher blood pressures associated with an endothermic metabolism; and the relatively high EQs of some dinosaurs.

Each of these indicators, however, is inconclusive: the erect stance argument has been criticized as being purely coincidental (and not causal) and the high EQs of some dinosaurs are matched by strikingly low EQs in others. Moreover, the absence in non-avian dinosaurs of respiratory turbinates has suggested that they perhaps didn't maintain the high rates of ventilation seen in living endotherms.

The presence in dinosaurs of Haversian canals appears to suggest endothermy. These, timed by LAGs, suggest rapid growth rates that today are known only in endotherms. These suggest the possibility that, as juveniles at least, many dinosaurs may have possessed endothermic metabolisms.

Because endothermy is, in terms of energy, quite costly to maintain, it was suggested that the ratio predators/prey in endothermic eocsystems ought to be significantly smaller than the ratio predators/prey in ectothermic ecosystems. Several attempts were made to calculate such ratios for dinosaurs. None ultimately proved definitive for a variety of reasons, including the unreliability of museum collections as accurate indicators of ancient communities, the fact that endothermic predators sometimes eat ectothermic prey (and vice versa), and the fact that the limiting factor on prey populations is not generally predation.

The existence of polar-dwelling dinosaurs has been interpreted as suggestive of an endothermic metabolism, since today large ectotherms don't get much above 20° N or below 20° S latitude. Yet, a high-latitude temnospondyl amphibian also preserved suggests that, for whatever the reasons, the past distribution(s) of ectotherms was greater than today.

With insulatory coatings potentially going as far back as Ornithodira, there is considerable evidence that endothermy may go back equally far within archosaurs. Regardless, as we saw in Chapter 7, *Archaeopteryx* and its avialan cohort enjoyed slower growth rates that we see in modern birds, suggesting that the insulation in basal ornithodires may be indicative of early flirtations with endothermy.

$^{18}O/^{16}O$ ratios are temperature-sensitive, and have been used in a variety of ways as paleothermometer in well-preserved fossil bone and teeth. The idea was that ectotherms would show a greater range of temperature fluctuations from core to extremities than endotherms. In fact, dinosaurs (and even some mammals) produced a somewhat mixed signal: while some dinosaurs, such as hadrosaurs, showed little temperature variation, ankylosaurs and two of the large theropods tested showed ectotherm-like variations. Reinforcing the point that endothermy and ectothermy are actually endpoints in a spectrum and that metabolisms among vertebrates are not easily predictable, an opossum, a living marsupial (mammal), also showed the kind of variation expected in an ectotherm. Latitudinal gradient studies suggest that dinosaurs maintained body temperatures that are significantly higher than ambient environmental temperatures. Clumped isotopic studies, although in the early phases, suggest similar conclusions in those dinosaurs in which they have been attempted.

While it is clear that the old "cold-blooded" lizard or crocodile model of dinosaur metabolism is defunct, the record suggests that dinosaurs potentially enjoyed a range of metabolic strategies, most or all of them rooted in some kind of at least primitive type of endothermy, perhaps intermediate between, and surely different from, the ectothermic poikilotherms and endothermic homeotherms of today.

SELECTED READINGS

Amiot, R., Lècuyer, C., Buffetaut, E., *et al.* 2006. Oxygen isotopes from biogenic apatites suggest widespread endothermy in Cretaceous dinosaurs. *Earth and Planetary Science Letters*, **246**, 41–54.

Bakker, R. T. 1975. Dinosaur renaissance. *Scientific American*, **232**, 58–78.

Bakker, R. T. 1986. *The Dinosaur Heresies*. William Morrow and Company, New York, 481 pp.

Bakker, R. T. and Galton, P. M. 1974. Dinosaur monophyly and a new class of Vertebrates. *Nature*, **248**, 168–172.

Barrick, R. E., Stoskopf, M. K., and Showers, W. J. 1997. Oxygen isotopes in dinosaur bone. In Farlow, J. O., and Brett-Surman, M. K. (eds.) *The Complete Dinosaur*. Indiana University Press, Bloomington, IN, pp. 474–490.

de Ricqlès, A., Horner, J. R., and Padian, K. 2006. The interpretation of dinosaur growth patterns. *Trends in Ecology and Evolution*, **21**, 596–597.

Desmond, A. 1975. *The Hot-Blooded Dinosaurs*. The Dial Press, New York, 238 pp.

Eagle, R. A., Tütken, T., Martin, T. S., *et al.* 2011. Dinosaur body temperatures determined from isotopic (^{13}C–^{18}O) ordering in fossil biominerals. *Science*, **333**, 443–445.

Erickson, G. M. 2005. Assessing dinosaur growth patterns: a microscopic revolution. *Trends in Ecology and Evolution*, **20**, 677–684.

Farlow, J. O. 1990. Dinosaur energetics and thermal biology. In Weishampel, D. B., Dodson, P., and Osmólska, H. (eds.) *The Dinosauria*. University of California Press, Berkeley, pp. 43–55.

Fricke, H. and Rogers, R. 2000. Multiple taxon–multiple locality approach to providing oxygen isotope evidence for warm-blooded theropod dinosaurs. *Geology*, 28, 799–802.

Grady, J. M., Enquist, B. J., Dettweiler-Robinson, E., Wright, N. A., and Smith, F. A. 2014. Evidence for mesothermy in dinosaurs. *Science*, 344, 1268–1272.

Hopson, J. A. 1980. Relative brain size: implications for dinosaurian endothermy. In Thomas, R. D. K. and Olson, E. D. (eds.) *A Cold Look at the Warm-Blooded Dinosaurs*. AAAS Selected Symposium 28, Westview Press, Boulder, CO, pp. 287–310.

Horner, J. R., de Ricqlès, A., and Padian, K. 2000. Long bone histology of the hadrosaurid dinosaur *Maiasaura peeblesorum*: growth dynamics and physiology based upon an ontogenetic series of skeletal elements. *Journal of Vertebrate Paleontology*, 20, 115–129.

Reid, R. E. H. 2012. "Intermediate" dinosaurs: the case updated. In Brett-Surman, M. K., Holtz, T. R., and Farlow, J. O. (eds.) *The Complete Dinosaur*, 2nd edn. Indiana University Press, Bloomington, IN, pp. 873–921.

Rich, P. V., Rich, T. H., and Wagstaff, B. E. 1988. Evidence for low temperatures and biologic diversity in Cretaceous high latitudes of Australia. *Science*, 242, 1403–1406.

Sellers, W. I., Manning, P. L., Lyson, T., Stevens, K., and Margetts, L., 2009. Virtual palaeontology: gait reconstruction of extinct vertebrates using high performance computing. *Palaeontologia Electronica*, 12, Issue 3; 11A: 26p; palaeoelectronica.org/2009_3/180/index.html

Spotila, J. R., O'Connor, M. P., Dodson, P. and Paladino, F. V. 1991. Hot and cold running dinosaurs: body size, metabolism, and migration. *Modern Geology*, 16, 203–227.

Thomas, R. D. K. and Olson, E. C. (eds.) 1980. *A Cold Look at the Warm-Blooded Dinosaurs*. AAAS Selected Symposium no. 28, 514 pp.

TOPIC QUESTIONS

1. What are meant by the terms endothermy, ectothermy, poikilothermy, and homeothermy?
2. What are remodeled bone, Haversian canals, LAGs, and respiratory turbinates?
3. Was a four-chambered heart evidence for endothermic dinosaurs? What was it used to support?
4. Give some anatomical evidence that dinosaurs were endothermic. Critique it.
5. What is a predator/prey biomass ratio? How were these used in assessing dinosaur metabolism?
6. Critique the predator/prey biomass ratios.
7. How would the distribution of animals on Earth have anything to do with their metabolisms? What kinds of strategies were available to dinosaurs for protection against long, cold winters?
8. What is ^{18}O? What is meant by the statement that "Fossil vertebrates carry around their own paleo-thermometers?"
9. Why would comparison of core temperatures to those of the extremities be useful in determining whether an animal had an endothermic or an ectothermic metabolism?
10. How are rates of growth calibrated in fossil dinosaurs?

The flowering of the Mesozoic

14

Dinosaurs in the Mesozoic Era

Throughout much of this book, we have considered dinosaurs as individuals; who they were, what they did, and how they did it. Now we'll step back and take a look at the large-scale ebb and flow of dinosaur evolution. Before we can do that, though, we need to think about the fossil record itself: what's there and what might be *missing*.

Preservation

Table 14.1 shows the distribution of dinosaurs among the continents through time. The paucity of dinosaur remains from Australia and Antarctica, however, is surely more a question of *local* preservation and inhospitable conditions today for finding and collecting fossils than defining where dinosaurs actually lived. Into this mix must be factored *geological* preservation; some time intervals simply contain more rocks than others. For example, the terrestrial Middle Jurassic is not well represented by rocks, with the result that it artificially appears to have been a time of very low tetrapod diversity (Box 14.1). The Late Cretaceous is rather the opposite, with the happy result that we have a rich record of Late Cretaceous dinosaurs. Several methods of estimating the completeness of fossil preservation have been developed to mitigate these problems (Box 14.2).

Dinosaurs through time

We can think of these geographical and temporal distributions as pages in a notebook, in which each succeeding page represents a new time interval, with new continental arrangements, and new and different assemblages of dinosaurs populating the continents. Considered in this way, the sequence of dinosaurs through time is like a grand pageant through Earth history, in which each interval of time has a characteristic fauna that gives that time a characteristic quality (Figure 14.1).

Now let's look at this information in a more quantitative way: the **diversity**, that is, the number of different *types*, of non-avian dinosaur genera over time (Figure 14.2), through the approximately 164 million years that they were on Earth.

Table 14.1. **Distribution of dinosaurs on continents during the Mesozoic Era. Crosses indicate dinosaurs known**

	Asia	Africa	South America	North America	Europe	Australia	Antarctica
Late Triassic	X	X	X	X	X	X	
Early Jurassic	X	X	X	X	X		X
Middle Jurassic	X	X	X	X	X	X	
Late Jurassic	X	X	X	X	X		
Early Cretaceous	X	X	X	X	X	X	
Late Cretaceous	X	X	X	X	X	X	X

Box 14.1   The shape of tetrapod diversity

For at least 30 years, the University of Bristol's M. J. Benton has been compiling a comprehensive list of the fates of tetrapod families through time. We see several interesing features of the curve that results from this compilation. Note the drop in families during Middle Jurassic time (Figure B14.1.1). This, as we have seen, is an **artifact**; that is, a specious result. This particular one comes from the lack of find localities more than from a true lack of families during the Middle Jurassic. Then, notice the huge rise in families during the Tertiary. Some of this may be real, and perhaps attributable in part to Tertiary birds and mammals (both of whom are very diverse groups), but some of it might be another artifact, due to what is called the "**pull of the Recent.**" The pull of the **Recent** is the inescapable fact that, as we get closer and closer to the Recent, fossil biotas become better and better known. This is because more sediments are preserved as we get closer and closer to the Recent, and a greater amount of sedimentary rocks preserved means more fossils. The big spike at the end of the Jurassic is the Morrison Formation of the US Western Interior, a unit that preserved an extraordinary wealth of fossils.

So a curve like Benton's requires skill to understand and to factor out the artifacts. Nonetheless, we can see that, generally, dinosaurian diversity increased throughout their

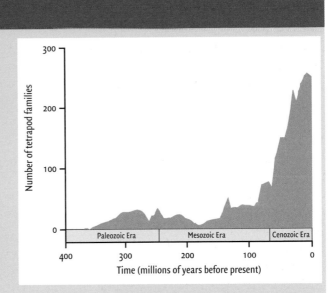

Figure B14.1.1. M. J. Benton's estimate of vertebrate diversity through time. On the x-axis is time; on the y-axis is diversity as measured in numbers of tetrapod families.

stay on Earth and, as they progressed through the Cretaceous, dinosaurs continued to diversify. The increase in diversity shown in Benton's diagram may reflect the increasing global endemism of the terrestrial biota, itself driven by the increasing isolation of the continental plates.

In the beginning... the Late Triassic (237–201.3 Ma)

Figure 14.2 shows that dinosaurs radiated quickly in the Late Triassic. Unambiguous dinosaurs first appear on Earth at ~231 Ma,[1] but exactly how dinosaurs came to be the dominant terrestrial vertebrates in the Late Triassic remains tantalizingly shrouded in misty antiquity. To cut to the chase, however, our best data now suggest that dinosaurs likely moved quickly into a world *abandoned* by other vertebrates; even the most ardent dinosaur advocates have failed to demonstrate that dinosaurs possessed a superior "secret weapon" that somehow allowed them to *out-compete* pre-existing tetrapods (mainly therapsids and primitive archosaurs; see Figures 14.3–14.7) and take charge. In the following section, we expand upon this interpretation.

The rise of dinosaurs – an ecological perspective

In Chapter 5, we discussed some of the basal dinosauromorphs closest to Dinosauria, and showed the transition, from non-dinosaur archosaurs to dinosaurs, from a phylogenetic perspective. Now we take a look at this transition from an ecological perspective.

---

[1] There are some preservational and dating uncertainties.

Figure 14.1. The great historical pageant of dinosaurs through time. The paleoenvironments of each time interval – and the dinosaurs that populated them – all have different qualities that characterized each successive ecosystem.

Apatosaurus

Ornitholestes

Diplodocus

Archaeopteryx

Cetiosaurus

CRETACEOUS

66.1 Ma

L A T E

Maiasaura

Edmontonia

Oviraptor

Parasauralophus

Triceratops

T. rex

Troodon

Euoplocephalus

Gallimimus

Protoceratops

Saltasaurus

Velociraptor

Centrosaurus

Pachycephalosaurus

Edmontosaurus

Therizinosaurus

Alamosaurus

Carnotaurus

100.5 Ma

E A R L Y

Carchardodontosaurus

Psittacosaurus

Suchomimus

Deinonychus

Utahraptor

Ouranosaurus

Protarchaeopteryx

Amargasaurus

Polacanthus

Baryonyx

Sinosauropteryx

Wuerhosaurus

Iguanodon

Hypsilophodon

Caudipteryx

Brachiosaurus

145 Ma

Box 14.2   Counting dinosaurs

How many dinosaurs have ever lived? What percentage of those that have lived do we actually know about? In this box, we introduce several of the ways that diversity can be estimated. We also look briefly at the inverse: how many are missing? We start with the idea of **completeness,** a word one sees often in this context, and an important concept in the geosciences: completeness is the measure of how much of a particular thing is preserved. Sedimentary rocks, as we have seen, record time, and so the completeness (or incompleteness) of the sedimentary rock record is related to the completeness (or incompleteness) of the temporal record. Fossils, as we also saw, are preserved in sedimentary rocks, and so the completeness of the sedimentary rock record is also related to the completeness of the fossil record. How to figure out what's there and what's missing?

In the past 30+ years, veteran sedimentary geologist P. M. Sadler and several colleagues, with the help of computers and piles of data, have taken on this problem. Because it all boils down to the completeness of the sedimentary record, Sadler and sedimentologist D. J. Strauss first constructed estimates of sedimentary completeness based upon a compilation of 25 000 rates of sediment accumulation (that is, rates of sediment deposition) in modern and ancient environments. Completeness, they determined, is ultimately a function of the age of the rocks, the long-term depositional rate, and the steadiness of the sedimentation over time. In a very general way, the larger the time scale under consideration, the more complete the sections. Later work by other authors refined some of these basic ideas. Sadler and paleontologist L. Dingus also attempted to apply the technique to fossils, estimating the completeness of particular intervals of time, such as blocks of time during which key evolutionary events, like mass extinctions, took place. They correctly reasoned that if not enough time was preserved, it would be rather difficult to learn about the events that took place (see Chapter 16).

Other ways

Sophisticated statistical treatments have opened up other ways to count dinosaurs as well. Dodson (1990) used a published compilation, and simply counted. Fastovsky *et al.* (2004) used an updated 2004 version of the compilation used by Dodson (1990), and applied a statistical technique called rarefaction to the data. This technique allowed them to compare different sized samples to determine whether the diversity of dinosaurs actually changed through time, or whether just the samples varied because of preservation.

Another interesting statistical approach was applied by Wang and Dodson (2006), who developed a method for estimating the number of fossils for particular groups *that have yet to be discovered*! They introduced a metric called "coverage," which statistically assesses exactly how closely the *known diversity* from a locality (that is, what has been collected) conforms to its *actual diversity* (that is, a complete inventory of what theoretically ought to be preserved in the locality). Using the coverage metric, Wang and Dodson were able to take the total number of currently known dinosaur genera – 527 genera as of the year 2006 – and estimate the total number of dinosaur genera that ever existed: approximately 1850 genera.

Cladistic estimates

Although we have heretofore emphasized the use of cladograms for reconstructing evolutionary relationships, they can be used to get a sense of the completeness of the fossil record of a particular group. Recall that, along with character distributions, cladograms also portray the relative sequence of the appearance of organisms; who comes before whom. For with a fossil record (like dinosaurs!), this relative sequence from the cladogram can be compared with the real sequence of appearance that comes from the geological record of the same dinosaurs. The cladogram-based sequence ought to compare well with the sequence of stratigraphic occurrence of the fossils themselves. Suppose, however, that the fossil record is not complete; even if an ancestor is not preserved, the cladogram allows us to *infer* when it must have existed. To understand this, we turn to an example.

From its outset 251 million years ago, the Triassic was dominated on land by therapsids. Among these, the sleek, dog-like cynodonts were the chief predators, while the rotund, beaked, and tusked dicynodonts were the most abundant and diverse of herbivores (see Figure 14.3a). From the middle and toward the end of the Triassic, therapsids shared the Earth with squat, armored, plant-eating, archosauromorphs called

Suppose Dinosaur X and Dinosaur Y were each other's closest relative; thus they share a unique common ancestor. If Dinosaur X is known from rocks dated at 100 Ma and Dinosaur Y came from 125 Ma rocks, then this ancestor had to be at least 125 million years old (that is, the age of the older of the two dinosaur species). And if this is true, then there must be some not-yet-sampled history between this ancestor and Dinosaur X – to the tune of 25 million years – all because of phylogenetic continuity calibrated through the use of stratigraphy. Such a 25 million year gap can be referred to as a **minimal divergence time (MDT)**; it can be calculated for any two taxa so long as their phylogenetic relationships and stratigraphic occurrences are known and is an estimate of the completeness of the fossil record. Lineages must have existed that have so far not left us a physical record (through fossils) of their existence; these are called **ghost lineages**. MDTs are measures of their duration (Figure B14.2.1).

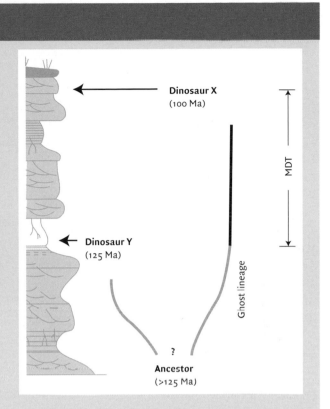

Figure B14.2.1. Ghost lineages and MDTs. Dinosaur X and Dinosaur Y are preserved 25 million years apart. If they are closely related, they both share a common ancestor that is at least as old as Dinosaur Y. Thus an estimate of the minimum divergence time (MDT) of the two lineages is as old as, or older than, Dinosaur Y (125 Ma). The record of that divergence is unpreserved and is therefore called a ghost lineage (curved gray lines on the figure).

Completeness of the ceratopsian fossil record

We exemplify how this method can give us an estimate of the quality of the fossil record of a particular group with Ceratopsia. It has sometimes been claimed that ceratopsians have one of the best fossil records among all dinosaur groups; that is, it is the most complete. How can we test this? There are about 32 ceratopsian species known, less than are found in theropods, sauropodomorphs, and ornithopods, yet more than in ankylosaurs, stegosaurs, and pachycephalosaurs. Averaged over their total time on Earth, ceratopsians apparently produced new species at the rate of one every 1.9 million years, as compared with a high of one new species per 1.4 million years for sauropodomorphs and a low of one per 5.6 million years for stegosaurs. By this reckoning, ceratopsians had relatively high rates of speciation.

To estimate the total diversity of a group, however, we turn to ghost lineages. For ceratopsians, MDT values range from 0 to nearly 30 million years, with an average of just over 5 million years. These are among the smallest MDTs for all Dinosauria, which suggests that the fossil record of this group is comparatively pretty well represented. Furthermore, actual ceratopsian species counts are nearly 70% of the total after ghost lineages have been added to the diversity total. On these measures, ceratopsians do indeed have one of the best records of all of the major dinosaur groups.

aetosaurs and some carnivorous crocodile-like archosaurs (see Figure 14.4). Dinosaurs appear not to have been important constituents of these Late Triassic ecosystems. Other organisms of the day included primitive turtles (Figure 14.5), some crocodile-like amphibians called temnospondyls (Figure 14.6), and pterosaurs (Figure 14.7). Oh yes, and the very earliest mammals, tiny, shrew-sized, insectivorous

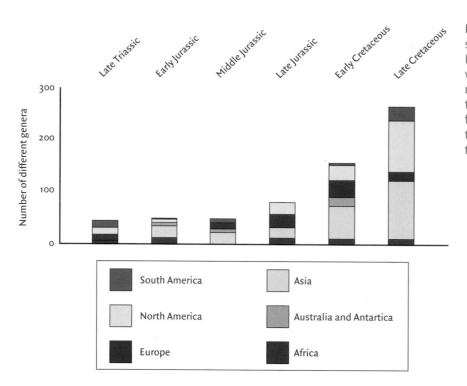

Figure 14.2. Changes in dinosaur diversity by continent measured through the Late Triassic–Late Cretaceous time interval. Each vertical bar shows the total number of different genera known from that particular time interval. Viewed from this perspective, dinosaurs appear to have steadily increased in diversity as the Mesozoic progressed.

creatures, were present (Figure 14.3d). As it turned out, their appearance on Earth was approximately coincident with – or even slightly preceded – that of dinosaurs. In short, Late Triassic terrestrial vertebrate faunas were not dinosaur-dominated; rather they were an eclectic mixture.

At the end of the Triassic, approximately 201 Ma, there was a great change in the fortunes of amniotes.[2] The majority of therapsids went extinct (one highly evolved group of therapsids, the mammals, of course survived). And it was only then that dinosaurs somehow rose to become the **dominant** terrestrial vertebrates, by which it is meant that they became the most abundant, diverse, and probably visible group of tetrapods. The pattern of waxing and waning in dominance, as one group supercedes another in evolutionary time, is called the **wedge** (Figure 14.8).

Puzzling, though, is *why* dinosaurs prevailed. Ideas have boiled down to two contradictory theories:

(1) Dinosaurs out-competed their contemporaries, earning the right, as it were, to be the dominant terrestrial vertebrates.

(2) Dinosaurs somehow survived, because their non-dinosaurian contemporaries went extinct, leaving a vacated planet open to dinosaur colonization.

Out-competition – wedge with edge

In the late 1960s and early 1970s, A. Charig, Curator of Lower Vertebrates at what was then called the British Museum of Natural History, argued that those archosaurs that had the new, "improved"[3]

---

[2] This is the notorious Triassic–Jurassic mass extinction, an event that caused a considerable perturbation in all life forms, not just land-dwelling amniotes. How long this extinction took and precisely what caused it remain subject of considerable contention; however, the data are starting to point to a geologically rapid event (lasting, perhaps, a few tens of thousand years), driven by a mid-Atlantic submarine volcanic outpouring of basaltic rock and the wholesale emission into the atmosphere of a variety of not-so-healthy greenhouse-inducing volcanic gases.

[3] Rather transparently, Charig's own preferred term to describe the parasaggital stance of dinosaurs.

Figure 14.3. Late Triassic therapsids ("mammal-like reptiles") and a very early mammal. (a) A large, 2.5 m herbivore (the dicynodont *Kannemeyeria*); (b, c) two carnivorous cynodontians (*Cynognathus*); and (d) an early mammal, the tiny (approximately 5 cm) *Eozostrodon*.

Figure 14.4. Assorted primitive archosaurs. (a) A phytosaur (*Rutiodon*); (b) an aetosaur (*Stagonolepis*); (c) a rauisuchian (*Postosuchus*); and (d) a primitive crocodile (*Protosuchus*).

Figure 14.5. The primitive Late Triassic turtle *Proganochelys*.

Figure 14.6. A temnospondyl (an archaic amphibian) grabbing a snack.

Figure 14.7. The primitive rhamphorynchid pterosaur *Dimorphodon*.

(A) The Edge (Competitive replacement of Synapsida by Diapsida)

Synapsid carnivores

Diapsid carnivores (including dinosaurs)

Synapsid herbivores

Diapsid herbivores (including dinosaurs)

(B) The Wedge (Diapsida enters ecospace abandoned by Synapsida)

Synapsid carnivores

Diapsid carnivores (including dinosaurs)

Diapsid herbivores (including dinosaurs)

Synapsid herbivores

| Early Triassic | Middle Triassic | Late Triassic |

Figure 14.8. The wedge: two views of the origin of dinosaurs during the Late Triassic. (a) *Wedge with the edge*: gradual competitive replacement of synapsids, primitive archosaurs, and rhynchosaurs (both herbivores and carnivores) by herbivorous and carnivorous dinosaurs. (b) *Wedge without the edge*: rapid opportunistic replacement facilitated by extinction.

parasaggital stance were then able to out-compete contemporary predatory, semi-sprawling therapsids for their food sources (see Figure 5.2). The immediate descendants of these flashy new parasaggital archosaurs were the dinosaurs. The inevitable consequence of such progressive improvements in limb posture, Charig argued, was the pattern of faunal succession at the end of the Triassic. We call this and any other evolutionary advantage a **competitive edge**; dinosaurs prevailed, according to Charig, by virtue of having better-designed limbs and therefore more efficient terrestrial locomotion.

At nearly the same time, R. T. Bakker was making similar arguments about the competitive superiority of endothermy in dinosaurs (see Chapter 13). He believed that, instead of limbs, it was the achievement of internally produced heat that gave dinosaurs (or their immediate ancestors) a competitive edge over contemporary and supposedly cold-blooded therapsids and rhynchosaurs. The same conclusions applied – dinosaurs won, therapsids lost, and the truth of the competitive superiority of endotherms over ectotherms could be read directly from the pattern of faunal succession at the end of the Triassic: the competitive edge of dinosaurs produced the wedge. This of course, presumed that the near-dinosaur antecedents – the non-dinosaurian dinosauromorphs we met in Chapter 5 – were *not* endotherms, something that, as we have seen from Chapter 13, is still something of an open question.

Wedge without edge?

M. J. Benton of the University of Bristol was not convinced that the edge produced the wedge in the Middle to Late Triassic fossil record of the earliest dinosaurs and their predecessors. In order for edges to lead to wedges, all of the players in the game have to be present to interact with each other. And, according to Benton, they were *not* (note Figure 14.8). Instead, he proposed that the fossil record of the last part of the Triassic is marked by not one but two mass extinctions. The earlier one, he argued, was the more extreme and ultimately most relevant to the rise of dinosaurs. As Benton sees it, this earlier Late Triassic extinction completely decimated rhynchosaurs and nearly obliterated dicynodont and cynodont therapsids, as well as several major groups of predatory archosaurs.

Likewise, there was a major extinction in the plant realm. The important seed-fern floras (the "*Dicroidium* flora," which contained not only seed-ferns, but also horsetails, ferns, cycadophytes, ginkgoes, and conifers; see below) all but went extinct as well, to be replaced by other conifers and **bennettitaleans** (see below, Figures 14.9 and 14.10). Dinosaurs appeared as the dominant land vertebrates only after the disappearance of therapsids, archosaurs, and rhynchosaurs. Thus the initial radiation of dinosaurs, according to Benton, occurred in an ecological near-vacuum, with the rapid loss of the dominant land-dwelling vertebrates setting the stage for the *opportunistic* evolution of dinosaurs. No competitive edge, because there was no competition.

The proposal of a double extinction remains controversial. The best biostratigraphic and geochronologic data available seem to be narrowing the window for the end-Triassic extinction, so that a single event with a single cause – submarine basaltic volcanism? – is strongly suggested. Regardless, Benton may still be right: and the dinosaurs and other organisms may have inherited a brave new world, largely depopulated by noxious events associated with the end-Triassic extinction. If so, then perhaps dinosaurs may have just squeaked by, survivors not because they were somehow superior to the presumed competition, but because they happened to inherit a deserted Earth. Instead of survival having been something intrinsic to dinosaur superiority, it may have been that they simply had better luck.

Ironically (as we shall see in Chapter 16), ~164 million years later the tables again turned, and mammals inherited an Earth this time deserted by the very dinosaurs who had taken it from them all those millions of years earlier. As we discuss in Chapter 16, the wedge this time was unequivocally produced without any edge.

Contintental distributions and the Late Triassic fauna

What kinds of evolutionary forces might have been driving the distinctive Late Triassic faunas? The very distributions of the continents likely play a role in the composition of global faunas.

Consider this example from the modern world: the large herbivore fauna of Africa is rather different from that of North America. And both differ from that of India. There are no land connections among these continents that would allow the fauna of one to spread to the other. Each of these faunas – in fact, the ecosystems of which they are a part – developed in relative isolation, and is therefore distinct. This type of distinctness is called **endemism**. A region that is populated by distinct faunas unique to it is said to show *high endemism*. High endemism is caused by evolution on widely separated continents, because there is no opportunity for faunal interchange.

Aternatively, if faunas of two continents appear very similar to each other, then it is likely that some land connection was present to allow the fauna of one continent to disperse to the other continent. Thus we can imagine a region characterized by *low endemism*; because the continents are closely allied with each other, and there are extensive opportunities for faunal interchange.

It is now clear that, during the Triassic and Early Jurassic, global terrestrial vertebrate faunas were characterized by unusually low endemism. The supercontinent of Pangaea still existed during this time, and land connections were more or less continuous among all of today's continents (see Figure 2.5). The interesting mixtures of faunas outlined here look similar on a global scale during these time intervals. While Pangaea remained united, land connections existed and endemism was low. Here, then, is an excellent example of large-scale, geological (abiotic) events driving and modifying large-scale patterns of biological evolution.

For all that, however, we should note that the first dinosaur-bearing faunas are rather patchily distributed, and somewhat enigmatic. The earliest dinosaurs, as we have seen, are known from South America. These include very primitive saurischians, whose affinities are likely primitive Theropoda and/or Sauropodamorpha (see Chapter 5). In those South American faunas, virtually no non-dinosaur dinosauromorphs occur, sending the very misleading signal that once the first dinosaurs appeared, they took charge. Yet, in somewhat younger (~7 myr) North American deposits, relatively advanced theropods are found (indicative of significant dinosaur evolution), happily co-habitating with non-dinosaur dinosauromorphs! Did these groups of animals co-exist as mixed faunas or did they not? In short, these very earliest records of dinosaurs do not give a definitive signal about the nature of early dinosaur ecosystems; here, then, the incompleteness of the early dinosaur record is surely hampering its interpretation.

Jurassic (201.3–145 Ma)

The Early Jurassic (201–174 Ma) was the first time on Earth when dinosaurs truly began to dominate terrestrial vertebrate faunas. Indeed, although dinosaurs appeared in the Late Triassic, UC Berkeley paleontologist K. Padian marks the Early Jurassic as the real beginning of the "Age of Dinosaurs." Most of the players in the terrestrial game were now dinosaurs, although they continued to share the limelight with some relict non-amniotes (see Chapter 4; Figure 14.6), a few of the very highly derived, mammal-like therapsids (including some puny mammals; Figure 14.3), turtles (who, by now, appeared very nearly like their modern descendants), pterosaurs (Figure 14.7), and the newly evolved crocodilians. Interestingly, the Early Jurassic faunas retained much of the low endemism that characterized the Late Triassic world. The unzipping of Pangaea was in its very earliest stages, and it had not gone on so long that the fragmentation of the continents was yet reflected through increased global endemism.

The Middle Jurassic (174–164 Ma) has historically been an enigmatic time in the history of terrestrial vertebrates. As noted earlier, Middle Jurassic terrestrial sediments are quite uncommon. When we look at the total diversity of tetrapods through time (Box 14.1), the curve all but bottoms out during the Middle Jurassic. Did vertebrates undergo massive extinctions at the end of the Early Jurassic? Almost certainly not. More likely the curve is simply reflecting the serendipitous absence of terrestrial Middle Jurassic sediments on Earth. Without a good sedimentary record to preserve them, we can have little knowledge of the faunas that came and went during that time interval.

Regardless, the Middle Jurassic must have been a very important time in the history of dinosaurs. With the dismemberment of Pangaea well underway by this time, dinosaurs had diversified, and endemism was on the rise. Many of the non-dinosaurian tetrapods that characterized earlier faunas – advanced therapsids, for example – were largely out of the picture. The insignificant exceptions to this, of course, were mammals, hanging on by the skin of their multi-cusped, tightly occluding teeth. The Middle Jurassic must have been a kind of pivot point in the history of dinosaurs, because it was then that most of the major dinosaur groups – sauropods, large theropods, thyreophorans, and ornithopods – assumed their familiar forms and consolidated their hold on terrestrial ecosystems. It's a shame that we cannot know more of this crucial time.

By the Late Jurassic (164–145 Ma; see Figure 2.6), global climates had stabilized and were generally warmer and more equable (less seasonal) than they presently are (see Chapter 2). Polar ice, if present, was reduced. Sea levels were higher than today. Dinosaur faunas were more endemic than ever before.

The Late Jurassic has been called the Golden Age of Dinosaurs.[4] With the Middle Jurassic as poorly known as it is, dinosaurs appear to have sprung into high diversity in the Late Jurassic. Many of the dinosaurs that we know and love were Late Jurassic in age. Somehow that special Late Jurassic blend of high sea levels, equable climates, small brains, and massive size epitomizes dinosaur stereotypes and exerts a fascination on humans. Many *were* large – gigantic sauropods (*Brachiosaurus*, *Diplodocus*, *Camarasaurus*, among others) as well as theropods that reached upward of 16 m – but many were not (for example, *Compsognathus*). It was during the Late Jurassic that (comparatively tiny) avialans like *Archaeopteryx* appeared. Moreover, this was the time of stegosaurs, early iguanodontians, avialans, and even a few ankylosaurs. By Late Jurassic time, dinosaurs had consolidated their dominance of terrestrial vertebrate faunas.

Cretaceous (145–66.1 Ma)

The Early Cretaceous (145–100.5 Ma) was a time of enhanced global tectonic activity. With this came increased continental separation, as well as greater amounts of CO_2 in the atmosphere, producing "greenhouse" climates. Climates from the Early through mid-Cretaceous (to about 96 Ma) were therefore warmer and more equable than today (see Chapter 2).

Who enjoyed these balmy conditions? Representatives of groups including all of our old friends from earlier times as well as a number of new groups of dinosaurs made their appearance. The Early Cretaceous marks the rise of the largest representatives of euornithopods. Ankylosaurs also became a significant presence among herbivores of the Early Cretaceous times, as did the earliest ceratopsians.

Moreover, the balance of the faunas seems to have changed. During the Late Jurassic, sauropods and stegosaurs were the major large herbivores, with euornithopods represented primarily by smaller members of the group. Now, in the Early Cretaceous (and, in fact, throughout the Cretaceous), euornithopods begin to make their mark. Sauropods continued their success story, and stegosaurs were still present (barely), but evidently on their way out. Was the spectacular Cretaceous ascendency of Euornithopoda due to the feeding innovations developed by the group? The parallel success of ceratopsians in Late Cretaceous time, and the independent invention by that group of similar feeding innovations, suggest that sophisticated chewing didn't hurt. Then, too, the Early Cretaceous witnessed a revolution in small carnivorous theropods, most notably the deinonychosaurids and troodontids of both North America and Asia.

During the Late Cretaceous (100.5–66.1 Ma; see Figure 2.7), never before in their history had Dinosauria been so diverse, so numerous, and so incredible. Something happened; there was a bump

---

[4] If only because Late Cretaceous dinosaurs weren't fully appreciated in the latter part of the 1800s when people began to imagine a "golden age" for dinosaurs!

in diversity without parallel in dinosaur history.[5] The Late Cretaceous boasted the beefiest terrestrial carnivores in the history of the world (tyrannosaurids), a host of sickle-clawed brainy (*and* brawny) killers worthy of any nightmare, and herds of horned herbivores, honking hadrosaurids, ankylosaurs, dome-heads, therezinosaurs, the truly titanic titanosaurs...

Climate seems not to have affected diversity. In fact, although diversity increased, from the mid-Cretaceous time onward, seasonality gently increased. This occurred at the same time as a marine regression, which undoubtedly played a role in the destabilization of climate. At the very end of the Mesozoic, there is no evidence for a sudden drop in temperatures, or any significant modification of climate; there was, however, a temporary increase in global temperatures just before the Cretaceous–Tertiary boundary. This has been attributed to the effects of Deccan volcanism (see Chapter 16).

Endemism

Southern continents tended to maintain the veteran Old Guard: a large variety of sauropods (some rather new versions of the Old Guard), some ornithopods and ankylosaurs, and, in South America, the abelisaurid theropods, a group representative of a somewhat primitive theropod stock. In northern continents, however, new, very different faunas appeared. Among herbivores, sauropods were still present, although oddly absent in northern North America. Lots of new creatures roamed, including potentially migrating herds of pachycephalosaurs, ceratopsids, and hadrosaurids. Europe, by contrast, consisted of an island archipelago, which dramatically increased endemism in the Old World.

Finally there is the magnificent diversity of Late Cretaceous theropods. Nothing shaped quite like tyrannosaurids had ever been seen, or has existed since. Yet, the Late Cretaceous theropod story might be better told in the diversity of smaller forms: oviraptorosaurs, alvarezsaurids, dromaeosaurids, ornithomimisaurs, troodontids, and therizinosaurs.

Across the Bering Straits?

North America and Asia share a rich Late Cretaceous record, including ceratopsians, tyrannosaurids, and ornithomimosaurs. Because of this, there may have been multiple migrations of herds of dinosaurs across a Bering **land bridge** throughout much of the Late Cretaceous (see Figure 11.36), just as humans and other Ice Age mammals are thought to have migrated to North America from Asia many tens of millions of years later.

And then, in the earliest Cenozoic, it was over. Just like that. While one of the great enigmas of dinosaur paleontology has historically been how the animals went extinct, the problem has begun to yield to concentrated study over the past 35 years. We'll save that story for Chapter 16.

After the ball is over

With the end of the Cretaceous, non-avian dinosaurs disappeared from Earth forever, and it definitely *was* the end of an Era. Mammals, well entrenched as the dominant terrestrial vertebrates in the Cenozoic, would be no more likely to give up their place in Tertiary ecosystems to dinosaurs than dinosaurs had been likely throughout the ~164 million years of their incumbency to give up *their* place to mammals!

---

[5] **Note added in proof.** This conclusion, that dinosaur diversity increased throughout the Late Cretaceous, was recently challenged in a controversial paper by Sakamoto and colleagues. They argued, in effect, that during the last 50 million years of the Cretaceous, the rate of origination of dinosaur groups was exceeded by the rate of their extinction; that, therefore, dinosaurs were a group aiming towards extinction from 50 million years before the end of the Cretaceous, until its conclusion, the Cretaceous–Paleogene mass extinction. One thing is certain: we have not heard the last of this subject! See: Sakamoto, M., Benton, M. J., and Venditti, C. 2016. Dinosaurs in decline tens of millions of years before their final extinction. *Proceedings of the National Academy of Sciences.* www.pnas.org/cgi/doi/10.1073/pnas.1521478113.

Or so the story is told by we mammals. But there is one small point for consideration: in today's world, some 66 million years after the non-avian dinosaurs supposedly became extinct, there are just under 5500 species of mammals. By contrast, there are somewhere around 10 000 species of avian dinosaurs (birds) alive today. Can it really be said that we've left the Age of Dinosaurs and entered the Age of Mammals?

Plants and dinosaurian herbivores

As in most extant terrestrial mammalian communities, the majority of dinosaurs were herbivorous. If dinosaurs were numerous enough, and their impact on terrestrial ecosystems was important enough, so the thinking goes, there ought to be some relationship between herbivorous dinosaur evolution and plants. To address this proposition, we begin by looking at plants.

Plants

Most **paleobotanists** – people who study extinct plants – recognize two major groupings of Mesozoic plants. The first is a non-monophyletic cluster of plants including ferns, lycopods, and sphenopsids (Figure 14.9). All of these plants tend to be low-growing and primitive, but, like most land plants,

Figure 14.9. Representative ferns, lycopods, and sphenopsids from the Mesozoic. Club moss: (1) *Pleuromeia* (Early Triassic). Ferns: (2) *Matonidium* (Jurassic–Cretaceous); (3) *Onychiopsis* (Jurassic–Early Cretaceous); (4) *Anomopteris* (Middle–Late Triassic); (5) Osmundaceae (Late Paleozoic–Recent), (6) Tree fern (Jurassic). Sphenopsids: (7) *Equisetum* (Late Paleozoic–Recent); (8) *Neocalamites* (Triassic–Lower Jurassic); (9) *Schizoneura* (Late Paleozoic–Jurassic).

they are **vascular**; that is, they possess specialized tissues that conduct water and nutrients through-out the plant.

The second major grouping of plants consists of gymnosperms and angiosperms. Together these two groups are united by the diagnostic character of possessing a **seed** (Figure 14.10). Seeds are ultimately nutrient-bearing pods apparently developed for the dissemination of gametes. Gymno-sperms are today best known as pines and cypress, and a lesser-known but Mesozoic-ly important group known as **cycadophytes**: plants with large, pineapple-like trunks and bunches of leaves springing out of their tops. **Angiosperms**, the flowering plants, today consist of magnolias, maples, grasses, roses, and orchids, among many other groups (Figure 14.10).

Several qualities distinguish these plant groups. In general, Mesozoic gymnosperms tended to be of three types: conifers, cycadophytes, and ginkgoes. Conifers – epitomized, for example, by pines – were very tall and woody plants. They had relatively little nutritive value pound for pound, possessing coarse thick bark and cellulose-rich leaves. The modern representatives of these plants tend to secrete a variety of ill-tasting or poisonous compounds as a strategy to discourage their consumption; there is no reason to suppose that their Mesozoic counterparts were any different.

Figure 14.10. Representative cycads, ginkgoes, gymnosperms, and angiosperms from the Mesozoic. (1) Cycadeoids or bennettitaleans, as they are sometimes called (Triassic–Cretaceous); (2 and 5) *Williamsoniella* spp. (Triassic–Jurassic); (3) *Wielandiella* (Jurassic); (4) *Williamsoniella sewardiana* (Jurassic). Ginkgo: (6) *Ginkgoites* (Triassic–Recent). Conifers: (7) *Sequoia* (mid Cretaceous–Recent); (8) *Araucaria* (Late Triassic–Recent); (9) *Pagiophyllum* (Triassic–Cretaceous). Angiosperms: (10) Magnoliaceae (magnolias; still small-flowered in the early days of their appearance on Earth; Cretaceous?–Recent); (11) Nymphaeaceae (water lilies; Late Cretaceous–Recent). *Inset*: Seed (dicot) in cross-section. The cotyledons, shoot apex, root apex, and suspensor are all parts of the embryonic plant. The endosperm is a food source for the embryo as it develops, and the seed coat protects the embryo and its food source.

Cycadophytes, on the other hand, tended to be fleshier and softer, with perhaps more nutritive value. Ginkgoes would also have been plants available for dinosaur consumption, and circumstantial evidence suggests they too were eaten by Mesozoic herbivorous dinosaurs (Figure 14.10).

Flowering plants evolved an entirely different approach to life from gymnosperms. Far from discouraging herbivores from consuming them, they evolved a variety of strategies to actively seduce herbivores into consuming them: bright tasty flowers, fruits with tough seeds that can survive a trip through a digestive tract. Consumption by herbivores in the case of angiosperms appears to be a strategy for seed dispersal, not the destruction of the plant.

Dinosaurs and plants

Figure 14.11 compares the record of Late Triassic through Late Cretaceous plant diversity with that of dinosaurian herbivores. The lower part of the figure gives approximations of the global composition of plants through the time of the dinosaurs. The upper part of the figure is divided into various groups of herbivorous dinosaurs.

Plants

In terms of plants, Figure 14.11 shows some key patterns. Lycopods, seed ferns, sphenopsids, and ferns decrease in global abundance during the Late Triassic interval. From then until the end of the Mesozoic, they constitute a roughly constant proportion of the world's floras. Not so with the gymnosperms, which dramatically increase their proportion of the total global flora during the Late Triassic. And it is clear that much of that increase is taken up by conifers, which consitute around 50 per cent of the world's total floras throughout the rest of the Mesozoic.

Our best guess is that angiosperms first evolved in the very early part of the Cretaceous; however, it was during mid-Cretaceous times that they underwent a tremendous evolutionary burst. The uniquely efficient angiosperm seed dispersal mechanisms afforded by flowers and their various dispersing agents – animal fur, insects, feces, you-name-it! – were (and are) unparalleled in the botanical world, and consequently flowering plants have literally blossomed as no other group of plants has.

Co-evolution

It cannot be purely by chance that the rise of tall coniferous forests is coincident with the appearance on Earth of the world's first tall herbivores: prosauropods (and later sauropods). Here we see the possibility of **co-evolution**, the evolution of one group affecting – and even effecting – the evolution of another. In this case, it's plants and dinosaurs: were those tall prosauropods favored by natural selection that could take advantage of comparatively succulent leaves at the tops of conifers? Alternatively, were conifers that were particularly tall favored by natural selection in response to the increasing height of prosauropods? Which is cause and which is effect is something we'll likely never know.

The figure also reveals another potentially compelling relationship. The rise of the angiosperms occurs at approximately the same time as several major radiations of dinosaur groups. Did these groups – ceratopsians, pachycephalosaurs, hadrosaurids, and late-evolved ankylosaurs – somehow take advantage of angiosperms as a food source and diversify? Is this a clue to what these dinosaurs were eating (Box 14.3)?

It appears that, during the middle of the Mesozoic, few vertebrates fed very *selectively* upon the relatively slow-growing conifers, cycads, and ginkgoes that formed the majority of terrestrial floras. Instead, it has been suggested, dinosaur feeding consisted of low browsing and was rather general-ized (similar to the way a lawn-mower "grazes" over whatever is in its path). Because so many of these Mesozoic herbivores were also very large and may have lived in large herds, they likely cleared

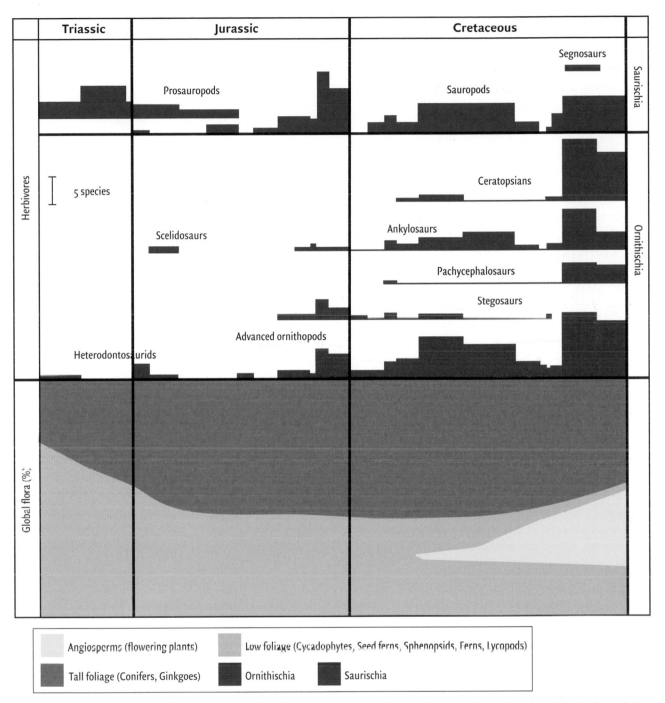

Figure 14.11. Comparison of changes in plant diversity and herbivorous dinosaur diversity during the Late Triassic through Late Cretaceous time interval. The upper part of the diagram shows diversity of major groups of herbivorous dinosaurs through time. Comparison between this diagram and that shown in Figure 14.2 suggests that one of the most important things driving dinosaur evolution and diversity was the development of new (or improved) ways to exploit the environments in which they lived (see the text).

expansive areas, trampling, mangling, uprooting, and otherwise disturbing areas that otherwise might be colonized by plants.

Such low-level, generalized feeding and disturbances of habitats tended to emphasize fast growth in plants, but discouraged the establishment of seed-dispersal relationships between plants and

Box 14.3 Dinosaurs invent flowering plants... or at least fuel their evolution?

In his 1986 popular book *Dinosaur Heresies*, R. T. Bakker proposed that dinosaurs "invented" flowering plants. The germ behind Bakker's hypothesis is that Late Jurassic herbivores, epitomized by sauropods, were essentially high-browsers, while Cretaceous herbivores, epitomized by ornithopods, ankylosaurs, and ceratopsians, were largely low-browsers. Bakker argued that Cretaceous low-browsers put tremendous selective pressures on existing plants, so that survival could occur only in those plants that could disseminate quickly, grow quickly, and reproduce quickly. Angiosperms, he argued, are uniquely equipped with those capabilities. In his scenario, Bakker has Cretaceous low-browsing dinosaurs eating virtually all the low shrubbery, and plants responding by developing a means by which animals simply couldn't keep up with the growth, reproduction, and dissemination of the plants. The idea, here, is that dinosaurs drove plant evolution.

Since Bakker's original proposal, comprehensive, quantitative treatments of this idea have been given by a variety of paleontologists in several publications.[6] These authors noted that although the coincidence of dinosaurs and angiosperms is impressive, actual data supporting the co-evolution of the two are more limited. We have seen that the most obvious direct evidence of dinosaurian diets – "mummies" and coprolites – all point to a gymnosperm-rich diet. In fact, because angiosperms were likely rare in the Early Cretaceous, it is only in the Late Cretaceous that angiosperms were numerous enough to have been an important constituent of dinosaur diets.

The issue is more nuanced. While the dinosaurs and angiosperms appear to have co-evolved in time, if one incorporates *space* as well as time, there appears to be a discordancy; where the angiosperms appear seems not to be where the dinosaurs are recorded as having radiated.

Other herbivores almost certainly had at least as great an impact on angiosperm evolution. Strikingly, insects underwent a major radiation in the mid- Late Cretaceous, and a variety of flower-loving pollinating forms evolved at that time, including bees, wasps, butterflies, and flies of many types. If coincidence of appearance is the evidence for the herbivorous dinosaur–angiosperm relationship, then surely an equal or stronger argument can be made for the appearance of all these flower-loving insects co-evolving with angiosperms. Other potentially important influences were mammals and birds; however, in all of these cases, evidence for direct interactions is sparse.

Another important potential influence on angiosperm evolution has also been proposed: CO_2 in the atmosphere. Recall that atmospheric CO_2 was high during the mid-Cretaceous; this also correlates strongly with the angiosperm radiation. As high levels of atmospheric CO_2 fuel plant growth and the turnover of generations, it is possible that atmospheric CO_2 was involved in the mix as well.

So, the simple relationship between herbivorous dinosaurs and flowering plants is in fact a mighty complex one; and as we have seen, the balance of the evidence appears to suggest that dinosaurs had very little to do with the rise and spread of flowering plants.

[6] Here are two samples:

Barrett, P. M. and Willis, K. J. 2001. Did dinosaurs invent flowers? Dinosaur-angiosperm coevolution revisted. Biological Reviews, 76, 411–447. Butler, R. J., Barrett, P. M., Penn, M. G., and Kenrick, P. 2010. Testing coevolutionary hypotheses over geological timescales: interactions between Cretaceous dinosaurs and plants. Biological Journal of the Linnaean Society, 100, 1–15.

animals. Thus the picture of Mesozoic plant–herbivore interactions appears to be one in which (a) plants produced vast quantities of offspring to ensure the survival of the family line into the next generation and (b) herbivores took advantage of the rapidly and abundantly reproducing resource base to maintain their large populations of large individuals. Plant–herbivore co-evolution during the Mesozoic appears to have been based on habitat disturbance, generalized feeding, and rapid growth and turnover among plants.

This still doesn't really demonstrate whether angiosperms and dinosaurs co-evolved; that is, each affected the trajectory of the other's evolution. Where we can actually track food sources, surprisingly perhaps, the evidence does not argue strongly for dinosaur–plant co-evolution (Box 14.3).

For example, "mummified" remains of hadrosaurids (*Edmontosaurus* and *Corythosaurus*; see Figure 12.9) do not show the remains of angiosperms in the digestive tract, but rather the remains of coniferous plants. Late Cretaceous coprolites, reliably attributed by size to either ceratopsids or hadrosaurids, contained conifer fragments as well. If angiosperms were fueling this dinosaur radiation, where are the angiosperm pieces that we might hope to find?

Dinosaur chewing efficiency increased markedly through the latter part of the Mesozoic. This is not to say that non-chewing dinosaurs were in a state of decline; as Figure 14.11 shows, sauropods – for whom, in most cases, chewing was a minimalist artform – were successful throughout the Cretaceous. Moreover, animals that indulged in rudimentary chewing – such as ankylosaurs and pachycephalosaurs – underwent strong evolutionary bursts during the latter part of the Cretaceous. Far less abundant were the sloth-like therizinosaurs (see Figure 9.31). Yet, was there something about the rise of angiosperms that fueled their unusual radiation too? By the chewing standards of their ornithopod brethren, these animals were mighty primitive.

Still, ceratopsids and hadrosaurids – groups that elevated chewing to new heights – are characteristic of the Late Cretaceous radiation. Did advanced chewing mechanisms allow hadrosaurids and ceratopsids to take advantage of food resources not heretofore available to other dinosaurs? Perhaps, but the evidence suggests that indiscriminate food selection and sophisticated chewing might have been ways to extract every last nutrient out of the nutrient-poor gymnosperms, rather than a means to access low-hanging, tasty, and nutritious angiosperm fruit.

To actively drive plant evolution would take a significant total biomass, perhaps more than the dinosaurs – for all their large size as individuals – could muster. Think *really* large total biomass: like perhaps insects! And even then, more might have been involved (see Box 14.3). So overall, despite the coincidence of a variety of herbivorous dinosaurs and the rise of angiosperms, it is by no means clear that dinosaurs fueled the angiosperm radiation – even if they nibbled a daily angiosperm or two.

The Late Cretaceous can rightly lay claim to being the "Golden Age of Dinosaurs"; as most dinosaurs of the time were herbivorous, it seems that, at a minimum, they themselves flowered during the flowering of the Mesozoic.

SUMMARY

Here we look at the overall sweep of non-avian dinosaur evolution. Factoring in time intervals of a poor geological record, in which preservation is artifically low, dinosaurs as a group increased markedly in number and diversity, particularly during the Late Jurassic-through-latest Cretaceous time interval. This increase is attributable to ceratopsian and ornithopod herbivores, and theropods.

The global pattern of dinosaur evolution from the Late Triassic to the Late Cretaceous is one of generally increasing endemism, likely attributable to the increasing separation of continental masses. Late Triassic and Early Jurassic dinosaur faunas shared their terrestrial world with a variety of other vertebrates; and global vertebrate faunas were relatively homogeneous. Distinct among all the Late Triassic and Early Jurassic vertebrates, however, dinosaurian herbivores were the first to be able to reach, and thus add to their diets, tall foliage.

By Middle Jurassic time, dinosaurs likely consolidated their dominance in the terrestrial realm, even though terrestrial deposits from this time interval are comparatively rare. This fact is especially unfortunate for understanding the details of dinosaur evolution, since many of the groups that became so abundant and diverse in the Cretaceous had their roots in the Middle Jurassic. The Late Jurassic has been called the "Golden Age of Dinosaurs," with the abundance of many familiar forms including very large theropods and sauropods.

Although the dinosaurs likely didn't drive Mesozoic plant evolution, it is probable that, as plants evolved effective methods for dispersal and colonization, dinosaurs hitched a ride, increasing markedly in number and diversity as they took advantage of the radiation of vascular plants.

The Cretaceous was a truly astounding time in dinosaur evolution. Aside from the wholesale dominance of new forms (in particular, ornithopods and ceratopsians, as well as a wide range of theropods), many of the spectacular adaptations that we've seen, such as advanced chewing, evolved in the Cretaceous. A driving force in all this evolutionary ferment may have been the rise of flowering plants; yet what we know of dinosaur diets suggests that the fibrous gymnosperms constituted the bulk of the nutrition.

SELECTED READINGS

Bakker, R. T. 1986. *The Dinosaur Heresies*. William Morrow and Company, New York, 481 pp.

Carrano, M. 2012. Dinosaurian faunas of the later Mesozoic. In Brett-Surman, M. K., Holtz, T. R., and Farlow, J. O. (eds.) *The Compete Dinosaur*, pp. 1003–1026.

Chin, K. 2012. What did dinosaurs eat: coprolites and other direct evidence of dinosaur diets. In Brett-Surman, M. K., Holtz, T. R., and Farlow, J. O. (eds.) *The Compete Dinosaur*, Indiana University Press, Bloomington, IN, pp. 589–601.

Dodson, P. 1990. Counting dinosaurs. *Proceedings of the National Academy of Sciences*, **87**, 7608–7612.

Fastovsky, D. E. 2000. Dinosaur architectural adaptations for a gymnosperm-dominated world. In Gastaldo, R. A. and DiMichele, W. A. (eds.) *Phanerozic Terrestrial Ecosystems*. The Paleontological Society Papers, vol. 6, pp. 183–207.

Fastovsky, D. E., Huang, Y., Hsu, J., *et al.* 2004. The shape of Mesozoic dinosaur richness. *Geology*, **32**, 877–880.

Sues, H.-D. 2012. Early Mesozoic continental tetrapods and faunal changes. In Brett-Surman, M. K., Holtz, T. R., and Farlow, J. O. (eds.) *The Compete Dinosaur*. Indiana University Press, Bloomington, IN, pp. 988–1002.

Tiffney, B. H. 1989. Plant life in the age of dinosaurs. *Short Courses in Paleontology*, **2**, 34–47.

Tiffney, B. 2012. Land plants as a source of food and environment in the Age of Dinosaurs. In Brett-Surman, M. K., Holtz, T. R., and Farlow, J. O. (eds.) *The Compete Dinosaur*. Indiana University Press, Bloomington, IN, pp. 569–587.

Wang, S. C. and Dodson, P. 2006. Estimating the diversity of dinosaurs. *Proceedings of the National Academy of Sciences*, **103**, 13 601–13 605.

Weishampel, D. B. and Norman, D. B. 1989. Vertebrate herbivory in the Mesozoic: jaws, plants, and evolutionary metrics. *Geological Society of America Special Paper* no. 238, pp. 87–100.

Wing, S. and Tiffney, B. H. 1987. The reciprocal interaction of angiosperm evolution and tetrapod herbivory. *Review of Palaeobotany and Palynology*, **50**, 179–210.

TOPIC QUESTIONS

1. What is meant by the words conifer, co-evolution, cycadophyte, angiosperm, diversity, and endemism?

2. What is the general pattern of dinosaur diversity through time? When were dinosaurs at their most diverse? When were they at their least diverse?

3. On what continents are the most dinosaurs found? Why might this be?

4. Describe climatic conditions throughout the Mesozoic.

5. Describe the degree to which the continents were separated throughout the Mesozoic. When were the continents most like they are now? When were they least like they are now?

6. What is the relationship between endemism and the distribution of continents? Why is this so?

7. Describe the general outlines of plant evolution through the Mesozoic.

8. What is the general relationship between dinosaur diversity and plant evolution through the Mesozoic?

9. What is the relationship between herbivore chewing specializations and plant diversity in the Mesozoic?

10. Drawing on material from other chapters, can you think of highly evolved behaviors that appear to be related to plant diversity?

11. What kinds of evolutionary changes characterize Theropoda in the context of plant diversity and increased ornithischian diversity in the Mesozoic?

A history of paleontology through ideas

15

CHAPTER OBJECTIVES

- Outline the history of paleontological thought
- Understand relationships to larger intellectual movements
- Introduce the stories of some famous paleontologists
- Provide a historical context for the subjects discussed in this book

Figure 15.1. Gideon Mantell (1790–1852), the man who first recognized non-avian dinosaurs for what they were.

The idea of ideas

Ernest Rutherford[1] once infamously remarked, "In science there is only physics; all the rest is stamp collecting." And what could be more like stamp collecting than paleontology, that ever-growing litany of obscure names, dates, and locations?

Paleontology *would* be stamp collecting, if it weren't for the ideas – the creativity – that grew with the field. The history of paleontology, therefore, is really the history of the ideas that forged the discipline.

In the beginning

Tradition usually identifies the beginning of dinosaur paleontology as 1822, when Mary Ann Mantell, wife of English physician Gideon Mantell, found large teeth along a Sussex country lane while her husband was busily tending patients (Figure 15.1). Like many physicians of the time, Gideon was something of a naturalist, and the discovery baffled him, because the teeth looked very much like those of the living herbivorous lizard *Iguana*, but were ominously much, much bigger (Figure 15.2).

---

[1] Nobel Prize-winning New Zealand physicist, 1871–1937; pioneer in radiation and radioactivity.

Figure 15.2. Mantell's *Iguanodon* teeth.

But of course the Mantells weren't the first humans to see dinosaur fossils; however, they may have been the first to interpret them meaningfully in a Western scientific context. Fossils of all types must have been remarked upon for as long as there have been humans.

For example, Adrienne Mayor, classical folklorist and historian of science, has reconstructed the origin of the legend of griffins, sharp-beaked, winged, four-legged creatures whose mythology was known across all of Europe and Asia (Figure 15.3). Her idea is that traders along ancient gold-trading caravan routes stretching from Europe through central Asia encountered abundant, beautifully preserved fossils of *Protoceratops* (see Chapter 11), whose strange (to them) combination of beak, frill, and limbs were explained as the mythical griffin's beak, wings, and legs. The richness of the Asian deposits was revealed more than a thousand years later in the American Museum of Natural History's Central Asiatic fossil Expeditions of the 1920s (Box 15.1) and ancient traders, Mayor

Figure 15.3. A griffin.

Box 15.1   Indiana Jones and the Central Asiatic Expeditions of the American Museum of Natural History

He stands in the middle of the remote, rugged, Mongolian desert: high leather riding boots, riding pants, broad-brimmed felt hat, leather-holstered sidearm hanging from a glittering ammunition belt. He carries a rifle and knows how to use it. Nobody else dresses like him, but then nobody else is the leader of the American Museum's Central Asiatic Expeditions to Mongolia (a place which, at the time of the expeditions, the 1920s, could have been the Moon). He is Roy Chapman Andrews, who 50 years later will be the inspiration, it is most plausibly rumored, for Indiana Jones (Figure B15.1.1).

Andrews always knew that he was a man with a destiny. Although he began his career at the American Museum of Natural History (AMNH) modestly (he scrubbed floors), training in mammalogy (an MA), sheer will, charisma, and a *very* good idea carried him the distance. He had traveled extensively, spoke several Asian languages more or less fluently (at a time when very few Westerners did), and constructed fabulous contacts in Beijing (then called Peking).

His idea was simple: to run an expedition to what was then known as Outer Mongolia and to see what he could find. Andrews' timing was superb: the Director of the AMNH, the powerful Henry Fairfield Osborn, had concluded that the cradle of humanity was located in Outer Mongolia, and so Andrews was effectively offering Osborn the opportunity to prove his thesis right (the possibility that Osborn could be *wrong* did not seem to be of concern). The logistics of the expedition were extravagant: Dodge cars, resupplied by a caravan of camels, would bear the brunt of the expedition (Figure 1.8). The expedition itself would consist of a range of Earth scientists – paleontologists, geologists, and geographers – to explore the Gobi Desert, the huge desert that forms the vast southern section of Mongolia (then called "Outer" Mongolia, as if to emphasize its remoteness) and northern China.

As it turned out, the Central Asiatic Expeditions were an unqualified success. Although Osborn's theory was not supported, Andrews brought back a wealth of fossils, including abundant dinosaur material, which made Osborn's error easy to forget. Among the most famous dinosaur finds of his expedition, for example, were *Protoceratops* (the species name of this famous dinosaur is *andrewsi*) and *Oviraptor* eggs – the first time that dinosaur

Figure B15.1.1. Roy Chapman Andrews (1884–1960), explorer, adventurer, and leader of what he revealingly called "The New Conquest of Central Asia."

eggs were ever found. Other incredible finds included *Velociraptor* and a group of tiny Mesozoic mammals (still the rarest of the rare). Andrews and his field parties also found the largest land mammal and the largest carnivorous land mammal of all time (both Cenozoic in age). Other fossils were obtained whose significance was not completely understood. For example, it was only in 1992 that a specimen of *Mononykus*, collected by Andrews' scientists in the 1920s, was finally correctly identified. All in all, it was quite a haul.

Andrews and his parties survived the Mongolian revolution of 1922, but eventually the expeditions came to an end when the political situation in China became too unstable and travel too dangerous. Andrews himself wrote popular accounts of his adventures and, having demonstrated exceptional fund-raising skills, eventually went on to get the job held much earlier by Osborn: Director of the AMNH. He assured his place in history, however, by leading the Central Asiatic Expeditions.

suggests, could hardly have failed to notice the bones of strange, articulated, bird-like creatures emerging from weathering desert sands. Mayor hypothesizes that the griffin legend spread from central Asia along trade routes to Europe.

Seventeenth and eighteenth centuries

The griffin legend, then, is one explanation, outside of a scientific context, for the observation of dinosaur material. But, with a few exceptions, the birth of the Western scientific tradition is generally reckoned to have occurred in association with the **Enlightenment**, the seventeenth-and eighteenth-century intellectual revolution devoted to the ability of reason and observations to reveal truth. The Enlightenment brought with it a number of scientific conclusions important to our story, including:

- The Earth is not static, that is, it has changed through time.
- The Earth of is of great antiquity (its age was not well understood until late in the twentieth century).
- The sequence of the rock record reveals the history of the Earth.
- Fossils are the remains of once-living organisms.
- Organisms on Earth were not static, they too had clearly changed through time, some in startling ways.

Figure 15.4. Robert Plot's drawing of the lower end of a *Megalosaurus* (?) thigh bone. Dr. Plot was the first Professor of Chemistry at the University of Oxford. For explanation of the Latin inscription, see footnote 2.

It was during this time that we have the earliest description of a dinosaur fossil, in this case the lower end of a theropod thigh (likely *Megalosaurus*) from Oxfordshire, England. As the bone was large, it was interpreted by the Reverend Dr. Robert Plot in 1677 to have been the end of a thigh bone of an **antediluvian** (pre-Biblical Flood) giant – man or beast (Figure 15.4).[2]

The nineteenth century through the mid-twentieth century

Mantell's discovery turned him into the world's first true dinosaur junkie. While there is no space to recount the details, his is the remarkable story of a man consumed by a passion for paleontology so great that he vanquished the skepticism of the greatest anatomist of his time (see Box 15.2), built a museum, and wrote the first description (and guided the first reconstruction) of a dinosaur – *Iguanodon*. It was a passion so all-consuming that it ultimately cost Mantell his livelihood and his marriage. Mantell's unexpected and dramatic life's journey is beautifully told in Deborah Cadbury's 2001 history, *Terrible Lizard*.

---

[2] Stranger still, in 1763, Richard Brooke re-drew the specimen in a publication on the uses of various natural objects (including fossils) in medicine. The specimen appeared to Brooke to preserve a giant's testicles; hence, his Latin description identified the fossil as "*scrotum humanum.*" It has been suggested that, tongue firmly in cheek, the rule of priority in the Linnaean classification (see Chapter 3) dictates that this first dinosaur bone should be referred to a genus *Scrotum*, species *humanum*.

Box 15.2 Sir Richard Owen: brilliance and darkness

Richard Owen (Figure B15.2.1) was the dean of natural historians in Victorian times, that iconic age of natural history. In his day, he was among the most powerful and influential scientists in England. His personality was at once brilliant, irascible, politically astute, ruthless, and condescending, and it would not be going too far to call him a liar. He was, to say the least, a man of contradictions.

Owen looked the part. He was tall and gaunt with high cheekbones and, as he grew older, strangely bulging eyes. Cloaked, hands resting gently upon a skull, he looked like he came directly from Central Casting for the part of a Victorian serial killer and his image was used, presumably to good effect, on posters for a 1980s punk rock band!

Owen was born in 1804, trained to be a physician, and early on demonstrated a penchant for anatomy. Bill Bryson inimitably describes a memorable event from the life of the young Owen as he copped corpses for dissection:

> Once while carrying the head of a black African sailor [that Owen had severed from the corpse for study], . . . Owen slipped on a wet cobble and watched in horror as the head bounced away from him down the lane and through the open doorway of a cottage, where it came to rest in the front parlor. What the occupants had to say upon finding an unattached head rolling to a halt at their feet can only be imagined.
>
> (Bryson, 2003, p. 87)

Figure B15.2.1. Sir Richard Owen (1804–1892), eventually of the British Museum of Natural History, the brilliant nineteenth-century English anatomist and the man who bequeathed to us the term "Dinosauria."

Dinosaurs in the Victorian Age

Perhaps it had to do with the Victorian penchant for collections and museums, perhaps it was just the novelty of the beasts being uncovered, but Victorian England was dino crazy. In 1824, the natural historian William Buckland (1784–1856) described a jaw fragment with a single recurved, serrated tooth as *Megalosaurus*. This was the first named dinosaur, now known to be a theropod, but at the time Buckland thought it was just a rather large lizard.

By 1842, enough of dinosaurs was known for the rising young English anatomist Richard Owen (Box 15.2) to invent a new term: Dinosauria (*deino* – terrible; *sauros* – lizard). The charter members of the group were *Iguanodon* (an ornithopod), *Megalosaurus* (a theropod), and *Hylaeosaurus* (an ankylosaur). Presciently, Owen's initial idea of Dinosauria was that its members were endotherms like mammals and birds, a conclusion based upon the now-discredited idea that Mesozoic air was somehow thinner than modern air: right idea, wrong reasons! It seems that Owen balked at the idea,

By the age of 21, Owen was hired by the Royal College of Surgeons in London to assist in the curation of the Hunterian Collection, a collection of biological oddities and medical curiosities amassed by John Hunter, a famous London surgeon. Hunter's notes had been destroyed in a fire, and so the daunting job was to organize, identify, and catalog disorganized drawers of biological detritus. Owen proved to be particularly adept, using clever inferences and his growing knowledge of comparative anatomy to identify and catalog specimens for which there was no recorded information. As his reputation grew, the medical career receded quietly into the distance.

Instead, he became a lecturer in comparative anatomy, and began to publish scholarly tomes on organisms ranging from the living chambered cephalopod *Nautilus* to the first description of the newly discovered *Archaeopteryx*. It was Owen who first described the exotic South American fossils that Charles Darwin brought back with him from his voyage on the *Beagle*, and, naturally enough, it was Owen who made the connection between the still-fragmental and isolated bits of fossil material that at the time constituted all of Dinosauria.

But along with the rise of Owen's powers and reputation came a rise in some unfortunate personality quirks. He was, not to put too fine a point upon it, unpleasant. He was arrogant and condescending to presumed inferiors; a very inclusive category. Charles Darwin, famously tolerant, disliked him enough to remark upon the fact in his autobiography. Owen dissembled, claiming for himself honors and positions that he didn't actually hold. He barred talented contemporaries from access to specimens that would have allowed them to carry out their science. Ever jealous of his place in history, he reserved some of his most finely honed vituperation for Gideon Mantell, who had the presumption to discover and describe the first dinosaurs, and then actually claim credit for his own accomplishments. Owen even went so far as to write an anonymous, scathing obituary of Mantell when the man finally had the good grace to die. The occasion of Owen's receiving the Royal Medal – the Royal Society's highest honor – was marred by the discovery that Owen claimed credit for someone else's discovery. Owen's very real, legitimate accomplishments juxtaposed against his personal nastiness appear almost pathological.

Eventually, people began to catch on to Owen's meaner side, and his star fell into eclipse. Owen was discredited at both the Zoological Society of London and the Royal Society, and eventually took a job as superintendent of the natural history collections in the British Museum. He had a stunning vision for the collections – that they would be available to any and all who cared to look, and advocated their separation from the rest of the British Museum. He died in 1892. The Natural History Museum, today a public museum much as Owen had envisioned it, finally separated from the British Museum in 1963.

which had currency in certain circles at that time, that organic evolution (as it was understood before Darwin) was a kind of linear process that ran from quite simple to more complex. Owen thought that by demonstrating that an ancient group of organisms had modern levels of complexity, he would successfully undermine the notion of evolution.

Victorians immortalized their conception of dinosaurs with a variety of images and sculptures. The dinosaurs were reconstructed as large, heavy-set quadrupeds, the most famous of which were created life-sized in plaster and tile by an English sculptor, Benjamin Waterhouse Hawkins (1807–1889), on the occasion of the reopening of the Crystal Palace in 1854.

The Crystal Palace, a very large glass and iron building, had been the brainchild of Prince Albert, Queen Victoria's Royal Consort, and was built to house the Great Exhibition, a monument to England's scientific and technological prowess. After the Great Exhibition closed the Crystal Palace was moved to Sydenham Hill (London) as part of a newly established public park with ornamental gardens. Waterhouse Hawkins posed his bestiary frolicking on the Crystal Palace

Figure 15.5. Sculptor Waterhouse Hawkins' life-sized dinosaurs on the Crystal Palace grounds, Sydenham, south London, England, still enchanting visitors after 160 years.

grounds (Figure 15.5), where, remarkably, they can still be seen.[3] It rarely gets more surreal than it did on New Year's Eve, 1853, when Owen and Waterhouse Hawkins hosted a dinner for England's best and brightest *inside* the unfinished sculpture of *Iguanodon* to celebrate the reopening (Figure 15.6).

From 1842 onward, membership of Owen's Dinosauria grew by leaps and bounds. Much of the attention was devoted to basic collecting and description, asking questions like, "What is this creature? A new genus? A species of an existing genus? Maybe even a new family?" This otherwise-healthy penchant for discovery, description, and naming reached absurd levels during the latter half of the century, fueled by the extraordinarily rich fossil beds of the North American West and the not-so-healthy competition between Yale's O. C. Marsh and the Philadelphia Academy's E. D. Cope (Box 15.3).

Within 30 years of the establishment of Dinosauria, there was a revolution in scientists' conceptions of how dinosaurs looked, driven by remarkable finds such as a complete hadrosaurid from New Jersey (1858) and 33 complete *Iguanodon* skeletons, recovered from coal seams outside of the town

---

[3] Not so Waterhouse Hawkins' efforts in Central Park, New York. There, the sculptor – or rather, the cost of his project – ran afoul of Tammany Hall, the "Boss" Tweed-led political machine that dominated New York City politics for much of the latter half of the nineteenth century. One spring day in 1871, Waterhouse Hawkins showed up at his studio to find his near-finished sculptures smashed beyond repair. The bits were buried in Central Park, where they are thought to remain to this day.

Figure 15.6. Contemporary lithograph of the New Year's Eve, 1853, dinner inside of Waterhouse Hawkins' model of *Iguanodon*.

of Bernissart, Belgium (1877–1878; Figure 12.4; Box 15.4). In the hands of imaginative, skilled paleontologists such as J. Leidy (the hadrosaurid) and L. Dollo (the *Iguanodon* specimens), dinosaurs were transformed from overfed, bear-like lumbering lizards to something more terrifying, unimaginable, and wonderful than anybody could have ever invented.

Dinosaurs divided

And they were *different*. Not just from living animals, but also from each other. This was duly noted by Harry Govier Seeley, vertebrate paleontologist at Cambridge University, and Friedrich von Huene, dean of German dinosaur paleontology at the University of Tübingen, both of whom recognized the fundamental division in Dinosauria between Ornithischia and Saurischia (Figure 15.7).

That dinosaurs had two different types of pelvis implied to Seeley that the ancestry of Ornithischia and Saurischia was to be found separately and more deeply among primitive archosaurs, within a now-abandoned group called "Thecodontia" (see below). Therefore, Seeley's dinosaurs were not, had he known the term, monophyletic (Figure 15.8).

The perception that dinosaurs were at least **diphyletic** (that is, having two separate origins) continued well past the middle of the twentieth century. Most paleontologists, until even the early 1980s, thought that dinosaurs had at least two and likely three or four, separate origins within "thecodonts." Certainly, saurischians and ornithischians must have had separate origins; after all, their hip structure was different. And among saurischians, surely sauropods and theropods had separate origins; after all, they *look* so different. And finally, among ornithischians, ankylosaur ancestry was also often sought separately within some thecodontian group.

And what about Owen's bold suggestion that these dinosaurs were endothermic? Within ten or so years – and despite some early advocacy of it by other natural scientists – it was largely forgotten, the victim of the "fact" that dinosaurs were reptiles, and reptiles are cold-blooded.[4] In 1953, Roy Chapman Andrews evocatively described *T. rex*'s meal in cold-blooded – literally and figuratively – terms, reflecting the prevailing ideas of the time:

[4] A few paleontologists, notably G. R. Wieland of Yale University, shared a vision of some kind of dinosaur homeothermy.

Box 15.3   Dinosaur wars in the nineteenth century: boxer versus puncher

One of the strangest episodes in the history of paleontology was the extraordinarily nasty and personal rivalry between late nineteenth-century paleontologists Edward Drinker Cope and Othniel Charles Marsh (Figure B15.3.1). In many respects, it was a boxer versus puncher confrontation: the mercurial, brilliant, highly strung Cope versus the steady, plodding, bureaucratic Marsh. Their rivalry resulted in what has been called the "Golden Age of Paleontology," a time when the richness of the dinosaur faunas from

western North America first became apparent – when the likes of *Allosaurus, Apatosaurus*, and *Stegosaurus* were first uncovered and brought to the world's attention. But the controversy had its down side too. Who were these men, and why were they at each other's throats?

Cope was a prodigy; one of the very few in the history of paleontology. By the age of 18, he had published a paper on salamander classification. By 24, he became a Professor of Zoology at Haverford College, Philadelphia. Blessed with

Figure B15.3.1. The two paleontologists responsible for the Great North American Dinosaur Rush of the late nineteenth century. (a) Edward Drinker Cope (1840–1897) of the Philadelphia Academy of Sciences; and (b) Othniel Charles Marsh (1831–1899) of the Yale Peabody Museum of Natural History.

Then it [*Tyrannosaurus*] settles to the feast. Huge chunks of warm flesh, torn from the Duckbill's body, slide down the cave-like throat . . . The King's stomach is full to bursting. Walking slowly to the jungle, he stretches out beneath a palm tree . . . For days, or perhaps a week, he lies motionless in a death-like sleep. When his stomach is empty, he gets to his feet and goes to kill again. That is his life – killing, eating, and sleeping.[5]

[5] Andrews, R.C. 1953. *All About Dinosaurs.* Random House, New York, pp. 64–67.

independent means, within four years he had moved into "retirement" (at the grand old age of 28) to be near Cretaceous fossil quarries in New Jersey. He quickly became closely associated with the Philadelphia Academy of Sciences, where he amassed a tremendous collection of fossil bones which he named and rushed into print at a phenomenal rate (during his life he published over 1400 works). He was capable of tremendous insight, made his share of mistakes, and was girded with the kind of pride that did not admit to errors.

Marsh, nine years Cope's senior, was rather the opposite, with the exception that he, too, eventually rushed his discoveries into print almost as fast as he made them (some thought faster) and that he, too, did not dwell upon his mistakes. Marsh's own career started off inauspiciously; with no particular direction, he reasoned that if he performed well at school he could obtain financial support from a rich uncle, George Peabody. This turned out to be perhaps the most significant insight in Marsh's life: Marsh persuaded Peabody to underwrite a natural history museum at Yale (which to this day exists as the Yale Peabody Museum of Natural History), and, while he (Peabody) was at it, an endowed chair for Marsh at the Museum.

The careers of the two paleontologists moved in parallel; Marsh slowly publishing but acquiring prestige and rank, while Cope frenetically published paper after paper. At first, there was no obvious acrimony, but this changed when Marsh apparently hijacked one of Cope's New Jersey collectors right out from under him. Suddenly, the fossils started going to Marsh instead of Cope. Then, in 1870, Cope showed Marsh a reconstruction of a plesiosaur, a long-necked, flippered, marine reptile. The fossil was unusual to say the least, and Cope proclaimed his findings in the *Transactions of the American Philosophical Society*. Marsh detected at least part of the reason why the fossil was so unusual: the head was on the wrong end (the vertebrae were reversed). Moreover, he had the bad manners to point this out. Cope, while admitting no error, attempted to buy up all the copies of the journal. Marsh kept his.

Cope sought revenge in the form of correcting something that Marsh had done. The rivalry ignited, and the battle between the two spilled out into the great western fossil deposits of the Morrison Formation. Both hired collectors to obtain fossils, the collectors ran armed camps (for protection against each other's poaching), and, between about 1870 and 1890, east-bound trains continually ran plaster jackets back to New Haven (Connecticut) and Philadelphia. There Marsh and Cope rushed their discoveries into print, usually with new names. The competition between the two was fierce, as each sought to out-science the other. Discoveries (and replies) were published in newspapers as well as scholarly journals, lending a carnival atmosphere to the debate. Because Philadelphia and New Haven were not that far apart by rail, it was possible for one of the men to hear the other lecture on a new discovery, and then rush home that night and describe it and claim it for himself. Because many of the fossils in their collections were similar, it was easy to do and each accused the other of it.

Both Cope and Marsh eventually aged and, in Cope's case, his private finances dwindled. Moreover, a new generation of paleontologists arose that rejected the Cope–Marsh approach, believing, not unreasonably, that it had caused more harm than good. Both men ended their lives with somewhat tarnished reputations. History has viewed the thing a bit more dispassionately, and it is fair to state that the result ultimately was an extraordinary number of spectacular finds and a nomenclature nightmare that has taken much of the past 100 years to disentangle (see Desmond, 1975). An interesting and unusual account of the Cope–Marsh feud was published as a graphic history (Ottaviani et al., 2005).

Ironically – for how our views have changed! – this description would have appeared stranger to a nineteeth-century paleontologist like Cope or Marsh than to most paleontologists 70 years later. In retrospect, it seems puzzling that thoughtful scientists could have so meticulously described the bones and studied the relationships, yet with hardly any thought assumed "reptilian" ectothermy for dinosaurs for so long. Yet a look at publications through this period suggests that dinosaur metabolism rarely crossed their minds. Such is the strength – or the inertia – of ideas.

Box 15.4   Louis Dollo and the beasts of Bernissart

Louis Antoine Marie Joseph Dollo, a Belgian paleontologist with a name almost as luxuriant as his moustache, gave us our first true picture of dinosaurs, through an incredible preservation of articulated *Iguanodon* skeletons in Belgium (Figure B15.4.1). Born in Lille, France, in 1857, Dollo first pursued a career in civil engineering, but soon was hired by the *Museé Royal d'Histoire Naturelle* in Brussels, Belgium. Here he was in charge of the study and museum exhibition of these specimens.

In 1878, commercial coal miners identified fossil bone some 322 m (1056 feet) below ground, which was immediately brought to the attention of the Brussels museum and to Dollo in particular. This occurrence of bone turned into a treasure trove of more than 30 articulated skeletons of the Early Cretaceous ornithopod called *Iguanodon* (Figure B15.4.2; see also Figure 12.4).

Dollo's research on *Iguanodon* was unlike contemporary approaches, which tended to ask questions about to which taxon the material belonged and how it would be classified. Instead, thanks in part to the exquisite preservation of the Bernissart material, he devoted himself to understanding the anatomy and function of these extinct forms in ways that had not been possible before. He sorted the *Iguanodon* material into two species by successively eliminating different sources of skeletal variation. He used the disparity between forelimb and hindlimb length, the development of ossified tendons across the back, and footprints to establish bipedality in *Iguanodon*. And he outlined new approaches to reconstructing the jaw systems of numerous dinosaurs including *Iguanodon*, putting them into their comparative context with living vertebrates. In doing so, Dollo turned paleontological attention to what he called "ethological paleontology" – the study of behavior and environment of extinct organisms – which Othenio Abel, a German paleontologist, termed **paleobiology** in 1912.

Dollo also worked on the other fossil forms from Bernissart, including its turtles, crocodilians, and amphibians.

Figure B15.4.1. Louis Dollo (1857–1931), the Belgian paleontologist of the Musée Royal de Sciences Naturelles Belgiques, who, along with Joseph Leidy, Professor of Anatomy at the University of Pennsylvania, first understood the shapes of dinosaurs.

When not busy with research on the riches of Bernissart, he conducted research on a number of new Late Cretaceous dinosaurs and mosasaurs from Belgium and elsewhere in the European lowlands, and on Antarctic fishes, among modern organisms.

Other than his work on *Iguanodon*, best known is Dollo's Law of Irreversible Evolution. This biological principle, which Dollo formulated in 1893, argued that evolution is not a reversible process. That is, structures eliminated during the course of evolution cannot themselves reappear in the same form within a given lineage of organisms.

Dollo died in his adopted home of Brussels in 1931.

Figure B15.4.2. Several death-posed *Iguanodon*, the great beast of Bernissart, Belgium as currently displayed at the Musée Royal d'Histoire Naturelle de Belgique. In Figure 12.4, we show one of the Bernissart *Iguanodons*, as originally mounted by Dollo.

(a)   (b)

Figure 15.7. (a) Cambridge University's Harry G. Seeley (1839–1909) and (b) Friedrich von Huene (1875–1969), University of Tübingen.

Dinosaurs in the first half of the twentieth century

The first 60 years of the twentieth century brought about an expansion and consolidation of our basic understanding of dinosaurs and their diversity. Collecting, describing, and naming were the game, and our understanding of fundamental dinosaur morphology and diversity was dragged into a modern framework. In North America in the early years of the twentieth century, spectacular collections were made by Charles H. Sternberg along the Red Deer River in Alberta, Canada (Box 15.5). There he and his crews floated along the river in a mobile field camp, swatting mosquitoes and harvesting Upper Cretaceous dinosaurs from the sandstones and mudstones exposed in its banks. No less impressive were the efforts of the American Museum of Natural History's redoubtable Barnum Brown (Box 15.6), who, one summer in 1902, unearthed a large theropod that his sponsor, H. F. Osborne, dubbed *Tyrannosaurus rex*. For exotic and ill-fated, however, German scientists seem to have had particularly poignant tales; in spite of this, their work led to the discovery of legendary beasts like *Spinosaurus* and *Brachiosaurus* (Box 15.7).

Dinosaur discoveries continued at a rapid rate, new names proliferated, skilled descriptions of the new material were written, but, from the standpoint of *ideas*, the field had largely stagnated. These discoveries were carefully collected and described by a host of extraordinarily fine, dedicated

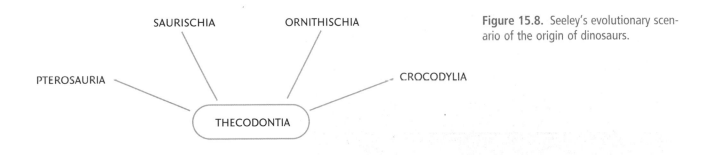

Figure 15.8. Seeley's evolutionary scenario of the origin of dinosaurs.

Box 15.5   Rollin' on the river

The greatest, most long-lived dynasty of fossil collectors was surely the Sternbergs, *père et fils*. Of these, the father, Charles Hazelius (1850–1943 (C. H.)), is perhaps the most highly regarded; yet, between him and his three sons, George F. (1883–1969), Charles M. (1885–1981 (C. M.)); and Levi (1894–1976), much of the last third of the nineteenth and the first half of the twentieth centuries was occupied with fossil collecting.

C. H., whose life-long piety translated into an interest in natural history and fossils, got his start in 1876, when he attempted to join O. C. Marsh's collecting teams in the west. When that didn't materialize, he turned to E. D. Cope, who sent him $300 and put him to work. Sternberg worked for Cope until 1897 (Cope died while Sternberg was in the field); but, during his long and productive life,

C. H. systematically mined the Pierre Shale in Montana and Wyoming (Upper Cretaceous mosasaurs, plesiosaurs, and ammonites; see Figure 16.11), the Niobrara Chalk in Kansas (Upper Cretaceous marine creatures such as turtles, mosasaurs, plesiosaurs, fish, and even pterosaurs), Tertiary-aged strata of Kansas and Oregon (a variety of mammals), and the Permian of Texas (recovering the first specimens of the primitive synapsid *Dimetrodon* (see Figure 4.10) and the temnospondyl *Eryops*). It was a stunning haul.

But C. H.'s reputation – and those of his sons, who accompanied him on these expeditions – was cemented by his raft-based explorations along the Red Deer River of Alberta, Canada (Figure B15.5.1). He writes matter-of-a-factly in his account of this work:

Figure B15.5.1. Charles H. Sternberg (1850–1943), professional dinosaur collector, and his crews floating along the banks of the Red Deer River, Alberta, Canada.

paleontologists, including (in addition to those mentioned above) W. Granger, C. W. Gilmore, J. B. Hatcher, L. M. Lambe, A. F. de Lapparent, R. S. Lull, W. D. Matthew, A. K. Rozhdestvensky, R. M. Sternberg, and C. C. Young. Each of these remarkable men made important contributions, and the full story of each would fill a book as long as this. It's really a shame that space keeps us from highlighting their lives and work. Yet no account of dinosaur paleontologists should omit the brilliant Franz Baron Nopcsa – paleontologist, Albanian nationalist, polyglot, and spy for the Austro-Hungarian Empire in World War I (Box 15.8).

The second part of the twentieth century to today

The 1960s and early 1970s are rightly known for social revolution, but they also spawned a revolution in paleontology. This was when the field of **paleobiology** was invented, a field that attempted to unravel the *biology* of fossil ecosystems.

We reached Drumheller [*Alberta, Canada*], where we purchased . . . a five-horsepower motor boat; we also built a flat boat 12' by 28', upon the deck of which we pitched two tents, one for sleeping purposes, the other for a kitchen . . . We threw a rope to Charlie [*C.M.*] in his motor boat, which he fastened to a post on the small stern deck. A kindly hand pushed us off into the stream, where Charlie got up power and towed us into the current . . . Our motor boat, under Charlie's management, went 'chug, chug' down the river at the rate of 5 miles per hour.

(Sternberg, 1985, p. 45)

Therein began one of the legendary collecting expeditions of all time. As the flatboat floated/motored down the river, they periodically put in at various promising exposures along the banks, set up camps on land, and excavated. The collection was formidable: the ceratopsians *Centrosaurus*, *Styracosaurus*, *Chasmosaurus*; the theropod *Gorgosaurus*; and the hadrosaurid *Lambeosaurus* (called "*Stephanosaurus*" by Sternberg). Along the way he met Barnum Brown (see Box 15.6), himself rafting and collecting.

C. H. was never an academic, and published relatively few scientific papers given the magnitude and importance of the fossils he collected. Yet he had a literary bent, and published an autobiography as well as an account of his fossil expeditions (see Selected readings). The latter, although originally self-published in 1917, became something of a classic and went to three editions.

The Sternberg family developed an important relationship with Canadian paleontology. Their collections remained in Canada and, in the end, while George became the curator of what was eventually named the Sternberg Museum at Fort Hays State University (Kansas), C. M. was hired on the staff of the Geological Survey of Canada, eventually ending up at the National Museum in Ottawa, while Levi was on the staff of the Royal Ontario Museum in Toronto. All taken, the Sternbergs represented two generations of lives devoted to paleontology.

C. H., as he aged, returned to Canada to be with his sons. His reaction to his adopted home sums the man up:

Though the ties of nearly a lifetime, that bound me to many a dear friend . . . must be severed, though I must leave the protecting folds of my father's flag and mine, and live under a flag that has waved a thousand years – under a monarch, in fact – I, a republican of republicans! Think of it! After three years' of residence in the beautiful city of Ottawa . . . after four seasons of work among buried dinosaurs and three winters spent in the laboratory of the Victoria Memorial Museum, I am free to confess . . . so far as personal liberty is concerned, that I was under the employ of his Royal Majesty George the Fifth of England . . . I have learned . . . that a man is as much a man amidst the snows of the Lady of the North, under the Union Jack, as under my own beloved Stars and Stripes.

(Sternberg, 1985, pp. 7–8)

Revolutions in dinosaur paleobiology

In retrospect, Yale University paleontologist J. H. Ostrom's 1969 description of *Deinonychus anthirropus* seems to have been the catalyzing event for the paleobiological revolution (Figure 15.9). Here was a predatory dinosaur (see Figure 6.6) obviously built for extremely high levels of activity. Its skeletal design simply made no sense considered otherwise. Ostrom doubted that such levels of activity were likely in an animal with the metabolism of a crocodile, and he argued for the possibility that *Deinonychus* might have been an endotherm.

Five years later, Ostrom published an exacting study of the earliest-known bird, *Archaeopteryx*. Reviving the ideas of T. H. Huxley, a contemporary of Charles Darwin, Ostrom concluded that a close relationship between dinosaurs and birds was inescapable. "All available evidence," he wrote, "indicates unequivocally that *Archaeopteryx* evolved from a small coelurosaurian dinosaur and that modern birds are surviving dinosaur descendants." This also suggested that dinosaur physiology

Box 15.6 "Mr Bones"

There have been dinosaur collectors; there have even been extraordinary dinosaur collectors, and then, in a league quite by himself, there is the legendary Barnum Brown (Figure B15.6.1). Born in 1873, and named after the then-popular circus showman P. T. Barnum, Brown had an extraordinarily long and stunningly productive career. He virtually single-handedly turned the American Museum of Natural History (AMNH) from a place with not a dinosaur on the premises to perhaps the world's greatest dinosaur collection. Its great hall of Cretaceous dinosaurs has been described as a monument to his accomplishments. Brown's lifetime spans Victorian to modern paleontology, and it is fair to say that Brown did his fair share to contribute to that growth.

Brown was a collector and, in the end, his most memorable contribution was collection and not scientific description (although he published many scientific papers). But what a collector! He began his collections in the fossil grounds of Wyoming originally prospected by Marsh. Initially he met

with little success (Marsh's collectors had done their jobs very well indeed), but toward the end of the season he discovered the still-productive Bone Cabin Quarry, a site so rich that local ranchers built an entire cabin out of fossil bones. After three years of collecting, 35 tons of fossil bones were sent back to the AMNH, including what eventually became the largest mounted specimen of its time, the AMNH's magnificent "brontosaurus" mount (see Figure B9.2.1a).

Meanwhile, in the early 1900s, Brown began to prospect in the now-legendary Hell Creek badlands of eastern Montana. There in 1902 he found the first of two magnificently preserved *Tyrannosaurus rex*. The thing was preserved in a calcite-hardened sandstone concretion, and, aside from the dynamite necessary to free it from the hillside in which it was found, he had to cut a road to carry the massive blocks out to the nearest railroad for shipment to New York (Figure B15.6.2).

Figure B15.6.1. Barnum Brown (1873–1963), collector with the American Museum of Natural History, with his most famous discovery: *Tyrannosaurus rex*. *Tyrannosaurus* is the fossil laid out behind Brown.

should be considered more along bird than crocodilian lines. With these two papers, Ostrom rewrote the book on both dinosaur physiology *and* bird origins!

That same year (as we have seen), Robert T. Bakker, Ostrom's student, and Peter M. Galton, a paleontologist from the University of Bridgeport, Connecticut, proposed to remove dinosaurs from "Reptilia" and establish them and birds as a new Linnaean class of vertebrates, Dinosauria (Figure 15.10). The basis for this proposal was the "key advancements of endothermy and high exercise metabolism." This radical proposal fired imaginations and incensed detractors. In the end,

Brown's third major venue for fossil collecting was the Red Deer River in Alberta, Canada. There, following the style of C. H. Sternberg (see Box 15.5), he fitted a barge with a canvas tent and prospected along the shores of the river. Here he collected trainloads of fossils, including the beautiful hadrosaur specimens for which the AMNH is justifiably renowned.

With these collections, Brown's reputation was thoroughly cemented. Besides possessing a remarkable intuition for finding spectacular fossils (he was said to be able to "smell" them), Brown was a fossil collector with *style*: he always stayed in the best accommodations when traveling, dressed impeccably even while carrying out the dirtiest field-work, and was often seen in tailored suits and in an opulent full-length fur-trimmed coat. The result was that his reputation preceded him, and often crowds pressed him as he arrived in the field – especially when he arrived by Waco monoplane. Indeed, he is plausibly rumored to have been quite the ladies' man.

By the early 1930s, Brown had cut a funding deal with the Sinclair Oil Company (whose logo, not coincidently, was, and is, a green sauropod) and expanded his collecting efforts to fossils other than dinosaurs and locales other than western North America. He traveled by virtually any conveyance available, and eventually ended up everywhere in the world, save Japan, Australia, Madagascar, and the South Sea Islands. His finds were numerous and varied: mummified musk ox, fossils of every stripe, and even the first Folsom projectile point, a find that indicated that humans were in North America far earlier than predicted. His discovery in 1934 of the Jurassic Howe Quarry bonebed was another highwater mark in a career full of them; more than 20 dinosaurs represented by 4000 bones.

Brown's many accomplishments and extraordinary life were immortalized by paleontologists Lowell Dingus and Mark Norell in a fine, well-deserved 2010 biography: see "Selected Readings."

Figure B15.6.2. A 1985 photograph of the site of Barnum Brown's first *T. rex* discovery. The arrow points to the remains of the wagon trail cut by Barnum Brown into the hillside to remove the massive blocks containing the fossil.

however, it didn't stick because the authors failed to sufficiently demonstrate any relationship between dinosaurs and birds other than the assertion that they shared an endothermic metabolism – a point that was heatedly debated, but ultimately left undemonstrated. Time has proven this idea resilient, and although perhaps the details of the proposal were flawed, the recognition that dinosaurs weren't overblown reptiles (in the Linnaean sense), that birds are dinosaurs, and that dinosaurs as a group are far more central to tetrapod phylogeny than had been previously recognized, are all ideas that have come to be widely accepted; if now 40 years after the concept was first proposed!

Box 15.7   Tails of Two Germans

Germany, from the latter part of the nineteenth century up until the end of the Third Reich (1945), was a country in which scientific accomplishment was highly prized and encouraged. So it is no surprise that some very distinguished German paleontological work came out during this time, carried out by very distinguished paleontologists. Two among these were Ernst Stromer and Werner Janensch.

Saharan dinosaurs

Ernst Stromer (Figure B15.7.1) was a member of Bavarian nobility,[a] and a professor at the University of Munich, who had built a fine reputation describing fossils he had collected after a trip to rich Saharan localities Wadi el Natrum and the Fayum Oasis (Egypt) after trips there in 1901 and 1902.

In the latter months of 1910 and into 1911, still bitten by the Sahara Bug, Stromer organized yet another camel-supported expedition to the western Sahara Desert, ultimately aiming at the Baharia Depression, a bone-dry, large (~1100 miles^2) desolate swath of the western Sahara with abundant terrestrial Mesozoic outcrops. The expedition was successful and, although Stromer returned to Munich in 1911, his crews continued to collect in the Baharia Depression region until 1914 and the outbreak of World War I.

Global events overtook paleontology, however, and getting the fossils out of Egypt (for study) became a nightmare. Egypt was a British "protectorate" (read, "colony"); Stromer was German; politics and the World War, of course, shut down both the expeditions and the removal of the fossils from Egypt. It took eight years but, after innumerable delays and negotiations, in 1922 Stromer was sent the specimens that he and his crews had collected. The fossils came battered and in bad shape, but at least they came. Stromer began a publishing streak that ultimately led to descriptions of *Aegyptosaurus* (a sauropod), *Bahariasaurus* (a large theropod), *Carcharodontosaurus* (a very large theropod; see Figure 7.5c), and *Spinosaurus* (the largest carnivorous dinosaur that ever lived; see Figures 6.26 and 7.4d), and in so doing, he solidified his reputation as one of the world's most careful, accomplished, paleontologists. Rapidly he obtained a professorship at the University of

Figure B15.7.1. Ernst Freiherr Stromer von Reichenbach (1870–1952), aristocratic Professor of Paleontology at the University of Munich, and discoverer of a many dinosaurs, including *Spinosaurus*, and other vertebrates in the Cretaceous beds of the Baharia Depression, western Sahara, Egypt.

Munich as well as a directorship of the Bavarian State Collection of Paleontology and Historical Geology, which he maintained until his mandatory retirement (because of age) in 1937.

But by then national events were overtaking paleontology in Germany. Stromer was old-school nobility and refused to join the rising Nazi Party (whose *heimat* was in Munich); worse yet, he denounced Hitler's regime and visibly maintained his Jewish friendships. As a member of the aristocracy, he was largely untouchable, but as Hitler plunged Germany (and the rest of the world) into World War II, the Nazis found a way to extract their literal pound of flesh: Stromer's three boys were sent to war; two to the Russian front. Nobody viewed the Russian front as a recipe for longevity, and in fact was one killed in 1941 and the second disappeared into Russia in 1944. The third son was killed on the western front in 1945, two weeks before Germany capitulated. Stromer paid dearly for his principles.

Meanwhile, in 1940, the State Collection was taken over by Karl Beurlen, a paleontologist with the correct political views. Along with being a rising paleontologist, Beurlen was an ardent support of the Führer, and simply refused to believe that the Third Reich could be defeated. And because he was sure of its invincibility, he also refused – despite Stormer's repeated imprecations – to protect the valuable

---

[a] The Germany of much of the nineteenth century was a collection of separate states, having only been united in the latter part of the century by Otto von Bismarck.

treasures of the State Collection by removing them (like so much else of value) to underground mines, such as the salt mines in nearby Salzburg.

And so it was that as April 24 turned to April 25, 1944, 16 speedy Royal Air Force deHavilland Mosquitoes with marker flares, and 244 bomb-laden Avro Lancasters, turned the Munich train station and surrounding buildings into utter rubble. Unfortunately, housed in one of those surrounding buildings was the Bavarian State Collection of Paleontology and Historical Geology. Stromer's fossils, most famously, *Spinosaurus*, were obliterated. Effectively all that was left were Stromer's meticulous notes and some detailed black and white photographs.

The end was bittersweet. On May 5, 1950, Stromer's middle son miraculously reappeared: having refused to make poison gas for Stalin, he spent the intervening six years since his disappearance at the Russian front in Siberian gulags. Two years later, Stromer died.[b]

Spinosaurus itself had to wait until well into the 2000s to be rediscovered. In 2008, then University of Chicago paleontology graduate student Nizar Ibrahim, working in the western Sahara (inspired in part by Stromer's work!) was approached by a man with some fossil bones, one of which Ibrahim recognized as possibly belonging to *Spinosaurus*. Five years later, he heard of a newly acquired specimen at the Natural History Museum of Milan that someone thought might be *Spinosaurus*. The bones in Milan looked promising and, strikingly, had the same coloration as the ones Ibrahim had been shown by the man in the western Sahara. Unfortunately, the origin of these bones in Milan was uncertain; they were thought to come from somewhere (but where?) in Morocco. Incredibly – and serendipitously – Ibrahim managed to track down the man who in 2008 had first shown him the bones, and persuaded him to take Ibrahim and a crew to the site where he had found the bones. The bones of 2008 and the Milanese fossils actually were from the same *specimen*! *Spinosaurus* had been rediscovered at last.

Tendaguru!

Tendaguru, located in the hinterland of Tanzania on the eastern coast of Africa and today monotonously formed of broad plateaus blanketed by dense torn trees and tall grass

thick with tse-tse flies, was formerly the site of perhaps the greatest paleontological expedition ever assembled, and much – thousands of millennia – before that it was the place where dinosaurs came to die.

Let's go back to 1907, when Tanzania was part of German East Africa. This was the era of massive western European colonialism in Africa. With the widespread colonialism came scientists. And to then German East Africa came paleontologists in search of fossils.

The fossil wealth of Tendaguru was first discovered in 1907 by an engineer working for the Lindi Prospecting Company (established 1903). Word spread quickly, ultimately to Professor Eberhard Fraas, a vertebrate paleontologist from the Staatliches Museum für Naturkunde in Stuttgart, who happened to be visiting the region. So excited was he at the prospect of collecting dinosaurs after his visit to Tendaguru that he took specimens back to Stuttgart (including what was eventually to be called *Janenschia*) and more especially started drumming up interest among other German researchers to continue field work in the area.

Wilhelm von Branca, director of the Humboldt Museum für Naturkunde in Berlin, first seized upon the opportunity presented to him by Fraas. Yet before mounting an expedition of the kind demanded by Tendaguru, Branca had to tackle the problem of its financial backing. By seeking support from a great many sources, he received more than 200 000 deutschmarks – a fortune for the time – from the Akademie der Wissenschaften in Berlin, the Gesellschaft Naturforschender Freunde, the city of Berlin, the German Imperial Government, and almost a hundred private citizens.

With money, material, and supplies in hand, the Humboldt Museum expedition set off for Tendaguru in 1909. For the next four field seasons, it was bonanza time. Under the leadership of moustached and jaunty Werner Janensch (Figure B15.7.2) for three of these seasons (Hans Reck took charge in the fourth season), these years were possibly the greatest dinosaur collecting effort in the history of paleontology. The first season involved nearly 200 workers, mostly natives, laboring in the hot sun as they dug huge bones out of the ground. During the second season, there were 400 workers and in the third and fourth seasons 500 workers. By the end of the expedition's efforts, some 10 km^2 of area was covered with huge pits, attesting to the diligence and hard work of these laborers.

[b] The full story of Ernst Stromer is given in Nothdurft, W. (with J. Smith) 2002. *The Lost Dinosaurs of Egypt*. Random House, NY, 242 pp.

cont.

But there's more. Many of these native workers brought their families with them, transforming the dinosaur quarries at Tendaguru into a populous village of upward of 900 people. With all these people, water and food was a severe problem. Not available locally, water had to be brought in, carried on the heads and backs of porters. And with the vast quantities of food that had to be obtained for workers and their families, and the pay for work carried out in the field, it is not surprising that the funds amassed by Branca disappeared at a great rate.

Still, the rewards were great indeed. Over the first three seasons, some 4300 jackets were carried back to the sea-port of Lindi – a four-day walk away and a trip made 5400 times there and back by native workers, each with the fossils balanced on his or her head – to be shipped from there to Berlin.

Overall, work at Tendaguru involved 225 000 person-days and yielded nearly 100 articulated skeletons and hundreds of isolated bones. When finally unpacked and studied, what a treasure-trove: in addition to ornithischians (*Kentrosaurus*, *Dryosaurus*) and theropods (*Elaphrosaurus*), and a pterosaur as well, the Tendaguru expeditions claimed not only two new kinds of sauropod (*Tornieria* and *Dicraeosaurus*), but also new material of *Barosaurus* and the finest specimen of *Brachiosaurus* ever found – now mounted and peering into the fourth floor balcony of the Humboldt Museum für Naturkunde.

The Humboldt Museum never went back to Tendaguru after 1912. In 1914, World War I erupted and, with the Treaty of Versailles, German East Africa became British East Africa. This shift in the continuation of European colonialism brought British paleontologists to Tendaguru in 1924, under the direction of W. E. Cutler. This team from the British Museum (Natural History) hoped to enlarge the

Figure B15.7.2. Werner Janensch (1878–1969), paleontologist at the Humboldt Museum, and the driving force behind the extraordinarily successful excavations at Tendagaru, Tanzania.

quarried area and retrieve some of the left-over spoils from the German effort. From 1924 to 1929, the British expedition had its ups and downs, finding more of the kinds of dinosaurs discovered earlier, but suffering some severe health problems including malaria, from which Cutler died in 1925. There has been no significant paleontological effort at Tendaguru since.

Endothermy

The debate over dinosaur endothermy climaxed with the 1980 publication of an American Association for the Advancement of Science special volume that covered the 1978 proceedings of a symposium devoted solely to dinosaur thermoregulation (please see "Selected Readings" here and in Chapter 13). In general, the contributions in the book struck a compromise between Bakker's concept of dinosaurs having bird/mammalian levels of endothermy and the then-prevalent view of dinosaurs as oversized, uninspired crocodiles. Still, most authors seemed to lean toward, at a minimum, some kind of homeothermy (see Chapter 12).

While modern research on dinosaur metabolism has lost some of its contentiousness, it continues apace. Areas of study proving to provide important insights are dinosaur developmental biology as tracked through histology, and stable isotope geochemistry (Chapter 13).

Box 15.8 Franz Baron Nopcsa: Transylvanian dinosaurs and espionage

If you visit Schönbrunn Palace, the Versailles-like summer home of the Austrian monarchy back in the heady days of the *fin-de-siècle* Austro-Hungarian Empire, and you go to the apartments of Emperor Franz Josef (1830–1916), you will find there a sign referring to a brilliant paleontologist: Franz Baron Nopcsa (Figure B15.8.1). Only he is there in another role: diplomat; his career as a paleontologist was doubtless beneath regal notice (and certainly not mentioned on the sign!).

Yet, there was never anyone quite like Franz Baron Nopcsa before and it is very unlikely that his kind will be seen again. He was one of the first paleontologists who saw to it that dinosaurs were interpreted in their full biological context. For this, he is generally regarded as the founder of the field of paleobiology. From him, we've learned about the unusual dinosaur fauna from Transylvania, that part of western Romania where his noble family's estate was located. This skeletal material formed the mainstay of Nopcsa's research, including soft tissue reconstruction and its relevance to jaw mechanisms, paleoecological reconstructions of the region as a Late Cretaceous island, and the evolution of the dwarf dinosaurs that lived on the

island. He also published extensively on the early evolution of birds from small predatory dinosaurs, the origin of new evolutionary conditions due to disease, the pituitary gland and large size in dinosaurs, and the relationship between bone histology, growth rates, and thermoregulation among dinosaurs. A polyglot, Nopcsa published not only in German, but also regularly in English, Hungarian, and French.

Remarkable though these achievements were, they were conducted against a background of scientific and political involvement in the founding of the state of Albania. Nopcsa was captivated by the geography and people of this stark, yet beautiful land of the western Balkans. He began working there in 1906 and by the end of his career had published some still-current monographs on the geography, geology, and ethnography of Albania and its people.

By 1912, Austria–Hungary was exceedingly worried about safe travel to the Mediterranean and saw Nopcsa's knowledge of the geography of the area to be tactically important to them. So Nopcsa became a spy during the first and second Balkan wars, the first of which was to establish the new country of Albania from the dying Ottoman Empire in Europe, and the second to prevent neighboring states from absorbing its territory. At that time, the new country needed a king, so Nopcsa volunteered for the job, suggesting to the Austro-Hungarian army chief of staff that he would fund the war effort with money he would obtain from marrying the daughter of some American millionaire and would pledge Albania as an ally to the Empire in exchange for recognition as King. As far as is known, his proposal received no response, although he continued his spy work in Romania during World War I.

Despite his international activities, Nopcsa was a private man. He lived most of his life in Vienna, except for two years as Director of the Hungarian Geological Survey in Budapest. Living with him was his secretary, friend, and lover, Bajazid Elmas Doda, an Albanian he met in 1906. Transylvania was ceded to Romania after World War I and the Nopcsa estate was lost. Thereafter, Nopcsa's mental health declined and early in the morning of April 15, 1933, he dosed Bajazid's tea with sleeping powder and then shot him. Going into his work room, Nopcsa wrote a suicide note and then killed himself.

Figure B15.8.1. Franz Baron Nopcsa (1877–1933), Transylvanian nobleman, patriot, spy, and brilliant paleontologist.

Phylogenetic systematics enters the fray

Amid all of this intellectual ferment, yet another revolution was not-so-quietly taking place. This was the *cladistic revolution* (see Chapter 3). The idea was not so new (although not nearly as old as that of endothermic dinosurs); the basics had first been articulated by a German entymologist, Willi Hennig, in 1950 (Figure 15.11). English translations of Hennig's ideas appeared in 1966 and again in 1979. Hennig's great insight was, as we've seen in Chapter 3, to develop a scientific (testable) method whereby relationship can be inferred from anatomy.

The method was initially not widely appreciated, and the results were a bit shocking to people trained in the traditional Linnaean classification system (see, for example, Box 4.1), with the result that between 1966 and 1990 (or thereabouts), this approach engendered considerable controversy. Nonetheless, when computer algorithms were developed that allowed cladograms to be generated from large and complex datasets, cladograms became a ubiquitous and powerful

Figure 15.9. John H. Ostrom (1928–2005), Yale University, the brilliant paleontologist whose ideas ignited the modern era of dinosaur research.

tool for deciphering the relationships of both living and extinct organisms (including dinosaurs). Starting in about 1984, cladistic analysis exploded onto the dinosaurian systematic scene. The results were four-fold:

(1) The disbanding of the venerable "Thecodontia."
(2) The origin of dinosaurs.
(3) The internal pattern of relationships within ornithischian and saurischian clades.
(4) The relationships of birds to dinosaurs.

In all four aspects, the changes wrought by cladistic analyses in our understanding of archosaurs in general, and dinosaurs in particular, were nothing short of revolutionary.

Thecodontia

"Thecodontia" (*theco* – socket; *dont* – tooth; teeth set in sockets) was a term invented in 1859 by Richard Owen to group primitive archosaurs. Thinking cladistically, however, "Thecodontia" was diagnosed using the same diagnostic characters as Archosauria; therefore, it must include *all* archosaurs – not just primitive ones. So the group "Thecodontia" was redundant when there was already a group called Archosauria. Cladistic analysis, therefore, propelled the abandonment of what had for 120 years been considered a very important group of animals.

Dinosaur origins

Recall that Seeley divided dinosaurs into two groups – ornithischians and saurischians (on the basis of their pelvic anatomy) – and that he viewed these as having separate origins within "Thecodontia." The elegant 1986 cladistic work of Jaques A. Gauthier (Figure 15.12), now at Yale University, however, provided ample corroboration of a monophyletic Dinosauria, identifying upward of ten derived features uniting all dinosaurs with each other (see Chapter 4). Since then, numerous cladistic analyses of both new and old taxa have affirmed that *dinosaurs share a single, most recent common ancestor, itself a dinosaur*. Ornithischia and Saurischia – each monophyletic – are more closely related to each other than they are to anything else.

Ornithischian and saurischian relationships

There are some other differences between the precladistic view of dinosaurs and a more modern one. For example, before cladograms, all large, carnivorous dinosaurs were grouped together within the evocatively named Carnosauria; Coelurosauria was kept for the smaller forms. Cladistic analysis, however, paints a very different picture: one in which several different lineages of large-sized theropods, with their short arms and huge heads, evolved *independently*. The large theropod *Carnotaurus* is as far from other large (tetanuran) theropods as it is possible for one theropod to be from another. And Coelurosauria, once the exclusive domain of small, light-bodied forms, is now home to the mighty *Tyrannosaurus rex* as well. The cladograms tell us that when theropods grow very large, they commonly, and independently, tend to take a similar form: large head, small arms, typically with reduced numbers of fingers, and long tails. What was it about their behavior or basic design that produced that shape?

Cladograms also revealed the important links between ceratopsians and pachycephalosaurs (once thought to be a kind of ornithopod), confirmed suspected links between ankylosaurs and stegosaurs, and situated the enigmatic therizinosaurs (once thought to be troodontids, or ornithischians, or who knows what else!) within Theropoda.

Birds as dinosaurs

Ostrom, as we have seen, constructed a compelling anatomical case for birds as dinosaurs in his 1974 paper on *Archaeopteryx*. In 1986, Gauthier published a now-classic paper on saurischian monophyly, in which he addressed Ostrom's observations from a cladistic viewpoint. Gauthier's analysis and much subsequent work form the backbone of our treatment of bird origins in Chapter 8 and, as we've seen, overwhelmingly confirm the fundamental dinosaurian ancestry of birds. With so many important evolutionary insights afforded by the use of cladograms, there

Figure 15.10. (a) Robert T. Bakker, then at Yale University, and (b) Peter M. Galton, University of Bridgeport, who first proposed that Linnaeus' venerable "Class" Aves be subsumed within a larger, new "Class Dinosauria."

should be no wonder why we have emphasized a cladistic approach throughout this book.

Our discussion of birds as dinosaurs would not be complete without at least a few words about feathers. Once the possibility of dinosaur endothermy became a question (as we have seen, this was in the late 1960s and early 1970s), it seemed that dinosaurs might need insulation, and it seemed obvious to many paleontologists that such insulation might come from feathers or proto-feathers. Thus, paleontologists predicted, we ought to find feathers on non-flying dinosaurs, on the assumption that feathers were initially more about endothermy than about flight. But as we have seen, feathers probably weren't initially about endothermy, but most likely about display or earlier, touch; however, we also have considerable evidence that they were co-opted for insulation before they were modified and co-opted for flight. We know this because of the extraordinary discoveries in Liaoning Province, China, and, as much as anybody, because of the work of Xu Xing, Chinese paleontologist *extraordinaire* at the Institute of Vertebrate Palaeontology and Palaeoanthropology, Beijing, China (see Box 15.9.8). Xu has been associated with formal descriptions of many new genera of feathered dinosaurs. These constitute one of the great revolutions in vertebrate paleontology (see Chapters 7 and 8).

Figure 15.11. Willi Hennig (1913–1976) of the Deutsches Entomologisches Institut, the German entomologist who was the father of cladistic analysis (phylogenetic systematics).

The extinction of (non-avian) Dinosauria

Our trip through the most recent ideas about dinosaur paleontology would not be complete without another thoroughly radical idea: extinction by asteroid impact (see Chapter 16). Since H. F. Osborn's time, the extinction of the dinosaurs was viewed by paleontologists as a gradual process of dwindling diversity, beginning well before the end of the Cretaceous. The prevailing view before 1980 was cleverly and succinctly summarized by University of California (Berkeley) paleontologist W. A. Clemens and colleagues who wrote (with apologies to T. S. Eliot):

This is the way Cretaceous life ended
This is the way Cretaceous life ended
This is the way Cretaceous life ended
Not abruptly but extended.[6]

Clemens and colleagues' article was aptly entitled "Out with a whimper, not a bang."

The revolution came in the form of the 1980 hypothesis that an asteroid came from outer space, smashed into the planet, and ultimately reset global ecosystems for all time

Figure 15.12. Jacques A. Gauthier (1951–present), then at the California Academy of Sciences, who carried out the seminal cladistic studies of bird origins and dinosaur monophyly.

(Chapter 16). What a strange, wonderful, and terrifying idea – that extraterrestrial events are critical forces shaping the history of life on Earth! The conceptual revolution provoked by this vision, however,

---

[6] Clemens, W. A., Jr, Archibald, J. D., and Hickey, L. J. 1981. Out with a whimper, not a bang. *Paleobiology*, 7, 297–298.

extended far beyond the deaths of a few dinosaurs, and reverberated throughout the geosciences.[7] For one thing, it took a global-system-based view of events: that the dinosaur extinction was not an isolated event pertaining only to dinosaurs (and thus dinosaur-based explanations for the extinction were sufficient), but rather it was a global biotic catastrophe with global physical ramifications. As such, it shined a (for some, unwelcome) spotlight into vertebrate paleontology, heretofore, a small, select community of scientists in large part talking exclusively to each other rather than to the scientific community as a whole.

At a deeper level, it provoked a fundamental revolution in the geosciences. Whereas the prevailing view in the geosciences, dating back to early nineteenth-century geologist Charles Lyell, had been that events on Earth occurred gradually over potentially immense time scales, the asteroid hypothesis catalyzed the widespread recognition that many Earth-shaping processes – as significant as asteroids, or as miniscule as a local flood – really occur on relatively short time scales; indeed, they might be said to occur, in a geological sense, instantaneously. Ultimately, this view has revolutionized the fields of paleontology and sedimentology. The asteroid and its aftermath are the subjects of our last chapter.

Today

There are more active, professional paleontologists working today than ever before in the history of the discipline (Box 15.9), and the insights derived from the field have become particularly important, not just in the field of evolutionary biology, but also as they relate to climate science. Yet, paleontology is today at a crossroads, with new skill sets required that would have been utterly foreign to paleontologists of a generation ago.

In evolutionary biology, it has become very clear, thanks to the work of scientists like Jacques Gauthier and Timothy Rowe (see Gauthier, Kluge, and Rowe, 1988, in "Selected Readings"), that the present-day biota does not give us a complete enough picture of the past record of evolution. Deeper insights are afforded by the fossil record; indeed, how would we ever know that birds are dinosaurs if we only knew the living representatives of Dinosauria?

Yet, modern evolutionary biology has acquired powerful new tools for unraveling the course of evolution, including: (a) molecular evolution, in which the molecular difference between two organisms is measured to determine how distantly or closely they are related (see Box 8.3); (b) evolution and development (or **evo-devo**), in which sophisticated genetic and embryological studies are revealing the way the organisms evolve new features and produce diversity; and (c) **molecular clocks**, in which the rate of molecular evolution can be used to date the time that two living organisms first diverged from a single (long-extinct) ancestor (see Box 8.3). In each of these cases, deep insights into the fossil record are necessary to optimize the results. Mary Schweitzer's work with *T. rex* soft tissues and blood cells would hardly have been what paleontologists expected to do 50 years ago; but we would surely call her work paleontology! So paleontology's contributions continue to be on the forefront; yet, modern paleontologists need a far greater sophistication and specialized training – more than was ever even imagined by the old timers – in the biosciences generally and in molecular genetics and embryology in particular. In recognition of this, a new term was invented to describe what we do: paleobiology.

We have also seen how many modern paleontologists now use geochemical tools unthinkable a generation ago: stable and unstable isotopes, as well as trace element compositions and geochemical signatures.

---

[7] See Powell, J. L. 1998. *Night Comes to the Cretaceous*. W.H. Freeman and Company, New York, 250 pp.

Another important tool widely employed by paleontologists only since the 1980s is statistics. Paleontologists deal with sparse datasets; incomplete datasets; data that are of variable quantity and quality; and all of these kinds of data require somewhat specialized mathematical treatments. An important part of a well-trained paleontologist's problem-solving arsenal, therefore, is statistical training.

All of these skills – molecular biology; geochemistry; statistics – are a far cry from the old-style collect-and-describe game of a generation ago. Of course, they haven't eclipsed it; they're just another addition to the professional toolbox.

Paleontologists have a unique perspective to offer the world about global climate change. The Earth, of course, has gone through many episodes of global warming associated with greenhouse conditions, not the least of which was, as we've seen, during the mid Cretaceous. What was that world like? More importantly, how did the biota respond to that climate? Since the Earth has experimented with greenhouse conditions in the past, it behooves us to consider how it and its biota responded to these events – information that may give us insights about what we may expect in the future. In this case, advanced expertise in geochemistry and computer-based climate modeling could serve as an important part of the preparation of future paleontologists.

So the days of pulling dinosaurs from the ground, describing, and naming them, are certainly still with us; it is clear from Benton's data (Figure I.1) that there are lots of dinosaurs to find. Yet, the significance of these finds – in terms of evolutionary biology, paleogeography, paleoecology, functional morphology, or some other paleobiological issue – can no longer be ignored. But in a way, the questions are more interesting, and the stakes far greater. It's a brave, new, and exciting world out there.

Box 15.9   Young Turks and old turkeys

"It's funny," University of Oregon paleopedologist Greg Retallack once said, "how quickly today's 'Young Turks' become tomorrow's old turkeys." The young, fire-breathing revolutionaries themselves become advocates of established dogma, as, even in their own lifetimes, their ideas are challenged and swept aside by a righteous, new, young, aggressive generation of scientists. Since the generation of paleontologists active in the 1950s and early to mid-1960s, at least two generations of scientists have made their marks: the "Baby Boom" generation, most now nearing retirement, and the Gen-X generation, currently at the height of their careers.

Our choices are, in a sense, arbitrary; there is a rather large number of paleontologists, specializing in dinosaurs, that we lack the space to mention in both of these generations, as well as many others who have worked with dinosaurs, as well as with other groups of organisms. The scientists we describe here must essentially represent – stand in for – the many, highly accomplished vertebrate paleontologists produced by these two generations.

Here we start by highlighting the careers of four members of the Baby Boomer generation of vertebrate paleontologists; revolutionary, creative thinkers, all, but perhaps no longer so young!

As we have seen, the 1970s were a time of intellectual ferment in paleontology, and for the dinosaur-loving public, nobody embodied those fresh winds of change more than Robert T. Bakker (b. 1945; Figure 15.10a). In a field historically known for bookish indifference to fame, Bakker has been a ubiquitous media presence; so much so that he was recognizably caricatured as Dr Robert Burke in *Jurassic Park II*. Bearded, long-haired, and dressed for battle in his field-ready best[a] (belying degrees from Yale and Harvard), Bakker was filled with amazing ideas about birds, dinosaurs and their world. A highly competent prose stylist (described in *Harper's Magazine* as "by far the most gifted writer in his profession"[b]) and a talented illustrator, Bakker was distinctive, articulate and out to change ideas.

Equally informal of manner and no less intellectually endowed, is Bakker's contemporary, John R. "Jack" Horner

[a] Including an iconic battered hat that pre-dated that of Indiana Jones – but not that of Roy Chapman Andrews!
[b] Silverberg, R. 1981. Beastly debates. *Harper's Magazine*, October, 1981, pp. 68–78.

Figure B15.9.1. Paleontologist *extraordinaire* Jack Horner, doing what he does best.

(b. 1946; Figure B15.9.1). For much of his career, Horner lacked formal advanced degrees (although he now has an honorary doctorate); yet, it would be hard to find a more creative (he is a MacArthur "genius award" recipient) and accomplished paleontologist. As we have seen, it was Horner who first recognized extended parental care in *Maiasaura*; it has been Horner who has, most recently, led the field in understanding bone histology and its relationship to dinosaur growth and metabolism; it was in Horner's laboratory that proteins from *T. rex* were first extracted (**see** Boxes 6.3 and 8.3).

Horner is a superb field paleontologist. His laconic, plainspoken manner, reputed to be the inspiration for Dr. Alan Grant in the film *Jurassic Park*, belies a canny, sophisticated, approach to almost all he undertakes, whether it be building the Museum of the Rockies (Bozeman, Montana, USA) into a world-class paleontological museum and an important center for cutting-edge histological research (both living and fossil), or publishing well over 100 research papers on virtually every subject pertaining to Dinosauria. Jack Horner continues to be a dominant and creative force in the field of vertebrate paleontology, using his own successes to make opportunities for other younger scientists.

Somewhat younger than Horner and Bakker is the Curator of the Division of Paleontology at the American Museum of Natural History (New York), Mark A. Norell (b. 1957; Figure B15.9.2). Norell's studies have been wideranging, including contributing to the development of the

Figure B15.9.2. Mark Norell, Curator of Paleontology at the American Museum of Natural History (New York).

concept of "ghost taxa" (see Box 14.2), the discovery and description of the unusual theropod *Mononykus* (see Figure 7.14), the discovery of the first embryo of a theropod dinosaur, and the first clear "proof" that dinosaurs nested on their eggs (*Oviraptor*; see Figure 6.32). Norell has led expeditions to the Gobi Desert for the past 18 years, during the course of which he discovered and described the dinosaurs *Shuvuuia, Apsaravis, Byronosaurus*, and *Achillonychus*, among other vertebrate fossils. He has even co-written several award-winning books, including *Discovering Dinosaurs* (1995) and *Unearthing the Dragon* (2005).

Our brief sampling of some of dinosaur paleontology's not-so-young Young Turks would surely be incomplete without the University of Chicago's redoubtable Paul Sereno (b. 1957; see Figure B15.9.3). Sereno belies the myth that they don't make paleontologists like they used to.

cont.

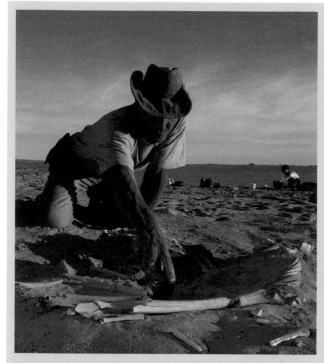

Figure B15.9.3. Paul Sereno in the waning sun of the Sahara Desert, digging up something not too dinosaur-like. He writes of this picture, "I [fell] prey to digging up 100 fossil humans in the Sahara – they got in the way of my dinosaurs!"

Figure B15.9.4. Smithsonian curator and dinosaur specialist Matt Carrano, framed by a toothy Late Cretaceous friend.

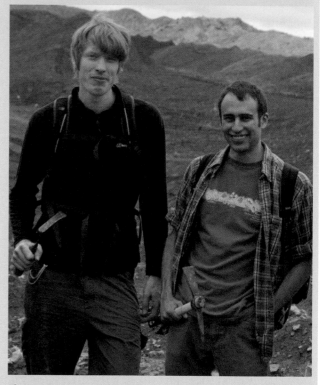

Figure B15.9.5. Paleontologists Richard J. Butler (left) and Stephen Brusatte (right) deep in the red Triassic clays of Lithuania.

Handsome and dashing, he was, after all, one of *People* magazine's "50 Most Beautiful People," *Newsweek*'s "100 People to Watch for the Next Millennium," and *Esquire* magazine's "100 Best People in the World." Reflecting his vigor/age ratio, did we not see him recently gracing the cover of the magazine of the American Association of Retired People [AARP]? – though he can hardly be said to have retired! Sereno has made it his business to travel to exotic locales (Egypt, Niger, Morocco, Argentina, China) and collect exotic fossils. Therein lies quite the list, including *Afrovenator abakensis* (a large theropod), *Carcharodontosaurus saharicus* (another large theropod, possibly larger than *Tyrannosaurus*), *Deltadromeus agilis and Rugops primus* (two more large theropods!), *Rajasaurus narmadensis* (a large, crested theropod), *Herrerasaurus ischigualastensis* and *Eoraptor lunensis* (two of the most primitive dinosaurs known; Figure 5.7), *Jobaria tiguidensis* (a 20+ m sauropod), the 13 m dinosaur-munching Saharan crocodile *Sarcosuchus imperator*, and the piscivorous large theropod *Suchomimus tenerensis*. Sereno, unsurprisingly, is a co-author on the full description of the newly reconstructed (and surely rethought) *Spinosaurus*.

The Gen-X generation has had no shortage of exciting talented paleontologists, either. We'll meet several, again with the understanding that their selection here does not imply greater importance for them than for other productive, active members of their generation.

We start with the US National Museum of Natural History's (Smithsonian Institution) Curator of Dinosauria Matthew T. Carrano (Figure B15.9.4). Carrano has wide-ranging interests, from collecting dinosaurs in the grand tradition, to large-scale evolutionary (macroevolutionary) patterns in the fossil record (such as we discuss in Chapter 14). Like many vertebrate paleontologists, he is actively involved with reconstructing phylogeny, particularly that of theropods – these days, a very active field indeed. He also has deep interests in functional morphology, and is working to reconstruct selective pressures on dinosaur evolution based upon function. Finally, Carrano has been involved in generating a comprehensive, computerized database for vertebrates, to be used as a resource for understanding the long-term, large-scale ebb and flow of vertebrate evolution.

Carrano is no stranger to exotic fieldwork, and has tromped around the world, including the wilds of Nigeria, Madagascar, and South America. His publications reflect his broad-ranging interests, touching phylogeny, biogeography, functional anatomy, and evolution, and extinction.

A second highly productive member of this generation, although slightly younger, is the University of Edinburgh's affable Stephen Brusatte (Figure B15.9.5). Brusatte is an American; trained in the United States, although he obtained his MS (and his wife!), in Bristol, UK, working with Baby Boomer M. J. Benton (another prodigious dinosaur paleontologist whose work has graced our pages). Brusatte seems to be everywhere and to have done everything: before he had even received his PhD (his advisor was Mark Norell; see above), he had written an excellent book for paleontology and zoology graduate students about dinosaurs (*Dinosaur Paleobiology*). He has published widely at an almost scary pace (he currently has ~86 publications). Brusatte is a dinosaur specialist; and while his nominal expertise is theropods (particularly coelurosaurs including tyrannosaurids), he has in fact made important contributions in dinosaur paleoecology, behavior, dinosaur origins *and* extinction, Mesozoic mammal evolution, functional morphology, phylogeny, all kinds of non-dinosaurian archosaurs (phytosaurs, pterosaurs, crocodiylians. . .); it would in fact be hard to list the full breadth of his interests and publication record.

Like most paleontologists, Brusatte likes the field and spends one to two months each year doing fieldwork. Among the places he has maintained active research are the Triassic of Portugal, the Jurassic of Scotland, the Paleocene of New Mexico (USA), and the Mesozoic of central Europe, in particular, Poland, Romania, and Lithuania.

Britain has produced many extraordinary paleontologists. Two representatives of Generation X that come to mind are the Natural History Museum's (formerly, the British Museum of Natural History; of Richard Owen fame) Paul Barrett and the University of Birmingham's Richard J. Butler. Both are extraordinarily accomplished, productive paleontologists. Moreover, both have creatively and successfully paired their research problems with sophisticated statistical approaches to the data.

Figure B15.9.6. Paul Barrett, getting his pants dirty in the field, excavating the partial skeleton of an early sauropodomorph in the Late Triassic Lower Elliot Formation near Lady Grey, Eastern Cape, South Africa.

cont.

Richard J. Butler, like Barrett and Brusatte, is an extraordinarily productive researcher (Figure B15.9.5). His interests involve the early evolution of dinosaurs, as well as the vertebrate recovery following Permo-Triassic mass extinction (the largest extinction in the history of Earth), in unidirectional breathing, both fossil and recent, in archosaur systematics and functional anatomy, and in the nature of – and patterns in – the fossil record. Butler has carried out fieldwork in the Triassic of Argentina, South Africa, Portugal, Lithuania, and Poland.

In the case of Paul Barrett (Figure B15.9.6), along with the standard taxonomic revisions and descriptions that are the bread and butter of the field, he is interested in macroevolutionary patterns and mechanisms; that is, large-scale controls on dinosaur diversity and morphology. Barrett also has a long-standing interest in terrestrial Mesozoic ecosystems and taphonomic issues that affect the fossil record. Along with dinosaurs, Barrett is interested in lizards, plesiosaurs, and ichthyosaurs. He has carried out extensive fieldwork in the Jurassic and Cretaceous of the UK, in the Early Cretaceous of China, in the Jurassic of South Africa, and in the Cretaceous of Australia and Antarctica.

Our next paleontologist is the University of Texas' highly accomplished Julia Clarke (Figure B15.9.7). A graduate of Yale University with Jacques Gauthier (see above), Clarke's career has been built around early bird evolution; a timely choice, given the revolution that our understanding of bird evolution has undergone in the past. Clarke is interested in bird diversity, systematics, and function – in particular, the modifications of the front limb from a grasping hand, to a combination of flight and propulsive mechanism (e.g., a wing), and then, in diving birds such as loons and penguins (and hesperornithiforms!), a swimming and diving mechanism. She has also studied the reconstruction of plumage coloration in non-avian dinosaurs. Her work on early birds (and their progenitors) as taken her afar, and she has carried out fieldwork in New Zealand, Antarctica, Mongolia, Argentina, the United States, and China. It was she that found the first representative of modern birds from the Mesozoic, the Cretaceous *Vegavis* (Figure 8.17). Clarke has published very extensively, and has become one of the world's authorities for primitive bird – and derived theropod! – evolution.

No discussion of today's best, most active paleontologists should be without some mention of Xu Xing, the prolific

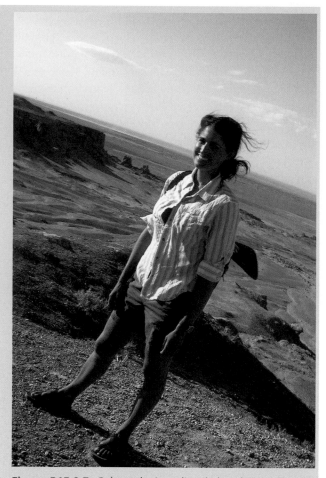

Figure B15.9.7. Paleontologist Julia Clarke, doing what she loves in the wilds of Argentina.

Chinese vertebrate paleontologist, and a researcher at the Institute for Vertebrate Paleontology and Paleoanthropology (IVPP), Beijing, China (Figure B15.9.8). Xu has worked all over the world – although it turns out that the riches that he has mined are largely to be found in China's famous Liaoning Province, ground zero for the feathered dinosaur revolution (see Chapters 6–8, and the many images of feathered dinosaurs that grace our pages). To understand the significance of these animals – not just for the origin of birds, but also for the origin of flight and the significance of feathers – it took a paleontologist with extraordinary breadth of vision, creativity, and just plain raw energy. Xu fits the bill, describing and analyzing more than 32 new, valid *genera* of fossil dinosaurs; doubtless a record for his generation; perhaps for all generations? The newspaper USA Today encapsulated his successes perfectly in its 2008 post: "Forget Indiana Jones: Dinosaur hunter Xu digs it."

Figure B15.9.8. Dr. Xu Xing, of the IVPP, perhaps the most prolific living paleontologist, and the man without whom feathered theropods would likely still be just a neat hypothesis.

All of these workers – from Bakker to Xu – are distinguished for the breadth of their interests, for their energy, and for the creativity that they have applied to the questions they are asking. They all seek and have obtained extensive independent funding, and have ultimately built their careers on the love of, and enthusiasm for, paleontology and Earth history.

And like most other active professional paleontologists, they know the history of their chosen field and, to a person, each has consciously inherited the mantle from the grand (and not so grand) old paleontologists who preceded them, beginning a long time ago with Gideon Mantell. They would all agree with Isaac Newton's famous epigram, "If I have seen further than others, it is by standing upon the shoulders of giants."

SUMMARY

Paleontology is a human endeavor, and like all human endeavors, ideas have changed as the context in which those ideas developed has changed. Our earliest suggestion that humans may have seen and taken note of dinosaur fossils comes from the recognition that mythical creatures may have been inspired by observations of very large or unfamiliar-looking dinosaur fossils.

Paleontology as a science began in the Enlightenment with the recognition that observation in combination with logic and rational thinking could reveal truths about the natural world. The earliest dinosaur fossil explicitly identified as something quite unlike anything alive today was found in 1822; within 40 years, not only was a variety of extinct animals recognized, but a relative geological time scale had been constructed. It remains valid in its essentials to this very day. In 1842, English anatomist Richard Owen established the word "Dinosauria" for an extinct group of reptiles, partly to demonstrate that organisms had not evolved. His concept of dinosaurs was one of bulky, elephantine quadrupeds.

Charles Darwin's idea of evolution by natural selection was published in 1859, and, with it, the burgeoning fossil record came to be seen as making sense in an evolutionary context. The idea that other worlds had existed that were very different from our own gained broad currency. At the same time, Dinosauria became far better known, and its members seemed so disparate that they were divided into two groups (Ornithischia and Saurischia) and thought to have had separate origins within early archosaurs (which were all lumped together as a group called "Thecodontia"). The rise of dinosaurs was interpreted in Darwinian terms as the competitive success of the superior forms (dinosaurs) over inferior forms (primitive archosaurs and advanced, non-mammalian synapsids).

The first 70 or so years of the twentieth century were largely about collecting, describing, and enhancing knowledge of the different forms of dinosaurs. This abruptly changed with John Ostrom's 1969–1970 interpretation of *Deinonychus* as endothermic and his 1974 re-evaluation of

Archaeopteryx as a theropod. These revolutionary views came as the field of paleontology was revitalized as paleobiology, and as phylogenetic systematics came to be recognized as a truly scientific way of inferring relationships among even extinct forms. Phylogenetic systematics demonstrated the fundamental monophyly of Dinosauria (also affirming, parenthetically, the monophyly of Saurischia and Ornithischia), destroyed "Thecodontia," and clearly showed that birds are dinosaurs.

When, in 1980, it was postulated that an asteroid caused the end-Cretaceous extinction (which obliterated the non-avian dinosaurs), paleontologists came to recognize that extraterrestrial events can have a profound effect on earthly events; that extinctions can occur regardless of how well adapted a particular group is. It was at this time that a re-evaluation of the rise of Dinosauria was carried out, in which the success of the group was no longer ascribed to competitive superiority, but rather to extinctions that had liberated ecospace for dinosaurs to colonize.

Today, dinosaurs continue to be studied in a variety of ways (see Box 15.9): through the discovery and description of new forms; as biological entities functioning in ancient ecosystems; via analyses of the large-scale evolution of the group; through histological and even molecular analyses; and from the standpoint of evolution and development.

SELECTED READINGS

Andrews, R. C. 1929. *Ends of the Earth*. G. P. Putnam's Sons, New York, 355 pp.

Bakker, R. T. 1986. *Dinosaur Heresies*. William Morrow, New York, 481 pp.

Bakker, R. T. and Galton, P. M. 1974. Dinosaur monophyly and a new class of vertebrates. *Nature*, 248, 168–172.

Benton, M. J. 1984. Dinosaurs' lucky break. *Natural History*, 93(6), 54–59.

Bryson, B. 2003. *A Short History of Nearly Everything*. Broadway Books, New York, 544 pp.

Cadbury, D. 2001. *Terrible Lizard*. Henry Holt and Company, New York, 384 pp.

Desmond, A. 1975. *The Hot-Blooded Dinosaurs*. The Dial Press, New York, 238 pp.

Desmond, A. 1982. *Archetypes and Ancestors*. University of Chicago Press, Chicago, IL, 287 pp.

Dingus, L. and Norell, M. A. 2010. *Barnum Brown. The Man who Discovered Tyrannosaurus rex*. University of California Press, 368 pp.

Gauthier, J. A. 1986. Saurischian monophyly and the origin of birds. In Padian, K. (ed.) *The Origin of Birds and the Evolution of Flight*. Memoirs of the California Academy of Sciences no. 8, San Francisco, pp. 1–56.

Gauthier, J. A., Kluge, A. G., and Rowe, T. 1988. Amniote phylogeny and the importance of fossils. *Cladistics*, 4, 105–209.

Hennig, W. 1979. *Phylogenetic Systematics*, translation by D. D. Davis and R. Zangerl. University of Illinois Press, Urbana, IL, 263 pp.

Lessem, D. 1992. *The Kings of Creation*. Simon and Schuster, New York, 367 pp.

Mayor, A. 2000. *The First Fossil Hunters*. Princeton University Press, Princeton, NJ, 361 pp.

Nothdurft, W. (with Smith, J.) 2002. *The Lost Dinosaurs of Egypt*. Random House, New York, p. 242.

Ottaviani, J., Cannon, Z., Petosky, S., Cannon, C., and Schultz, M. 2005. *Bone Sharps, Cowboys, and Thunder Lizards: A Tale of Edward Drinker Cope, Othniel Charles Marsh, and the Gilded Age of Paleontology*. GT Labs, New York, 168 pp.

Preston, D. J. 1986. *Dinosaurs in the Attic: An Excursion into the American Museum of Natural History*. St. Martin's Press, New York, 244 pp.

Shubin, N. 2008. *Your Inner Fish: A Journey Into the 3.5-Billion-Year History of the Human Body.* Pantheon Books, New York, 230 pp.

Sternberg, C. H. 1985. *Hunting Dinosaurs in the Bad Lands of the Red Deer River, Alberta, Canada.* NeWest Press, Edmonton, Alberta, 235 pp.

Thomas, R. D. K. and Olson, E. C. (eds.) 1980. *A Cold Look at the Warm-Blooded Dinosaurs.* AAAS Selected Symposium no. 28, Washington, D.C., 514 pp.

TOPIC QUESTIONS

1. Who invented the term "Dinosauria?" What was his idea about dinosaur metabolism?

2. In relation to paleontology, what is meant by "the edge?" and "the wedge?"

3. Decribe a dinosaur as imagined by Victorian paleontologists. What changed our ideas from the Victorian conception?

4. Describe the kinds of vertebrate faunas that existed just before dinosaurs became the most abundant and diverse terrestrial vertebrates.

5. What were the dramatic "revolutions" that changed the face of paleontology after the 1960s?

6. Cite an example of how early twentieth-century imperialism moved the science of paleontology forward.

7. Give two examples of how cladograms changed *paleontological* thinking.

8. In what fundamental ways does the legend of the griffin differ from our understanding of *Protoceratops*?

9. Contrast what you know of the Triassic–Jurassic extinction with the Cretaceous–Tertiary extinction.

10. Explain the difference between a cladistic view of dinosaur origins and H. G. Seeley's understanding of the origins of dinosaurs.

The Cretaceous–Tertiary extinction
The frill is gone

16

CHAPTER OBJECTIVES

- Learn about the Cretaceous–Paleogene boundary
- Learn about the dinosaur extinction
- Evaluate scientific hypotheses in the light of incomplete data

How important were the deaths of a few dinosaurs?

The mass extinction to which the non-avian dinosaurs finally succumbed after thriving for about 164 million years on Earth is called the **Cretaceous–Paleogene extinction**, commonly abbreviated **K-Pg**.[1] The K-Pg mass extinction (Box 16.1) involved much more than just dinosaurs. Among the "highlights" were:

- a ~10 km asteroid collided with Earth;
- the great cycles of nutrients that formed the complex food webs in the world's oceans temporarily shut down;
- many marine and terrestrial animals, as well as many plants, went extinct;
- landscapes were deforested;
- epic tsunamis rolled across ocean basins, and
- wildfires likely raged.

By comparison with that, how important were the deaths of a few dinosaurs?

Asteroid impact!

Whatever else is truth, the Cretaceous ended – and the Paleogene began – with a real bang. Here's how we learned about it. In the late 1970s, geologist Walter Alvarez and a team of co-workers (Figure 16.1) were studying K-Pg marine outcrops now exposed on land near a town called Gubbio, in Italy. They were struck by the fact that the lower half of the Gubbio rock exposure is composed of a

Figure 16.1. The team of University of California (Berkeley) scientists who first successfully proposed the theory of an asteroid impact at the K-Pg boundary. Left to right: geochemists Helen V. Michel and Frank Asaro, geologist Walter Alvarez, and physicist Luis Alvarez.

---

[1] The extinction is thus said to have occurred at the Cretaceous–Paleogene (or K-Pg) boundary. The "Pg" in K-Pg obviously stands for the Paleogene Period, the first period of the Cenozoic Era. The "K" the German word *Kreide*, or chalk, because the Cretaceous was first identified at the chalk cliffs of Dover (England). In older nomenclature, the boundary is sometimes called "K/T" for Cretaceous/Tertiary. The Tertiary is an outdated term that included what we now call the Paleogene.

rock made up entirely of thin beds of the microscopically sized shells of *Cretaceous* marine organisms. The upper half of the exposure was almost exclusively of thin beds of the microscopic shells of *Paleogene* marine organisms. Between the two was a thin (2–3 cm) layer of clay, obviously the K-Pg boundary.

Analyses showed that the clay layer contained unusually high concentrations of **iridium**, a rare, platinum-group metal.[2] Instead of the expected amount at the Earth's surface, about 0.3 parts per billion (ppb), the iridium content was a whopping 10 ppb at Gubbio. So the **iridium anomaly**, as it came to be called, contained about 30 times as much iridium as Alvarez and his co-workers had expected to find (Figure 16.2).

Iridium is normally found at the Earth's surface in very low concentrations, but it is found in higher concentrations in the core of the Earth and from **extraterrestrial** sources; that is, from outer space. Given that, the Alvarez group determined that the source of the iridium had to be extraterrestrial. The deal was sealed when they found iridium anomalies at two other K-Pg sites, one in Denmark and the other in New Zealand. Science often proceeds via brilliant intuition, and with exactly that (and little else), they concluded that, at 66.1 Ma,[3] a large **bolide**, that is, a fiery impacting body such as an asteroid, had to have smacked into Earth, delivering the iridium and, coincidentally, causing the K-Pg mass extinction. Luis Alvarez, Nobel Prize-winning physicist, and a member of the team, described the relationship between an asteroid impact and the iridium layer in this way:

> When the asteroid hit, it threw up a great cloud of dust that quickly encircled the globe. It is now seen worldwide, typically as a clay layer a few centimeters thick in which we see a relatively high concentration of the element iridium – this element is very abundant in meteorites, and presumably in asteroids, but is very rare on earth. The evidence that we have is largely from chemical analyses of the material in this clay layer. In fact, meteoric iridium content is more than that of crustal material by nearly a factor of 10^4. So, if something does hit the earth from the outside, you can detect it because of this great enhancement. Iridium is depleted in the earth's crust relative to normal solar system material because when the earth heated up [during its formation] and the molten iron sank to form the core, it "scrubbed out" [i.e., removed] the platinum group elements in an alloying process and took them "downstairs" [to the core].
>
> (Alvarez, 1983, p. 627)

Because the three sites are distributed around the globe, Alvarez and co-workers calculated that the asteroid had to have been about 10 km (about 6 miles) in diameter to spread an iridium dust layer globally. Really, it was more of a good idea than much else; but at least one geoscientist has since called it "the most important idea in geology."[4]

In the intervening years, a tremendous amount of work has been done to explore the possibility of an asteroid impact at 66.1 Ma. Most importantly, the number of K-Pg sites with anomalous concentrations of iridium at the boundary has reached well over 100 (Figure 16.3). Moreover, the iridium anomaly was discovered on land (Figure 16.4) as well as in ocean sediments, affirming that it is a global phenomenon.

---

[2] It is a common misconception that iridium metal is toxic and deadly. In fact, like its chemical relatives gold and platinum, it is quite unreactive. For those with significant disposable incomes, boutique fountain pens and watches made with iridium are available. Your bank account might go extinct if you buy them, however.

[3] The precise date was not actually known at the time; people thought it was around 65 Ma. It took almost 30 years before the high-precision date we report here – 66.1 Ma – was obtained.

[4] UC Berkeley plate tectonicist M. A. Richards. A lot of people would disagree; certainly plate tectonics, which explains so much more than continental drift, is a contender for the prize. Our personal preference would perhaps be the gift of geological time. Nonetheless, the 1980 Alvarez paper provoked a revolution in the geosciences that extended far past the K-Pg boundary. A user-friendly treatment of this is found in Powell, J. L. 1998. *Night Comes to the Cretaceous*. W. H. Freeman and Company, New York, 250pp.

Age (Myr)

Height above or below K/T (cm)

Iridium abundance (ppb)

Figure 16.2. The iridium (Ir) anomaly at Gubbio, Italy. The amount of Ir increases dramatically at the clay layer to 9 parts per billion (ppb), and then decreases gradually above it, returning to a background count of about 1 ppb. On the right are numbers representing the thickness of the rock outcrop; on the left the time intervals (in millions of years (Myr)) and rock types are identified. Note that the vertical scale is *linear* close to the K-Pg boundary, but *logarithmic* away from the boundary, to show results well above and well below the boundary.

Shocked quartz and **microtektites** also came to be recognized as part of the fingerprint left by the asteroid. "Shocked quartz" is the name given to quartz that has been placed under such pressure that its molecular structure becomes deformed (Figure 16.5). It is now recognized that the kind of pressure that can cause such deformation could only be generated by impacts; indeed, shocked quartz is now known from many different impact sites, and has come to be recognized as a diagnostic criterion for meteor impacts.

Microtektites are small, droplet-shaped blobs of silica-rich glass. They represent material thrown up into the atmosphere in a molten state due to the tremendous energy released when a meteor strikes the Earth. Quick cooling occurs while they're still airborne and then they plummet down on Earth as a rain of solid, glassy blobs (Figure 16.6).

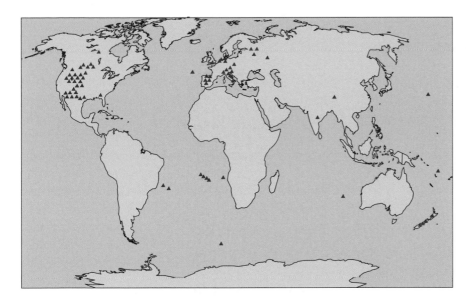

Figure 16.3. More than 300 K-Pg iridium anomalies are known around the world, about a third of which are shown here.

Figure 16.4. The iridium-bearing clay layer in Montana; one of the first localities on land where anomalous concentrations of iridium were discovered. Lines bracket thin Cretaecous–Paleogene boundary claystone. The sediments below the boundary claystone preserve dinosaurs; none has been found above the boundary claystone. To the left of the image are visible deposits of a small channel that carved into and eroded the claystone.

Box 16.1   Extinction

Paleontologists generally divide extinctions into two categories. The first are the so-called **background extinctions**, isolated extinctions of species that occur in an ongoing fashion. The second type is called **mass extinctions**. The latter certainly have caught media and the public's attention, and they appear to be something *qualitatively* as well as quantitatively different than background extinctions.

Background extinctions

Although background extinctions are less glamorous than mass extinctions, they are essential to biotic turnover: University of Tennessee paleobiologist M. L. McKinney has estimated that as much as 95 per cent of all extinctions can be accounted for by background extinctions. Isolated species disappear from a variety of causes, including out-competition (the edge), depletion of resources in a habitat, changes in climate, the growth or weathering of a mountain range, river channel migration, the eruption of a volcano, the drying of a lake, the spraying of a pesticide, or the destruction of a forest, grassland, or wetland habitat.

Dinosaur populations had a species-turnover rate of around 2 million years per species. This means that each species lasted about 2 million years, before a new one appeared and the old one disappeared.[a] Although some dinosaur extinctions coincided with earlier mass extinction events (such as those at the Triassic–Jurassic and Cretaceous–Paleogene boundaries), most dinosaurs fell prey to background extinctions. By far the majority of favorite and famous dinosaurs – *Maiasaura, Dilophosaurus, Protoceratops, Deinocheirus, Styracosaurus, Velociraptor, Iguanodon, Ouranosaurus, Allosaurus* (to name a tiny fraction) – were the victims of background extinctions. The ultimate dinosaur extinction didn't wipe out the total number of species accumulated over ~164 million years, it killed only the latest-evolved representatives of the group (see Figure 14.1).

Mass extinctions

Mass extinctions involve *large numbers* of *species* and *many types* of species undergoing *global extinction* in a *geologically short* period of time. None of these has a truly precise definition, because there are no fixed rules for mass extinctions. Indeed, how do we know that there even were mass extinction "events" and how can we recognize them? A compilation of invertebrate extinctions through time (Figure B16.1.1) shows that, although extinctions

[a] This is a simple statistical average for all of Dinosauria. It was not necessary for an older species to disappear before its descendant appeared.

The "smoking gun"

As early as 1981, a bowl-shaped structure 180 km in diameter, buried under many meters of more recent sediments, was reported from the Yucatan Peninsula of Mexico, in the region near the town of Chicxulub (translated approximately as "devil's tail"; Figure 16.7). Ten years later, drill cores taken through the structure revealed shocked quartz: Chicxulub was a buried impact structure.

At about the same time, an approximately 1 m thick sequence of glass was discovered in Haiti, suggesting that the source of the glass had to be somewhere relatively nearby. Its chemical composition was shown to be the same as the composition of the rocks that make up the Chicxulub structure.

The pieces really started falling into place. Several years earlier (1988) evidence of a tsunami in K-Pg deposits in the Gulf Coast region of Texas had been reported. The Chicxulub site was well situated to produce the tidal wave deposits recognized in the sedimentary record. Finally, the Chicxulub structure has been dated at 66.1 Ma, the time of the K-Pg boundary.

Further study of Chicxulub below the surface of the Earth, using sophisticated geophysical techniques, showed a bulls-eye pattern with a circular peak and large concentric rings around it, representing topography preserved in buried rocks below the surface (Figure 16.8). Interestingly characterize all periods (it is these that are termed background extinctions), there are intervals of time in which extinction levels are significantly elevated above background levels. Such intervals are said to contain the mass extinctions. Fifteen such intervals are recognized, of which five clearly towered above the others (Figure B16.1.1).

The 15 mass extinctions are classified into "minor," "intermediate," and "major" mass extinctions, on the basis of the amount of extinction that took place above background. In the entire history of life, only one extinction qualifies as "major"; that is, the **Permian–Triassic** (commonly called **Permo-Triassic**) extinction. The remaining four of the Big Five – including dinosaur extinction – are considered to have been "intermediate." The rest are considered "minor," although undoubtedly not to the organisms that succumbed during them.

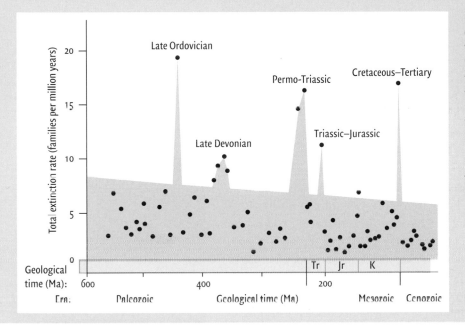

Figure B16.1.1. A compilation of extinction through time, taken from the work of D. M. Raup and J. J. Sepkoski. The most significant were the Late Ordovician (445–444 Ma), the Late Devonian (~373–359) Ma, the Permo-Triassic (251 Ma), the Triassic–Jurassic (201 Ma), and the Cretaceous–Paleogene (66.1 Ma). A sixth mass extinction may soon join the list of the "most significant": the one we're in the middle of right now! Tr, Triassic; J, Jurassic, K, Cretaceous.

enough, the northwest part of the outermost ring is broken through. The distinctive ring pattern suggests that a large asteroid, 10–15 km in diameter,[5] approached Earth from the southeast at a low angle of about 30°. The distribution of iridium, shocked quartz, and microtektites across the Western Interior of North America (north and west of the crater) reinforce the idea of a low-angle, directional impact (Figure 16.9).

What did the asteroid do to Earth when it struck? Numerous scenarios were proposed, most of them inspired by post-nuclear apocalyptic visions. Of these, only a few remain current.

- *Blockage of sunlight.* It was initially hypothesized that sunlight would have been blocked from the Earth for about three to four months. This would have caused a cessation of photosynthesis and a short-term temperature decrease (now called an **impact winter**).
- *Infra-red radiation pulse.* An 8 km asteroid, travelling at many times the speed of sound, and then being instantaneously brought to a complete stop would have to release the tremendous

[5] Recall that previous estimates of size were based upon the global distribution of the ejecta; this estimate was based upon the morphology of the impact site.

Figure 16.5. Shocked quartz from the terrestrial K-Pg boundary in eastern Montana. The etched angled lines across the face of a grain of quartz represent a failure of the crystal lattice along known crystallographic directions within the mineral. Grain is 70 μm across (1 μm = 10^{-6} m).

kinetic energy it possessed.[6] It has been proposed that tremendous amounts of energy in the form of infra-red radiation and heat must have been released immediately upon impact. The initial global heat release at ground zero might have been 50 to 150 times as much as the energy of the sun as it normally strikes Earth. One group of scientists likened this radiation at the Earth's surface to an oven left on the broil.

- *Global wildfires.* With so much instantaneous heat production, fires might have broken out spontaneously around the globe. Soot-rich horizons from K-Pg sites in Europe, North America, and New Zealand have been identified, in which the amount of the element carbon (the soot) was enriched between 100 and 10 000 times over background. The soot has been attributed to wildfires, perhaps the result of the infra-red heat pulse.

All of these catastrophic effects are short term, which means that they affected the globe for days, months, or at most a few years. In a longer-term sense, that is, on ecological time scales (10^4–10^6 years), climates were little affected by the asteroid impact. What we know of climates in the latest Cretaceous suggests that, considered overall, they did not differ significantly from those in the early Paleogene.

Figure 16.6. A microtektite.

Volcanic eruptions

The Earth wasn't completely peaceful – it never is – before the asteroid impact. In fact, in India, there was an unusual and very powerful volcanic episode going on – the **Deccan Traps**.

---

[6] D. S. Robertson and colleagues (2004) succinctly described it this way: "an asteroid 10–15 km in diameter, having a mass of ~1–4 × 10^{15} kg, arriving at tens of kilometers per second at an angle of perhaps 45° to Earth's surface, that produced a collapsed transient cavity ~80–100 km in diameter and a multi-ring basin 170–200 km in diameter on the Yucatán peninsula of Mexico." (Robertson, D. S. *et al.* 2004. *GSA Bull.*, **116**, 761.)

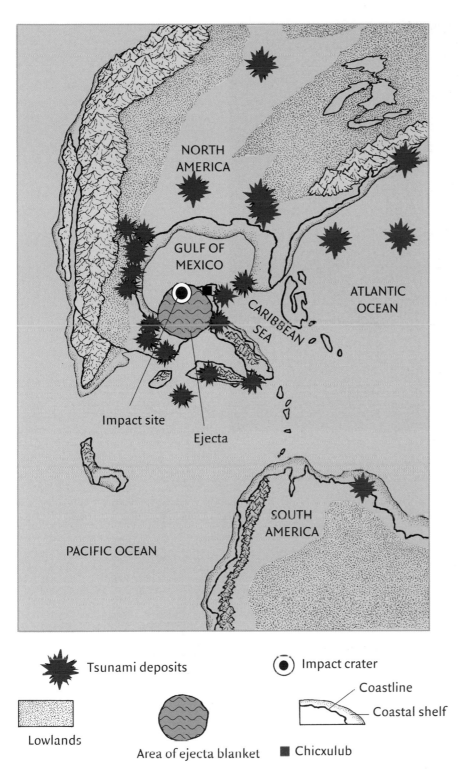

Figure 16.7. Paleogeographic map of the ground-zero region for the K-Pg asteroid, Yucatan Peninsula, Mexico. The geography of the region as we know it today is superimposed over the geography of the region at 65.5 Ma.

Tsunami deposits

Impact crater

Coastline

Coastal shelf

Lowlands

Area of ejecta blanket

Chicxulub

Deccan volcanism wasn't your basic Mt. Vesuvius–Krakatoa experience – short-term explosive eruption sending a geyser-like column of hot ash miles into the sky, followed by red-tinged sunsets from all the ash particles floating in the atmosphere – no, it was something potentially far deadlier.

Figure 16.8. Three-dimensional geophysical reconstruction of the remnants of the Chicxulub crater. A gravimeter measures subsurface changes in gravitational attraction of rocks under the town of Chicxulub. These variations in gravitational attraction show a large-scale bull-seye pattern of concentric rings, diagnostic of a meteor impact. North is toward the top of the page.

Crater showing oblique strike

Figure 16.9. Reconstruction of an asteroid impact with Earth. Planetary geologist P. H. Schultz and geobiologist S. L. D'Hondt suggest that the asteroid struck Earth at an angle of about 30°, coming from the southeast.

The Deccan Traps were a series of continental flood basalts: hot basaltic lava pouring through fissures in the Earth's crust. No big explosion, no huge dust cloud, but as the molten lava poured out over the Earth's surface, the pressure that the lavas were under when buried deep within the Earth, was released, and gases – certainly CO_2, SO_2, NO_2, and methane, among many others – were emitted

Figure 16.10. Latest Cretaceous basalt flows of the Deccan Traps, exposed on the western part of the Indian subcontinent.

into the atmosphere. Ultimately the Deccan Traps produced about >1.3 million km^3 of basalt, leaving a very palpable record on the Indian subcontinent (Figure 16.10).

The Deccan Traps are composed of three separate, closely timed (in a geological sense) events, the most important of which contributed a basaltic plateau some 3000 km thick on the western Indian subcontinent, right at the K–Pg boundary. That was a mighty pile of lava. Might this have affected the dinosaurs, too?

Biological record of the latest Cretaceous

Like any good murder mystery, no amount of volcanoes; bolides; hot climates; cold climates; natural (or unnatural) catastrophes – or any other causes – can explain *any* extinction until we understand the anatomy of the extinction itself. What went extinct? When did those extinction(s) occur? Was there one or were there many? Were they fast or were they slow? Who survived and who did not? These things were very poorly known at the time of the Alvarez hypothesis, which made their conclusion that much more radical. Now, with many years under our belts of research in this mass extinction, we have much better – but still incomplete – answers than we had in 1980.

Oceans

Continental seas and shelves

Because the shallow seas that covered large expanses of the continents receded before the K-Pg boundary, very few shallow marine deposits are preserved that record the last 2–3 million years of the Cretaceous. And because many groups of organisms lived and died in shallow continental seas and shelves, we lack data for such groups.

How well or badly fishes and sharks fared remains largely conjectural, although it is apparent that, where sharks, skates, and rays are understood (at this point, only in North America), there was a significant extinction.

The whale- and dolphin-like marine diapsids called **ichthyosaurs** (Figure 16.11a) are known to have disappeared well before the K-Pg boundary. Not so in the case of marine-adapted lizards called

Figure 16.11. Some of the better-known inhabitants of Cretaceous seas. Vertebrates are: (a) ichthyosaur (*Platypterygius*), (b) mosasaur (*Tylosaurus*), and (c) plesiosaur (*Elasmosaurus*). (d) The shelled, tentacled invertebrates toward the bottom of the drawing are cephalopod mollusks called ammonites.

mosasaurs (Figure 16.11b). Recent work suggests that these went extinct geologically abruptly, at the end of the Cretaceous. More equivocal is the record of **plesiosaurs**, the long-necked, Loch Ness-type, fish-eating diapsids of the Jurassic and Cretaceous (Figure 16.11c), for whom there are, Loch Ness notwithstanding, no credible post K-Pg records.

Among fossil invertebrates, perhaps the most famous group is the ammonites (Figure 16.11d). Ammonites lived right up to the K-Pg boundary, before finally going extinct. Another important group of invertebrates are the bivalves. Careful study has shown that, with one exception (which went extinct much earlier), 63 per cent of all bivalves went extinct sometime within the last 10 million years of the Cretaceous. The record is, unfortunately, not yet more precise than this, but it does show that the extinction took place without regard for latitude: bivalves in temperate regions were just as likely to go extinct as those in the tropics.

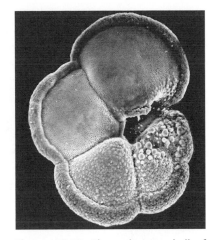

Figure 16.12. The carbonate shell of a modern planktonic (free-swimming) foraminifer, *Globorotalia menardii*. The long dimension is 0.75 mm.

Marine microorganisms

Because of the richness of their fossil record, **foraminifera**, marine microscopic, shelled, single-celled organisms that are either **planktonic** (living in the water column) or **benthic** (living within sediments) have dominated discussions of K-Pg boundary events (Figure 16.12). **Micropaleontologists** studying foraminifera have shown persuasively, since as early as the late 1970s (and in many studies thereafter, including the observations of the Alvarez team at Gubbio) that the planktonic foraminiferan extinction was abrupt, with only a few species crossing the boundary into the Paleocene.

An entirely different group of marine microorganisms, **calcareous nanofossils**, also shows an abrupt extinction. It is safe to say that most (but not all!) paleontologists working with marine microorganisms are inclined toward a catastrophic view of the extinction.

The "Strangelove" ocean

Some of the most exciting results came from a series of studies of K-Pg oceanic **primary production**; that is, the amount of organic matter synthesized by organisms from inorganic materials and sunlight.

At the K-Pg boundary there was observed a rapid and complete breakdown in nutrient cycling between surface and deep waters, to less than 10 per cent of what it had been. For some oceanographers, this signaled that the world's oceans were effectively all but dead. For the next 1.5 million years, **nutrient cycling** remained at levels well below those preceding the original drop.[7]

Obviously, nutrient cycling is fundamental to oceanic health. With oceans covering 75 per cent of the Earth's surface (or even more during the many high sea levels experienced during Earth history), it would not be an exaggeration to state that Earth's marine and terrestrial ecosystems are dependent upon these great cycles. In short, dead oceans could have threatened the whole biosphere.

Terrestrial record

For better or worse, virtually all of what we know of the K-Pg boundary on land also comes from the Western Interior of North America (Figure 16.13). There, several well-studied, complete sections have provided the best insights available into the dynamics of the extinction.

---

[7] The moribund K/T oceans were called "Strangelove" oceans after Dr. Strangelove, Peter George's brittle, grotesque character, played by Peter Sellers in the eponymous film, who was troubled by the possibility of a scorched, post-nuclear world.

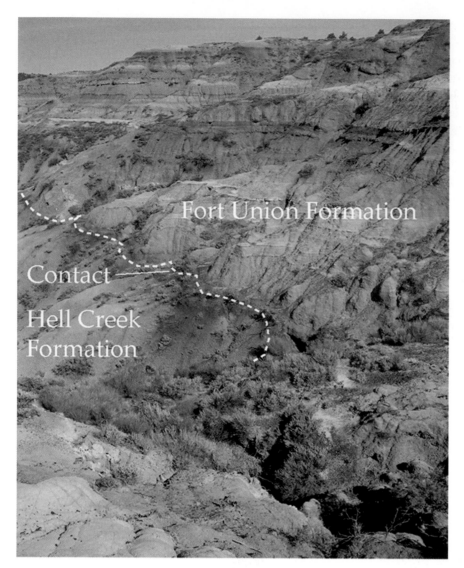

Fort Union Formation

Contact

Hell Creek
Formation

Figure 16.13. The K-Pg boundary in eastern Montana, USA. The boundary is midway up the butte, right at the dotted line. Below is the dinosaur-bearing latest Cretaceous Hell Creek Formation; above is the Paleogene Fort Union Formation. No dinosaurs have ever been found in the Fort Union Formation.

Plants

The plant fossil record in the Western Interior has two major components, a **palynoflora** (spores and pollen) and a **macroflora** (the visible remains of plants, especially leaves; Figure 16.14). After 15 years of intensive scrutiny, both records agree nicely with each other and both records indicate that a major extinction occurred geologically instantaneously at the K-Pg boundary.

Interestingly, pollen that is typical of early Paleocene time does not immediately follow the extinction of the Cretaceous pollen. Instead there is a high concentration of fern spores just after the iridium anomaly, suggesting that, immediately after the extinction of the Cretaceous plants, there was a "bloom" of fern growth, interpreted to be a pioneer community growing on a devastated post-impact landscape. In time, the fern flora gave way to a more diverse angiosperm flora characteristic of the early Paleocene. The North American plant record, however, is clear: the pollen show a 30 per cent extinction right at the K-Pg boundary.

Outside North America, an interesting southern high-latitude flora is known from New Zealand. There, an abrupt pollen and spore extinction, as well as the fern spike, are also known. In short, the pollen record suggests that the terrestrial K-Pg boundary was characterized by global deforestation.

Figure 16.14. Plant fossils. (a) Late Cretaceous leaf. The leaf is from an angiosperm that became extinct at the K-Pg boundary. The specimen is from just outside Marmarth, North Dakota, USA. (b) Pollen grains belonging to the genera *Proteacidites* (1) and *Aquilapollenites* (2), both important genera in measuring the moment of the terrestrial K-Pg extinction. *Proteacidites* is about 30 μm across; *Aquilapollenites* is about 50 μm.

The megafloral record based upon 25 000 plant specimens for the Western Interior of North America shows that while some environmental changes caused extinctions earlier than the K-Pg boundary, a 57 per cent extinction took place precisely at the boundary, exactly correlated with the pollen extinction and iridium anomaly. If slightly more time is included in the analysis, the extinction climbs to a whopping 78 per cent!

Which number is right? Probably neither, and the true extinction number is likely somewhere between the two. Although 57 per cent went extinct directly at the boundary, some of those that *appear* to have gone extinct before the boundary might have actually gone extinct right at the boundary as well, but were not preserved. Still, these are large collections from many localities, and so the expectation is that the real extinction was closer to 57 per cent than to 78 per cent.

The extinction of >57 per cent of the known Cretaceous angiosperms from the Western Interior of North America suggests that, as suspected, the fern "bloom" may have been a response to the absence of flowering plants that would normally have occupied and colonized the ecosystem.

Vertebrates

Vertebrates suffered their fair share of extinction at the K-Pg boundary: the poster children are of course dinosaurs. But as we've said, lots of animals went extinct, and some of these are actually far better suited than dinosaurs to understanding what might have happened.

Patterns of **survivorship**, that is, who made it across the boundary and who did not, obviously might shed light upon events 66.1 Ma. And indeed, these data were extracted from the K-Pg vertebrate fossil record of the US Western Interior in 1990 by paleontologists J. D. Archibald and L. J. Bryant. Overall, an important pattern emerged: organisms that lived in aquatic environments (that is, rivers and lakes) showed up to 90 per cent survival, whereas organisms living on land showed as little as a 10 per cent survivorship. Thus the extinction seems not to have drastically affected aquatic organisms such as fishes, turtles, and crocodiles, but apparently wreaked havoc among terrestrial organisms such as mammals and, of course, dinosaurs. Several other survivorship patterns, not as statistically robust, also appeared in the data: small vertebrates are favored over large vertebrates, ectotherms over endotherms, and non-amniotes over amniotes (Figure 16.15).

Two other recent studies have corroborated these results: a 2014 study of Western Interior amphibians, such as salamanders, showed a decline only in the last 400 kyr[8] of the Cretaceous. Snakes and lizards, fully terrestrial animals, evidently also underwent a major, abrupt extinction.

Mammals

Mammals were alive and well in the Late Cretaceous, having first appeared on Earth during the Late Triassic. Mammals turn out to be extremely useful fossils: their small teeth preserve well and show a lot of morphology, allowing us to precisely identify them, obtain insights into aspects of their diet and behavior, and use them biostratigraphically.

In general, as we have seen (Chapter 14), they were small and possibly nocturnal throughout the "first two-thirds of mammalian history," as paleomammalogist W. A. Clemens trenchantly put it. By the end of the Cretaceous, however, they were undergoing some important changes, as recorded in the Western Interior of North America. University of Washington paleontologist G.P. Wilson has reconstructed the pattern of their extinction.

Some ~500 kyr before the boundary, Wilson recognized a first episode of faunal change, involving decreases in body size and possible migration out of the region. A second wave of change occurred ~200 kyr before the boundary, which consisted of a series of apparent stepwise disappearances of fossil mammals. He correlates these extinctions with short-term temperature changes identified from the plant record, although he is unable to rule out the possibility that they are artifacts of the Signor–Lipps effect (Box 16.2). Whichever the case, the upshot was significant: mammals show in total a 75% extinction in this region. But the question of step-wise extinction vs. Signor–Lipps artifact looms over these data; thus, we will return to them a bit later in this chapter when we try to figure out what it all means.

---

[8] kyr = 1000 years.

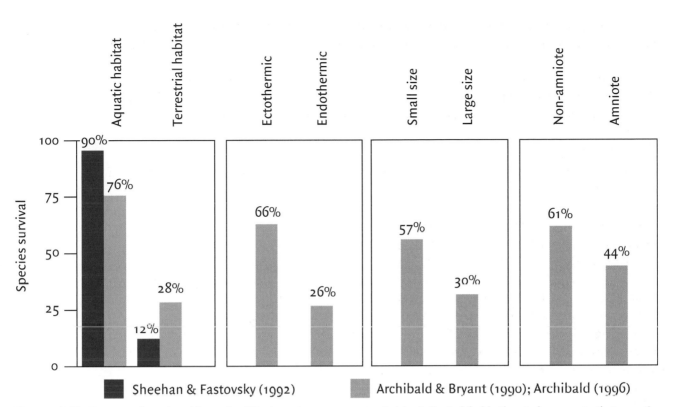

Figure 16.15. Patterns of survivorship at the K-Pg boundary as reconstructed by J. D. Archibald. The study suggests that aquatic habitat, ectothermy, small size, and the absence of an amnion were qualities that statistically facilitated survival across the K-Pg boundary. Of these, aquatic habitat may have been the most important; in a separate publication, D. E. Fastovsky and P. M. Sheehan reconstructed the aquatic and land dwelling survivorship pattern as even more extreme than that proposed by Archibald, with land-dwelling organisms showing only a 12 per cent survivorship, but aquatic organisms showing 90 per cent survivorship (shown in red).

Dinosaurs

Non-avian dinosaurs are difficult animals to study (Box 16.3) and for many years, no scientific study of dinosaurs at the K-Pg boundary was ever carried out. Inexplicably, although no data were ever published to show this, it was long thought that dinosaurs gradually died off about 10 million years before the boundary.

While that is clearly not correct, the exact pattern of how they went out remains controversial.[9] While the overall trajectory of dinosaur diversity through time is ever-increasing (see Figure 14.2), data obtained in 2012 suggest that, globally, dinosaurs may have experienced a decline in diversity several million years before the K-Pg boundary. Indeed, it may have depended upon who you were and what you did: theropods may not have declined, while ceratopsians and hadrosaurs may have. These fluctuations in their biodiversity, however, were no greater than many that had occurred in the preceding ~164 million years of dinosaurs on Earth. Nobody has identified anything that makes this particular fluctuation in dinosaur populations unique.

Decline in biodiversity or no, we can be sure that there were very much more than just a few dinosaurs hanging on right before the end of the Cretaceous. Indeed, non-avian dinosaurs were still a very diverse, abundant, and important terrestrial group of vertebrates within <500 kyr of the K-Pg boundary. How quickly, then, were they eliminated from Earth?

---

[9] See for example, Chapter 14, footnote 4.

Box 16.2. Getting fooled by the fossil record: the Signor–Lipps effect

A key question in any reconstruction of an extinction event is how fast or slowly it occurred. But it turns out that this can be very tricky to ascertain, in part due to a problem called the "Signor–Lipps effect," named after paleobiologists Phil Signor and Jere Lipps, who first articulated the problem.

We begin with a thought experiment. We take a handful of various coins, and distribute them on a table (Figure B16.2.1a). If they are randomly distributed, the diversity of coins (pennies, nickels, dimes, and quarters) is unchanging all over the table. Now, we draw an arbitrary line right through the center, as in Figure B16.2.1b. That line cuts across the full diversity of coins; that is, because the coins are randomly distributed on the table, anywhere one puts that line should reflect the same diversity (all coins represented).

Let's suppose the coins represent fossils, and the different types of coins represent different types of fossil animals. Now, think stratigraphically: imagine that the bottom of the pictures in Figure B16.2.1 is older, and the top of the pictures is younger; it can be said that the diversity of fossil animals does not change throughout the full amount of "time" represented by the table (from bottom to top). And, where the line is arbitrarily drawn, the diversity of fossil animals (represented by coins) is still constant (because nothing has changed).

Now, let's say that the arbitrary line actually represents an extinction boundary (Figure 16.2.2; E is of course the same line as in Figure 16.2.1b; here, however, it is labeled "E" and represents an "Extinction boundary" in the figure; thus the "extinct" coins above the boundary are grayed).

(a)

(b)

Arbitrary line

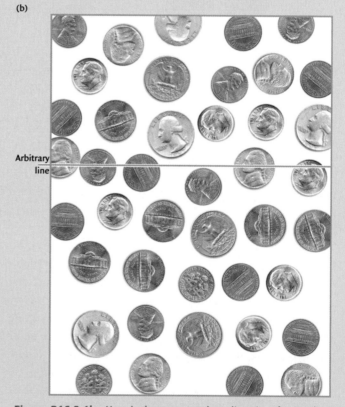

Figure B16.2.1a. A diversity of coins (pennies, nickels, dimes, and quarters) representing the diversity of a fauna or flora. The coins are randomly distributed.

Figure B16.2.1b. Here is the same random diversity of coins, but with an arbitrary line drawn across that same random distribution of coins/taxa.

So while there are no fossil animals above the boundary (they go extinct), the diversity of fossil animals leading up to the boundary is still unchanged and constant. The "extinction," therefore is *abrupt*; diversity is unchanged preceding the "extinction" (the arbitrary line), and then after it, nothing is present (e.g., survives).

You are the paleontologist, though, and you'd like to reconstruct this extinction – and if these coins were fossils, you would have no idea how they are distributed through time. So you ask the basic question: was the extinction abrupt or gradual?

If it is *gradual*, you correctly assume, the diversity of fossils (coins) will decrease as you approach the extinction boundary. So you look at successively smaller intervals approaching the boundary, and what you discover is that as you get closer and closer to the boundary, the diversity of fossils (coins) starts decreasing: first one goes "extinct," then another, and finally, as you get quite close to the boundary, only one type of fossil is left (Figure B16.2.2). The extinction of the fossils, you therefore conclude, is stepwise through time; while you started with four different types of coins (representing fossils) in the stratigraphically lowest parts of the section, by the time you get to the boundary, there is only one type of coin present. So that looks like a gradual extinction, with the fossils (again, the coins) going extinct in stepwise fashion.

This is the Signor–Lipps effect. You drew an arbitrary line across the full diversity of fossils (coins), and so you know that diversity did *not* decrease. Yet, as you attempted to reconstruct the pattern of extinction, diversity appeared to decrease as you approached the boundary – as you expected that it would if the extinction were not abrupt, but rather stepwise (gradual). You've been fooled!

So we return the question of the mammal extinction: as the K-Pg boundary is approached, mammal diversity decreases in the last few hundred thousand years. Should we take this as a stepwise (gradual) extinction, or is this just an artifact of the Signor–Lipps effect, and the extinction was actually abrupt? G. P. Wilson, studying the patterns of mammal extinction at the K-Pg boundary, and fully aware of the Signor–Lipps effect, was rightly unsure of which was the signal that he had observed.

Figure B16.2.2. Here, the arbitrary line from Figure 16.2.1b represents an extinction boundary (E) if we think about this stratigraphically. The coins in this exercise represent a fauna and/or floral and they are visible in the figure to show that it is the same image as Figure B16.2.1. However, they are greyed above E to indicate that they would no longer be there if there was actually an extinction. Between I_3 and the "extinction" E, the largest interval of time, the full diversity is represented. Between I_2 and E, a smaller interval, closer to the boundary, there is a slight decrease in diversity. Between I_1 and E, the smallest interval – and closest to the boundary – there is yet a further decrease in diversity. But, since we know that diversity is actually unchanged (we have not altered our original distribution of coins/taxa), the apparent decrease in diversity as the boundary is approached is an artifact of how we study, and not truly a gradual or stepwise extinction. As we saw in Figure 16.2.1b, diversity did not truly change as the "extinction boundary" is approached; the extinction, therefore, is abrupt (even it if looked gradual as an artifact of our flawed method of trying to understand it).

Box 16.3   Dinosaurs: all wrong for mass extinctions

What are some of the problems with reconstructing changes in dinosaur populations over time? For one thing, dinosaurs are, by comparison with foraminifera for example, large beasts and, more importantly, not particularly common.[a] For this reason, the possibility of developing a statistically meaningful database is impractical, and rigorous studies of dinosaur populations are very hard to carry out. Just counting dinosaurs can be difficult. Mostly, you don't find complete specimens, and adjustments have to be made. For example, if you happen to find three vertebrae at a particular site, they might be from one, or two, or three individuals. The only way to be sure that they belong to a single individual is to find them articulated. Suppose they are not; then one must speak of **minimum numbers of individuals**, in which case the three vertebrae would be said to represent one individual: that would be the minimum number of individual dinosaurs that could have produced the three vertebrae. On the other hand, if one found two left femora, then the minimum number of individuals represented would be two.

It would be nice to use all the specimens that have been collected in the past 170 years of dinosaur studies in a study of changes in dinosaur diversity. Unfortunately, dinosaur specimens have been collected in the past generally because they are either beautiful or rare; hardly criteria for ensuring that an accurate census of dinosaur populations has been performed. So any study that really is designed to get an accurate census of dinosaur abundance or diversity at the end of the Cretaceous must begin by counting specimens in the field, which is a labor-, time-, and cost-intensive proposition.

Then, of course, the taxonomic level at which to count dinosaurs can create problems. Suppose that two specimens are found; one is clearly a hadrosaurid and the other is an indeterminate ornithischian. The indeterminate ornithischian might be a hadrosaurid, in which case we should count two hadrosaurids. But then again it might not (because its identity is indeterminate), in which case calling it a hadrosaurid would give us more hadrosaurids in our survey than actually existed. On the other hand, calling both specimens "ornithischians" is quite correct, but not very informative, if we hope to track the survivorship patterns of *different types* of dinosaurs.

Finally, within the sediments themselves, problems of correlation exist. Suppose that, in Montana, we record the last (highest level) dinosaur in the Jordan area and then record the last (highest) dinosaur in the Glendive area, about 150 km away from Jordan. Can these two dinosaurs be said to have died at the same time? How could one possibly know? Suppose that in fact these dinosaurs died 200 years apart. An interval of 200 years, viewed from a vantage point of 65 million years, is literally a snap of the fingers. Yet 200 years is a long time when one is considering an instantaneous global catastrophe that ideally is measured in milliseconds.

[a] How rare are dinosaurs in this part of the world? Of course, we cannot know the density of dinosaurs within the rocks, but their surface density was calculated by sedimentologist P. White and colleagues, using the Sheehan et al. database (Fastovsky and Sheehan, 1997, p. 527), reported, "White and Fastovsky calculated that 0.000056 dinosaurs are preserved per m^2 of outcrop. Considered more realistically, in a statistical sense one must search a 5 m wide path of exposed rock that is 4 km long to find a single dinosaur fragment identifiable to family level (or lower)." (Fastovsky, D. E. and Sheehan, P. M. 1997. Demythicizing dinosaur extinctions at the Cretaceous–Tertiary boundary. In, Wolberg, D. L., Stump, E. and Rosenberg, G. D. (eds.) Dinofest International. Academy of Natural Sciences, Philadelphia, p. 527.)

In the late 1980s and through the 1990s, field-based studies were finally designed and carried out to determine the rate of dinosaur extinction. All of them took place in the American West: two on what had been a low-lying coastal plain in what is now eastern Montana and western North Dakota, and one in an intermontane basin in what is now Wyoming, which formed in the ancestors of the present-day Rocky Mountains, rising to the west (Figure 16.16) The three studies all concluded the same thing: the dinosaur extinction was geologically abrupt.

The two studies in the coastal plains of what is now eastern Montana and western North Dakota were quantitative censuses of dinosaur diversity during the last 1.5 million years of the Cretaceous. One looked at the **ecological diversity**; that is, the proportion of the total dinosaur

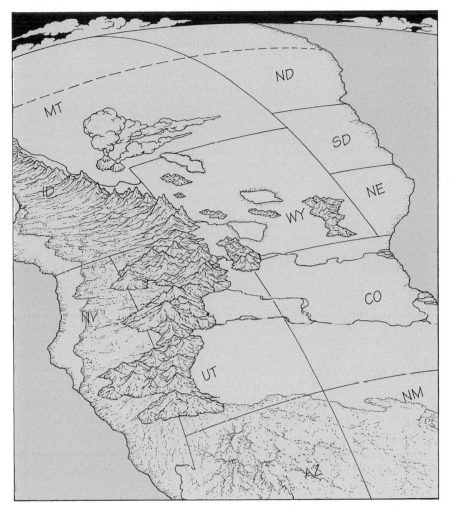

Figure 16.16. Paleogeography of the Western Interior of the USA, as it would have looked during Late Cretaceous time.

population taken by each of eight families of dinosaurs. The other counted genera of all vertebrates through the last 1.5 million years of the Cretaceous in that region, looking for changes in either abundance or diversity. Both demonstrated that, within about 160 000 years of the K-Pg boundary, neither ecological diversity nor abundance and generic diversity changed (Figure 16.17).

The last of the three studies, carried out in what was an ancestral Rocky Mountain intermontane basin (Figure 16.18), utilized an approach very similar to the coastal plain study of vertebrate genera described above. And the results from that study were much the same as those from the other studies: the extinction of the dinosaurs was geologically abrupt. Major extinctions occurred in most groups, but particularly in dinosaurs and mammals (Figure 16.18). A key point, however, is that none of these studies can distinguish whether the extinction took every day, or whether it took only the last minute, of that last 150 000 years of the Cretaceous.

In summary, the very limited data from the Western Interior of the USA strongly indicate an abrupt end for the non-avian dinosaurs. Only time and much further study will enable us to integrate other dinosaur-bearing localities from around the world into what is already known of North America.

The study of *avian* dinosaurs brings its own challenges, because bird fossils are so rare. We know that this group suffered very significant extinctions; all Cretaceous toothed birds – four major groups

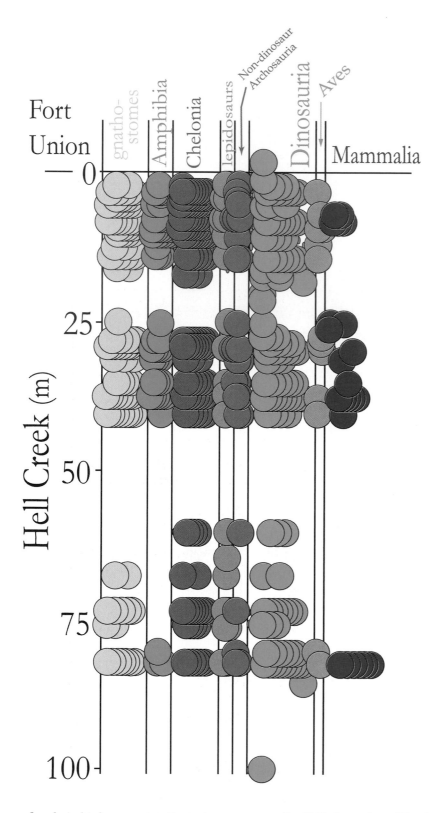

Figure 16.17. Evidence for sudden extinction of the dinosaurs. The vertical axis shows meters through the Hell Creek Formation, the uppermost unit in the Western Interior of the USA. "0" is the K-Pg boundary. The horizontal axis shows various vertebrate groups (including dinosaurs) that are found within the Hell Creek. Virtually all vertebrate groups are present throughout the thickness of the Hell Creek; there is no gradual decrease in the groups as the boundary is approached. The data indicate that the extinction of the dinosaurs and other vertebrates at 65.5 Ma was geologically instantaneous.

of archaic birds – went extinct at or very near the K-Pg boundary. We also know that some of the modern groups of birds – for example, ducks and their relatives – had begun to evolve before the close of the Cretaceous, and obviously survived whatever killed off their archaic relatives (see Figure 8.17). The survival of the modern groups may have been tied into an ability to fly and seek

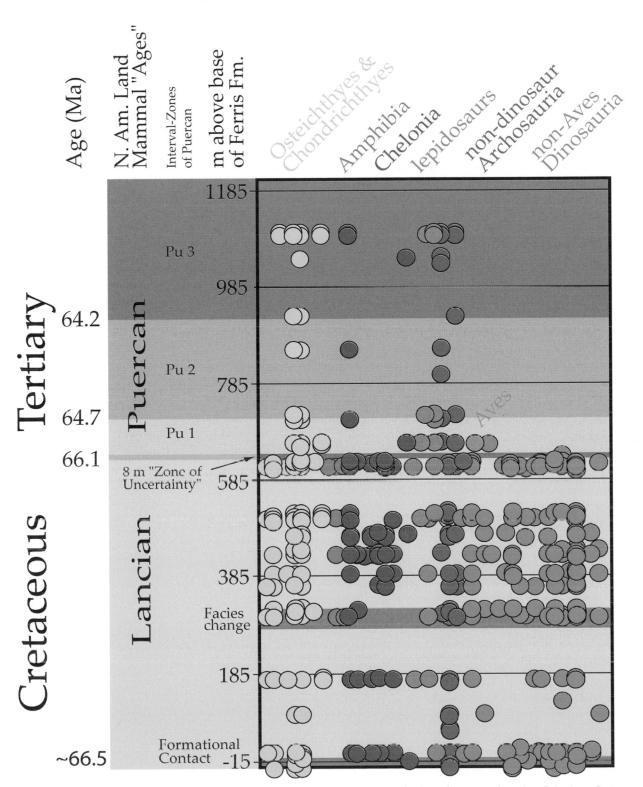

Figure 16.18. Evidence for sudden extinction of the dinosaurs in intermontane basin sedimentary deposits of the Late Cretaceous of the US Western Interior, as demonstrated by Lillegraven and Eberle (1999). The vertical axis shows meters through the K-Pg Ferris Formation and time (Ma). The horizontal axis shows various vertebrate groups (including dinosaurs) that were found in the study. Virtually all vertebrate groups are present below the boundary; there is no gradual decrease in the groups as the boundary is approached. After the boundary, diversity is drastically reduced. The data indicate that the extinction of the dinosaurs and other vertebrates 65.5 Ma was geologically instantaneous.

Table 16.1. Proposed causes for the extinction of the dinosaurs (based in part upon Benton, 1990)

I. Proposed biotic causes

A. *Medical problems*
  (a) Slipped disks in the vertebral column causing dinosaur debilitation
  (b) Hormone problems
    (1) Overactive pituitary glands leading to bizarre and non-adaptive growths
    (2) Hormonal problems leading to eggshells that were too thin, causing them to collapse in on themselves in a gooey mess
  (c) Decrease in sexual activity
  (d) Blindness due to cataracts
  (e) A variety of diseases, including arthritis, infections, and bone fractures
  (f) Biting insects carrying diseases that did dinosaurs in over hundreds of thousands to millions of years
  (g) Epidemics leaving no trace but wholesale destruction
  (h) Parasites leaving no trace but wholesale destruction
  (i) Change in ratio of DNA to cell nucleus causing scrambled genetics
  (j) General stupidity

B. *Racial senescence*
  This is the idea, no longer given much credence, that entire lineages grow old and become "senile," much as individuals do. Thus, in this way of thinking, late-appearing species would not be as robust and viable as species that appeared during the early and middle stages of a lineage. The idea behind this was that the dinosaurs as a lineage simply got old and the last-living members of the group were not competitive for this reason

C. *Biotic interactions*
  (a) Competition with other animals, especially mammals, which may have out-competed dinosaurs for niches, or perhaps ate their eggs
  (b) Overpredation by carnosaurs (who presumably ate themselves out of existence)
  (c) Floral changes
    (1) Loss of marsh vegetation (presumably the single most important source of food)
    (2) Increase in deforestation (leading to loss of dinosaur habitats)
    (3) General decrease in the availability of plants for food with subsequent dinosaur starvation
    (4) The evolution in plants of substances poisonous to dinosaurs
    (5) The loss from plants of minerals essential to dinosaur growth

II. Proposed physical causes

A. *Atmospheric causes*
  (a) Climate became too hot so they fried
  (b) Climate became too cold so they froze
  (c) Climate became too wet so they got waterlogged
  (d) Climate became too dry so they desiccated
  (e) Excessive amounts of oxygen in the atmosphere caused:
    (1) changes in atmospheric pressure and/or atmospheric composition that proved fatal; or
    (2) global wildfires that burned up the dinosaurs
  (f) Low levels of CO_2 removed the "breathing stimulus" of endothermic dinosaurs
  (g) High levels of CO_2 asphyxiated dinosaur embryos
  (h) Volcanic emissions (dust, CO_2, rare earth elements) poisoned dinosaurs one way or another

B. *Oceanic and geomorphic causes*

   (a) Marine regression produced loss of habitats

   (b) Swamp and lake habitats were drained

   (c) Stagnant oceans produced untenable conditions on land

   (d) Spillover into the world's oceans of Arctic waters that had formerly been restricted to polar regions, and subsequent climatic cooling

   (e) The separation of Antarctica and South America, causing cool waters to enter the world's oceans from the south, modifying world climates

   (f) Reduced topographical relief and loss of habitats

C. *Other*

   (a) Fluctuations in gravitational constants leading to indeterminate ills for the dinosaurs

   (b) Shift in the Earth's rotational poles leading to indeterminate ills for the dinosaurs

   (c) Extraction of the Moon from the Pacific Basin perturbing dinosaur life as it had been known for 140 million years (!)

   (d) Poisoning by uranium from Earth's soils

D. *Extraterrestrial causes*

   (a) Increasing entropy leading to loss of large life forms

   (h) Sunspots modifying climates in some destructive way

   (c) Cosmic radiation and high levels of ultraviolet radiation causing mutations

   (d) Destruction of the ozone layer, causing (c)

   (e) Ionizing radiation as in (c)

   (f) Electromagnetic radiation and cosmic rays from the explosion of a supernova

   (g) Interstellar dust cloud

   (h) Oscillations about the galactic plane leading to indeterminate ills for the dinosaurs

   (i) Impact of an asteroid (for mechanisms, see the text)

refuge, it may have been dumb luck, or it may have been something that we have not yet landed upon. Whatever the case, the sparse fossil record of birds makes understanding the dynamics of the K-Pg boundary as it relates to avian dinosaurs very difficult to interpret.

Extinction hypotheses
Testability and parsimony

Much – indeed, most – of what has been proposed to explain the extinction of dinosaurs does not even possess the basic prerequisites for a viable, scientific theory (see Part I and Chapter 3).

(1) *The hypothesis must be testable.* A scientifically meaningful hypothesis must be *testable*; that is, it must have predictable, observable consequences. Without testability, there is no way to falsify a hypothesis and, in the absence of falsifiability, we are then considering **belief systems** rather than scientific hypotheses. If an event occurred and left no traces that could be observed (by whatever means available), science is simply not an appropriate tool to investigate the event.

(2) *The hypothesis must be parsimonious.* By this we mean that it must explain as many of the events as possible. If each step of the event (or events) requires an additional *ad hoc* explanation, our hypotheses lose strength. They are strongest when that explanation which explains the most observations is used. If we can explain all that we observe at the K-Pg boundary with a single hypothesis, we have produced the most parsimonious hypothesis and it has a good chance of being correct.

Box 16.4 The real reason the dinosaurs became extinct

Not every published hypothesis has been serious. In 1964, for example, E. Baldwin suggested that the dinosaurs died of constipation. His reasoning went as follows. Toward the end of the Cretaceous, there was a restriction in the distribution of certain plants containing natural laxative oils necessary for dinosaur regularity. As the plants became geographically restricted, those unfortunate dinosaurs living in places where the necessary plants no longer existed acquired stopped plumbing and died hard deaths. The same year, humorist W. Cuppy noted that "the Age of Reptiles ended because it had gone on long enough and it was all a mistake in the first place," a view with which many characters in the *Jurassic Park* series would have probably agreed.

The November, 1981 issue of the *National Lampoon* offered its explanation, entitled "Sin in the Sediment." The Christian right was the target:

It's pretty obvious if you just examine the remains of the dinosaurs . . . Dig down into older sediments and you'll see that the dinosaurs were pretty well off until the end of the Mesozoic. They were decent, moral creatures, just going about their daily business. But look at the end of the Mesozoic and you begin to see evidence of stunning moral decline. Bones of wives and children all alone, with the philandering husband's bones nowhere in sight. Heaps of fossilized, unhatched, aborted dinosaur eggs. Males and females of different species living together in unnatural defiance of biblical law. Researchers have even excavated entire orgies – hundreds of animals with their bones intertwined in lewd positions. Immorality was rampant.

In 1983, sedimentary geologist R. H. Dott Jr. published a short note in which he vented his frustrations with the pollen season, suggesting that it was pollen in the atmosphere that killed the dinosaurs. He called his contribution "Itching Eyes and Dinosaur Demise."

The issues raised by the *National Lampoon* were compelling enough to again be raised in 1988 by the *Journal of Irreproducible Results*. There, L. J. Blincoe developed a then-new hypothesis about the "fighting dinosaurs" specimen (see Figure 6.25):

A thorough but cursory review of fossil specimens. . . has revealed a unique fossil found in the Cretaceous "beds" of Mongolia in 1971. The fossil featured two different species of dinosaur, one a saurischian carnivore (*Velociraptor*), the other an ornithischian herbivore [*sic*] (*Protoceratops*), in close association at the moment of their deaths. Prejudiced by their preconceived notions of dinosaur behavior, paleontologists have almost unanimously interpreted this find as evidence of a life and death struggle [*see Chapter 11*]

. . . However, an alternative theory has now been developed which not only explains this unusual fossil, but also answers the riddle of the dinosaurs' disappearance. Quite simply, when their lives were ended by sudden catastrophe, these two creatures were locked together . . . in a passionate embrace. They were, in fact, prehistoric lovers.

Extinction hypotheses

In Table 16.1 we present about 80 years of serious, published proposals designed to explain the extinction of the dinosaurs (although see Box 16.4). The majority was published within the past 40 years. Consider each; you don't need to be a professional paleontologist to reject most of them, for most fail to meet the twin criteria for science enumerated above.

Any theory that purports to explain K-Pg events in a meaningful way must also explain the other events associated with the boundary. With that in mind, the hypothesis that an asteroid impact caused the events at the K-Pg boundary becomes an interesting and plausible hypothesis.

Figure B16.4.1. O'Donnell's take on the cause of extinction of dinosaurs.

The implications of this startling interpretation are clear: dinosaurs engaged in trans-species sexual activity. In doing so they wasted their procreative energy on evolutionarily pointless copulation that resulted in either no offspring or, perhaps on rare occasions, in bizarre, sterile mutations (the fossil record is replete with candidates for this later category.[a]

For ultimate cause(s) of the extinction, however, we think O'Donnell's perspective published in the *New Yorker* says it all (Figure B16.4.1).

[a] **Blincoe, L. J. 1988**. Journal of Irreproducible Results, 33, 24.

Does the idea that an asteroid impact caused the K-Pg extinctions have predictable consequences?

Clearly, the answer to the above question is "Yes." Firstly, if the asteroid produced global consequences, evidence for it should be visible globally. After 35 years and hundreds of thousands of person-hours of research, the evidence for global influence of the K-Pg boundary asteroid impact is undeniable (see Figure 16.3). And it delivers on the promise of predictable consequences in terms of the extinction.

In the case of the bivalves and plants, the fact that the extinctions took place regardless of latitude is strong evidence that those extinctions were due to a global effect, which was apparently unrelated

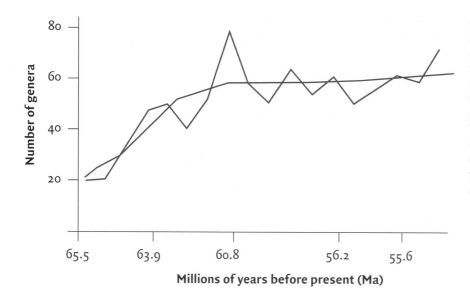

Figure 16.19. Radiation of mammals after the K-Pg boundary. The vertical axis shows species of mammals; the horizontal axis shows time. The dark blue line is the exact counts of genera at any particular time; the red line is in the inferred, general shape of mammalian diversity. The interpretation (red line) shows a rapid increase in the number of genera of mammals during the first 3–5 million years or so of the Paleogene, followed by a leveling off of diversity.

to climate. Had climate, for example, been involved as a causal agent, one might expect to see latitudinal changes in the patterns of extinction, but as we have seen such is not the case.

Other evidence comes from the rate at which the extinction took place. If the asteroid really caused the extinction, the event should have been what, in 1981, W. A. Clemens (after W. S. Gilbert) dubbed a "short, sharp shock." Patterns of *gradual* extinction would falsify the asteroid impact as a causal agent, whereas patterns of *abrupt* or *catastrophic* extinction would support the hypothesis.

Recovery

Catastrophic events tend to leave a distinctive mark: organisms that first colonize deserted **ecospace** tend to **speciate** rapidly, to be rather small, and to adopt **generalist** lifestyles (rather than developing a highly specialized behavior such as exclusively meat-eating or herbivorous behaviors). Such organisms are termed **disaster biotas** and are known in vertebrate, invertebrate, and plant communities.

At the K-Pg boundary, recall that in the plant realm, the initial colonizing flora was a short-lived growth of ferns. These have been interpreted as a disaster flora developing in a disrupted and unstable landscape.

The mammals that evolved throughout the recovery at the K-Pg boundary were extremely small generalists. They speciated rapidly, taking about 5 million years (Figure 16.19) to evolve a range of sizes and develop **specializations** (such as herbivores and carnivores). Theirs is the pattern of a disaster fauna that came through a catastrophic event and inherited deserted ecospace.

Recent phylogenies based upon the rates of molecular evolution have suggested that modern mammals' roots are to be found within the Cretaceous, showing that the mammalian radiation that characterized the Paleogene was actually well underway during the latest Cretaceous. In fact, the far-distant ancestors of modern mammals were likely scurrying around during the Late Cretaceous, but the rapid species' turnovers of the earliest Paleogene disaster faunas show the clear mark of a catastrophic event.

Does the asteroid impact hypothesis explain all the data?

In fact, there does appear to be a correlation between extinction selectivity and the asteroid as a causal agent in the extinctions. Those marine creatures that suffered the most extinctions were those that depended directly upon primary productivity for their food source. Such creatures included not

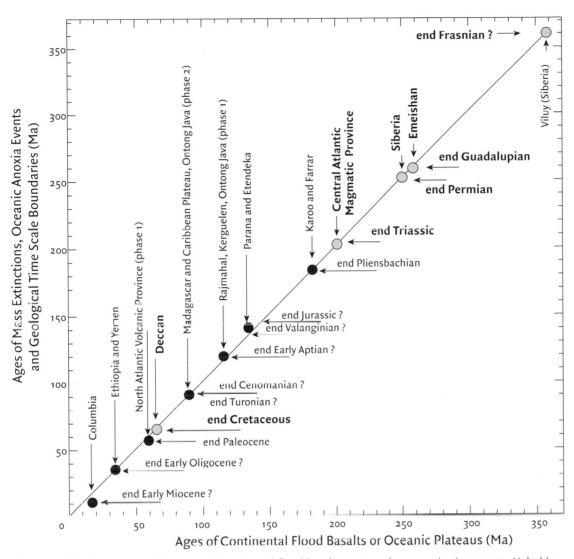

Figure 16.20. Comparison of the ages of continental flood basalt events and mass extinction events. Light-blue colored dots represent the most important mass extinction events. The straight line indicates excellent correlation between the ages of the basalts and the timing of the mass extinction events. Data compiled by Vincent Courtillot.

only the planktonic foraminifera and other planktonic marine microorganisms, but also ammonites, other cephalopods, and a variety of mollusks. On the other hand, organisms that not only depended on primary productivity but could also survive on **detritus**, that is, the scavenged remains of other organisms, fared consistently better. In marine deposits, detritus-feeders were apparently less affected by the extinction.

In the terrestrial realm, the strong selectivity between land-dwelling and aquatic tetrapod survival (see Figure 16.13) correlates with feeding strategy: aquatic vertebrates tend to utilize detritus as a major source of nutrients, while land-dwelling vertebrates are far more dependent upon primary productivity.

In this scenario, the tetrapods that survived the K-Pg boundary were primarily aquatic detritus-feeders. This is because river and lake systems can serve as a repository for detrital material, and organisms that live in such environments and can utilize this resource were protected against short-term drops in primary productivity. They may also have been protected from the strong infra-red radiation pulse, as well as the wildfires. In short, then, a lot of evidence points to an abrupt extinction.

Other hypotheses

So there is a striking coincidence between the timing of the asteroid impact, and these biological events. Moreover, recent high-precision numerical dating shows that, within a very few tens of thousands of years,[10] the asteroid impact and the biological extinction occurred at the same time. In fact, although it was not as well understood then as now, this remarkable coincidence of events is what first suggested to the Alvarezes that maybe the asteroid and the extinction were paired as cause and effect.

However, there is another remarkable coincidence that nags almost as compellingly as the asteroid. For many years, French volcanologist Vincent Courtillot has documented the relationship between episodes of continental flood basalt (CFB) volcanism and mass extinctions. The relationship that he observed is really quite striking; each continental flood basalt event since the Permian seems to have been associated with a mass extinction (Figure 16.20).

The K-Pg iconic CFB is the Deccan Traps, that humongoid outpouring of continental flood basalts in India described earlier in this chapter (see Figure 16.10). These bracketed the K-Pg boundary, beginning at about 67.5 Ma, and occurring in three major eruptive phases until about 64.7 Ma. Work done in 2015 suggests that the major eruptive phase (the middle one) took place within a period of <750 kyr (66.3–65.5 Ma), nicely bracketing the K-Pg boundary. Should Deccan volcanism and the extinction be paired as well? And if so, were its effects any less dire than the putative effects of the asteroid?

It is hard to say, "Absolutely not!" particularly as recent research suggests that the mass extinction at the Permo-Triassic boundary may have been associated with CFBs. Yet, you will note that the Deccan basalts began before the extinctions and continued after. In between each of the pulses of volcanism, life was restored, as shown by horizons containing abundant, well-preserved plant and animal fossils preserved between the flows. What, then, besides its size, was different about the second pulse? Or did its size exceed some key threshold and somehow cause an extinction that previous Deccan episodes did not?

Other aspects of the K-Pg extinction do not clearly implicate flood basalts: the global fires, the selectivity of the extinction, the evidence for physical disruption at the boundary globally, and chemical signals around the Gulf of Mexico region that match those of the impact crater, and of course the presence at the K-Pg boundary globally of an extraterrestrial signature, namely iridium and shocked quartz. Moreover, volcanic events implicated in other mass extinctions have left a global record. By contrast to this and to the asteroid, there is no significant global record of the Deccan volcanic event. For all these reasons, then, attractive as the continental flood basalt idea is, it does not fit best the known data profile that characterizes the K-Pg extinctions.

And yet... just a week or two before these words were written, tectonicist M. A. Richards (see Footnote 1 in this chapter) and a host of co-authors including Walter Alvarez of the 1980 hypothesis fame, proposed something of a compromise: perhaps, they suggested, the asteroid impact provoked the main episode of Deccan volcanism, and the two events together caused the extinction. The possibility of this occurring could be demonstrated on theoretical grounds; so, conceptually, it is not out of the question. This hybrid extinction mechanism takes us to the realm of multiple-cause hypotheses, which, although they may explain the data more mechanistically, run the risk of not being parsimonious. We consider these in the next section.

Multiple causes

It's been said that "complex extinctions require complex causes," and a number of workers have suggested that a unique suite of events fortuitously coalesced at the K-Pg boundary, to produce the

---

[10] An astounding achievement, considering that these events occurred tens of *millions* of years ago.

pattern of selectivity that we have observed in the extinctions. These events include the asteroid impact, the Deccan volcanism, sea-level fall, global climatic deterioration through warming just before the K-Pg boundary, competition among various groups... in fact, something of a laundry list of what is known about boundary events. In this case, every known physical event is imputed to have caused "weakened" ecosystems, and made them more susceptible to extinction. Does this better explain the data?

Well, not exactly. However, the reason why it does not provide a satisfactory explanation for what caused the extinction is at least in part a philosophical one. Ultimately in this way of thinking, each K-Pg event is ascribed a role in the extinctions; ultimately, therefore, any and all K-Pg events must have been involved in the extinctions. Because of this, the hypothesis is unfalsifiable: if all K-Pg events are involved in the extinction, there is no way to eliminate any particular event. If we can't know which event – because none can be absolutely eliminated – it really says nothing new about the K-Pg boundary: the hypothesis is so general that it ultimately explains nothing.

Ironically (and sadly), the multi-cause hypothesis might very well be correct: a unique suite of events uniquely occurring at the K-Pg boundary could have uniquely caused this mass extinction. The scenario could go like this: the asteroid impact, itself unimaginably destructive, could have provoked the Deccan Traps, which released various gases (CO_2; NO_2; SO_2) into the atmosphere, none of which likely improved life for the Earth's biota.

In the marine realm, the combination of volcanism and an asteroid would have darkened skies, leading to inhibition of photosynthesis and a drop in primary productivity of phytoplankton. Food chains, ultimately dependent upon that primary productivity, would have collapsed.

In the terrestrial realm, perhaps a few other things had already undermined existing ecosystems: some extra competition among mammalian species, some latest Cretaceous temperature increases; but for whatever the reasons, populations were not replenished, and dinosaur-based ecosystems became unusually susceptible to collapse. The heat pulse and forest fires induced by the asteroid, as well as the darkened skies produced by the asteroid and the Deccan Traps would have then destroyed the standing vegetation (the primary production on land), and shut down food chains that were primary production- and not detritus-based. Aquatic organisms would have been protected by their wet habitats, and their survival potential would thus have been enhanced.

Alternatively, the same scenario might have occurred *without* the Deccan Traps provoking the asteroid. In this case, the Deccan Traps themselves might have weakened marine and terrestrial ecosystems,[11] and the asteroid would have then provided the *coup de grâce* to these staggering ecosystems.

This idea, of weakened ecosystems made particularly vulnerable to an asteroid impact, has been called the "press-pulse" hypothesis; ecosystems were stressed (pressed) by a variety of potential stressors; and then, in a more susceptible state zapped (pulsed) by the asteroid impact. Here, G. P. Wlson's data are crucial – and ambiguous. If they represent a step-wise extinction, then perhaps they are reflecting the weakened, pre-asteroid ecosystems postulated in the press-pulse hypothesis. But, if they are just an artifact of the Signor–Lipps effect, then they support the idea of an abrupt extinction, such as one would predict from an asteroid impact.

So what happened at the K-Pg boundary?

As we have seen, the very nature of the multi-cause hypothesis precludes its testability, and therefore for us, its significance is diminished. Moreover, the Deccan Traps appear superfluous (and thus

---

[11] By any number of mechanisms: darkened skies decreasing photosynthesis; an atmosphere terminally polluted with various gases, many of which could have come back to Earth as acid rain; temperature increases precipitated by those same greenhouse gases in the atmosphere. Multi-cause hypotheses are truly multi-caused!

non-parsimonious): why invoke them when (a) the asteroid was sufficient to do the job; and (b) the pattern of extinction matches the expected pattern from an asteroid impact?

Scientists who view the asteroid impact as the cause of K-Pg events generally envision some kind of dramatic and short-term disturbance to the ecosystem. Such a disturbance was likely a dust cloud blocking sunlight for a few months, a pulse of infra red radiation, and global wildfires. It may never be possible to know exactly which factor(s) did which deed(s), and trying to reconstruct it in so exact a manner may be stretching the resolution of the fossil record well past its capability. Whatever the disturbance to the terrestrial ecosystem, it seems to have had, at the very least, a deadly effect on global primary productivity, which in turn seems to have decimated organisms solely dependent upon primary productivity. Aquatic habitats evidently provided some measure of protection, and detritus-feeders survived. Whatever else is true, the absence of non-avian dinosaurs after their ~164 million years of terrestrial importance was the event that allowed the mammals to diversify and occupy the place in the global ecosystem that they presently hold.

SUMMARY

Although there has been considerable speculation about what happened to the non-avian dinosaurs at 66.1 Ma, any meaningful hypothesis must operate within the bounds of science: it must be testable and it must explain as much of the data as possible. Given those constraints, the single-cause hypothesis matches what is known about the K-Pg extinction: the hypothesis that an asteroid hit the Earth and killed many organisms, including the non-avian dinosaurs. Multi-cause hypotheses are weaker because what is known of the extinction does not require further causes than an asteroid impact.

The evidence for a large (10 km in diameter) asteroid striking the Earth at 66.1 Ma in the Yucatan region of Mexico is now extensive, and virtually incontrovertible. Its immediate effects are likely to have been, globally, blockage of sunlight for 3–4 months and the propagation of an energy pulse and associated wildfires in the terrestrial realm as well as, locally, tidal waves and severe environmental disruption.

Key among the various biotic conseqences of these events is that the great nutrient cycles that characterize healthy oceans were severely disrupted.

The asteroid impact is consistent with the limited amount that is known of the dinosaur extinction, which is that in the Western Interior of North America, at least (the only place in the world in which rates of dinosaur extinction have been studied), the dinosaur extinction was as instantaneous as we are able to resolve at a distance of 66.1 Ma.

The patterns of terrestrial vertebrate survivorship are relatively well understood. Analyses show that vertebrates not in aquatic ecosystems were most susceptible to extinction; other inferences, less robust, suggest that larger animals, endotherms, and amniotes also were less likely to survive than smaller animals, ectotherms, and anamniotes. Non-avian dinosaurs were large organisms, and clearly dependent upon primary productivity.

One explanation for the survivorship of organisms living in aquatic food webs is that these are generally not as dependent upon primary productivity as those in land-based food chains. Because primary productivity was clearly perturbed at the K-Pg boundary, organisms in aquatic food webs were likely protected both by dietary preference and by the fact that the aquatic realm provided a refuge from the physical catastrophes caused by the asteroid impact.

The recovery took 2–3 million years in the oceans, and about 5 million years on land.

SELECTED READINGS

Alvarez, L. W. 1983. Experimental evidence that an asteroid impact led to the extinction of many species 65 Myr ago. *Proceedings of the National Academy of Sciences*, **80**, 627–642.

Alvarez, W. 1997. *T. rex and the Crater of Doom*. Princeton University Press, Princeton, NJ, 185 pp.

Alvarez, L. W., Alvarez, W., Asaro, F., and Michel, H. V. 1980. Extraterrestrial cause for the Cretaceous–Tertiary extinction. *Science*, **208**, 1095–1108.

Archibald, J. D. 1996. *Dinosaur Extinction and the End of an Era*. Columbia University Press, New York, 237 pp.

Archibald, J. D. 2011. *Extinction and Radiation: How the Fall of the Dinosaurs Led to the Rise of Mammals*. Johns Hopkins University Press, Baltimore, MD, 108 pp.

Benton, M. J. 1990. Scientific methodologies in collision the history of the study of the extinction of the dinosaurs. *Evolutionary Biology*, **24**, 371–400.

Fastovsky, D. E., and Bercovici, A. 2016. The Hell Creek Formation and its contribution to the Cretaceous–Paleogene extinction: A short primer. *Cretaceous Research*, **57**, 368–390.

Frankel, C. 1999. *The End of the Dinosaurs – Chicxulub Crater and Mass Extinctions*. Cambridge University Press, New York, 223 pp.

Hartman, J., Johnson, K. R., and Nichols, D. J. (eds.) 2002. *The Hell Creek Formation and the Cretaceous–Tertiary Boundary in the Northern Great Plains*. Geological Society of America Special Paper no. 361, 520 pp.

Koeberl, C. and MacLeod, K. G. (eds.) 2002. *Catastrophic Events and Mass Extinctions: Impacts and Beyond*. Geological Society of America Special Paper no. 356, 746 pp.

Lillegraven, J. A. and Eberle, J. J. 1999. Vertebrate faunal changes through Lanican and Puercan time in southern Wyoming. *Journal of Paleontology*, **73**, 691–710.

Maas, M. C. and Krause, D. W. 1994. Mammalian turnover and community structure in the Paleocene of North America. *Historical Biology*, **8**, 91–128.

Powell, J. L. 1998. *Night Comes to the Cretaceous*. W. H. Freeman and Company, New York, 250 pp.

Richards, M. A., Alvarez, W., Self, S., *et al.* 2015. Triggering of the largest Deccan eruptions by the Chicxulub impact. *Geological Society of America Bulletin*, published online on 30 April 2015 as doi:10.1130/B31167.1.

Robertson, D. S., McKenna, M. C., Toon, O. B., Hope, S., and Lillegraven, J. A. 2004. Survival in the first hours of the Cenozoic. *Geological Society of America Bulletin*, **116**, 760–768. doi: 10.1130/B25402.1.

Ryder, G., Fastovsky, D. E., and Gartner, S. (eds.) 1996. *The Cretaceous–Tertiary Event and Other Catastrophes in Earth History*. Geological Society of America Special Paper no. 307, 569 pp.

Sharpton, V. L. and Ward, P. D. (eds.) 1990. *Global Catastrophes in Earth History: An Interdisciplinary Conference on Impacts, Volcanism, and Mass Mortality*. Geological Society of America Special Paper no. 247, 631 pp.

Schoene, B., Samperton, K. M., Eddy, M. P., *et al.* 2015. U-Pb geochronology of the Deccan Traps and relation to the end-Cretaceous mass extinction. *Science*, **347**, 182–184. doi: 10.1126/science.aaa0118.

Silver, L. T. and Schultz, P. H. (eds.) 1982. *Geological Implications of Large Asteroid and Comets on Earth*. Geological Society of America Special Paper no. 190, 528 pp.

Wilson, G. P., Clemens, W. A., Horner, J. R., and Hartman, J. H. (eds.) 2014. *Through the End of the Cretaceous in the Type Locality of the Hell Creek Formation in Montana and Adjacent Areas*. Geological Society of America Special Paper 503, 392 pp.

TOPIC QUESTIONS

1. What do the words regression, detritus, iridium anomaly, shocked quartz, microtektites, Chicxulub, ichthyosaurs, plesiosaurs, ammonites, mosasaurs, foraminifera, planktonic, and benthic refer to?
2. What are primary production and disaster faunas?
3. What is meant by nutrient cycling? Why is that important?
4. What is the K-Pg boundary? When was it?
5. What kinds of physical events took place at the K-Pg boundary?
6. Describe the biotic extinctions that took place at the K-Pg boundary.
7. Describe the results of the studies that concluded that the dinosaurs died abruptly.
8. What is meant by "geologically abrupt" when speaking of the extinction of the dinosaurs?
9. Describe the kinds of physical events that might have occurred when the asteroid hit the Earth.
10. Choose four extinction hypotheses from Table 16.1 and evaluate them in terms of the criteria for a viable scientific theory.

GLOSSARY

Using this Glossary

The goal of this Glossary is to help to clarify language and images that may be unfamiliar to students who happen to live in the Recent (us!). Below, therefore, follows a complete listing of all the words highlighted in this textbook, as well as other words of relevance to the subject of dinosaur paleontology. Readers are provided with the chapter in which the word is highlighted; in some cases, however, readers are also referred to places where the concept(s) embodied by the word is treated. Finally, in a few cases, readers are referred to a figure as well. A relatively few words have no cross-reference, implying that the ideas they represent recur throughout the book.

A

ADP (adenosine diphosphate) A molecule involved in the energy production of a cell. It is produced when ATP breaks down to release energy.

ATP (adenosine triphosphate) A molecule synthesized by the body as a means to store energy. The energy is stored in the three phosphate bonds, one of which is then broken to release energy (ATP → ADP + inorganic phosphate + energy). *See* Chapter 12.

Acetabulum Hip socket. *See* Chapter 4.

Acromion (acromial) process A broad and plate-like flange on the forward surface of the shoulder blade. *See* Chapter 5.

Adenosine diphosphate *See* ADP. *See also* Chapter 12.

Adenosine triphosphate *See* ATP. *See also* Chapter 12.

Advanced In an evolutionary context, shared or derived (or specific), with reference to characters. *See* Chapter 3.

Aerobic A type of metabolism involving a complex series of oxidation steps through the citric acid cycle. *See* Box 12.2.

Akinetic *See* Kinetic. *See also* Chapter 7.

Allometry The condition in which, as the size of organisms changes, their proportions change as well. For example, if an ant were scaled up to the size of a 747 airplane, its features – body, head, legs, etc. – would no longer have the same proportions relative to each other. *See* Chapter 12.

Altricial Pertaining to organisms that are born relatively underdeveloped, requiring significant parental attention for survival. *See* Chapter 7.

Alvarezsauridae An unusual group of theropods equipped with stout carpometacarpus-like features. Their phylogenetic status is currently poorly understood. *See* Chapter 11.

Alveolus (pl. alveoli) A sac-like anatomical structure. *See* Chapter 8.

Amnion A membrane in some vertebrate eggs that contributes to the retention of fluids within the egg. *See* Chapter 4.

Amniote An organism bearing amniotic eggs. *See* Chapter 4.

Anaerobic Without oxygen. *See* Chapter 12.

Analog In anatomy, structures that perform in a similar fashion but have evolved independently. *See* Chapter 3.

Analogous Adjectival form of analog. *See* Chapter 3.

Anamniote(s) An organism whose egg has no amnion. *See* Chapter 4.

Anapsid(s) The group that contains all amniotes with a completely covered skull roof. *See* Chapter 4.

Ancestral In an evolutionary sense, relating to forebears. *See* Chapter 3.

Angiosperms Flowering plants. *See* Chapters 4, 5, 6, and 13.

Ankylosauria Armored quadrupedal thyreophorans (Ornithischia) whose bodies were encased in a bony pavement of osteoderms. *See* Chapter 5.

Ankylosauridae Along with Nodosauridae, one of the two clades that make up Ankylosauria. *See* Chapter 5.

Antediluvian Occurring before the Biblical flood. *See* Chapter 14.

Anterior Pertaining to the head-bearing end of an organism.

Antorbital fenestra An opening on the side of the skull, just ahead of the eye. This is a character that unites the clade Archosauria. *See* Chapter 4.

Arboreal Pertaining to trees; as the arboreal hypothesis, referring to the idea that bird flight evolved by birds jumping out of trees. *See* Chapter 10.

Archosauria A clade within Archosauromorpha. The living archosaurs include birds and crocodiles. *See* Chapters 4 and 10.

Archosauromorpha The large clade of diapsids that includes the common ancestor of rhynchosaurs and archosaurs, and all its descendants. *See* Chapter 4.

Ascending process of the astragalus A wedge-shaped splint of bone on the astragalus that lies flat against the shin (between the tibia and fibula) and points upward. Diagnostic character of Theropoda. *See* Chapter 4.

Assemblage In paleontology, a group of organisms. The term is used to refer to a collection of fossils in which it is not clear how accurately the collection reflects the complete, ancient formerly living community. *See* Chapters 2 and 12.

Asteroid A large extraterrestrial body. *See* Chapter 15.

Astragalus Along with the calcaneum, one of two upper bones in the vertebrate ankle. *See* Chapter 10.

Atom The smallest particle of any element that still retains the properties of that element. *See* Chapter 2.

Atomic number The number of protons (which equals the number of electrons) in an element. *See* Chapter 2.

B

Background extinctions Continually occurring, isolated extinctions of individual species. As distinct from mass extinctions. *See* Chapter 15.

Badlands Extremely rough country, generally carved by rivers and floods. *See* Chapter 1.

Barb Feather material radiating from the shaft of the feather. *See* Chapter 10.

Barbule A small hook that links barbs together along the shaft of the feather. *See* Chapter 10.

Barometric Adjectival form of barometer. Barometers measure atmospheric pressure; the atmospheric pressure at the sea level is 1 atmosphere (1 atmosphere = 14.7 lb/in^2). Atmospheric pressure increases dramatically with depth of submersion. *See* Chapter 8.

Beak Sheaths of keratinized material covering the ends of the jaws (synonym: rhamphotheca). *See* Chapters 4–7.

Belief system A conceptualization that is faith-based, knowledge that does not require the application of logic, reason, and/or observation. *See* Chapter 15.

Bennettitaleae A group of extinct cone-bearing plants. *See* Chapter 13.

Benthic With reference to the marine realm (oceans), living within sediments. *See* Chapter 16.

Biogeography Pertaining to the distribution of organisms in space. *See* Chapter 6.

Biomass The sum total of the weights of organisms in the assemblage or community being studied. *See* Chapter 12.

Biostratigraphy The study of the relationships in time among groups of organisms. *See* Chapter 2.

Biota The sum total of all organisms that have populated the Earth.

Body fossil The type of fossil in which a part of an organism becomes buried and fossilized as opposed to trace fossil. *See* Chapter 1.

Bone histology The study of bone tissue. *See* Chapter 12.

Bonebeds Relatively dense accumulation of bones of many individuals, generally composed of a very few kinds of organisms. *See* Chapter 6.

Brain endocasts Internal casts of braincases. *See* Chapter 12.

Braincase Hollow bony box that houses the brain; located toward the upper, back part of the skull. *See* Chapter 4.

C

Cadence In locomotion, the rate at which the feet hit the ground. *See* Chapter 5.

Calcareous nanofossils Fossils of extremely small planktonic microorganisms. *See* Chapter 16.

Carbohydrates A family of 5- and 6-carbon organic molecules whose chemical bonds, when broken, release energy. *See* Chapters 5 and 12.

Carpal Wrist bone. *See* Chapter 4.

Carpometacarpus (pl. carpometacarpi) Unique structure in all living, and in most ancient, birds, in which bones in the wrist and hand are fused. *See* Chapter 10.

Cast Material filling up a mold. *See* Chapter 1.

Cellular respiration The breakdown of carbohydrates through a regulated series of oxidizing reactions. *See* Chapter 12.

Cenozoic An Era, lasting from 65.5 Ma to present. *See* Chapters 2, 13, and 15.

Centrum The spool-shaped, lower portion of a vertebra, upon which the spinal cord and neural arch rest. *See* Chapter 4.

Cerapoda The ornithischian clade of Ceratopsia + Pachycephalosauria + Ornithopoda. *See* Part II.

Ceratopsia Beaked dinosaurs of Asia and North America who, together with Pachycephalosauria, make up Marginocephalia. *See* Chapter 6.

Character An isolated or abstracted feature or characteristic of an organism. *See* Chapter 3.

Cheek The muscular, fleshy organs along the side of the jaw that retain food in the mouth. *See* Part II.

Cheek teeth Teeth that lie against the cheeks; in mammals, the cheek teeth consist of the premolars and molars. *See* Part II.

Chronostratigraphy The study of geological time. *See* Chapter 2.

Clade Group of organisms in which all members are more closely related to each other than they are to anything else. All members of a clade share a most recent common ancestor that is itself the most basal member of that clade. Synonymous with "monophyletic group" and "natural group." *See* Chapter 3.

Cladogram A hierarchical, branching diagram that shows the distribution of shared, derived characters among selected organisms. *See* Chapter 3.

Clavicle Collarbone. *See* Chapter 4.

Coelophysoidea The theropod clade containing *Coelophysis* and its close relatives. *See* Chapter 9.

Co-evolution The idea that two organisms or groups of organisms may have evolved in response to one another. *See* Chapter 13.

Collect To obtain fossils from the Earth. *See* Chapter 1.

Computed tomography (CT) A medical scanning technique using computers to construct a two-dimensional image using X-ray imaging around an axis of rotation. It has proven to be an extremely successful means of resolving incompletely prepared fossils as well. *See* Chapter 12.

Continental effects The effect on climate exerted by continental masses. *See* Chapter 2.

Convergent In anatomy, pertaining to the independent invention (and thus, duplication) of a structure or feature in two lineages. The streamlined shape of whales, fish, and ichthyosaurs is a famous example of convergent evolution. *See* Chapter 9.

Coprolite Fossilized feces. *See* Chapters 1 and 9.

Coracoid The lower (and more central) of two elements of the shoulder girdle (the upper being the scapula). *See* Chapter 4.

Coronoid process A bony enlargement or process at the back of the lower jaw for the attachment of jaw-closing musculature. *See* Part II.

Cretaceous A Period lasting from 146 Ma to 65.5 Ma.

Cretaceous–Tertiary boundary That moment in time, 65.5 million years ago, between the Cretaceous Period and the Tertiary Period. *See* Figure 2.4 and Chapter 15.

Cretaceous–Tertiary extinction The event, 65.5 million years ago at the Cretaceous–Tertiary boundary, in which all the non-avian dinosaurs, as well as many other terrestrial vertebrates, became extinct. *See* Chapter 16.

Crop To cut short; in anatomy, to bite off. *See* Part II.

Crurotarsi A clade of archosaurs including crocodilians and their close relatives. *See* Chapter 4.

Curate In paleontology, to incorporate, preserve, and catalog specimens into museum collections. *See* Chapter 1.

Cursorial Pertaining to running; as the **cursorial hypothesis**, referring to the idea that bird flight evolved by birds running along the ground. *See* Chapters 4 and 10.

Cycadophyte A bulbous, fleshy type of gymnosperm. *See* Chapter 13.

D

Deccan Traps Interbedded volcanic and sedimentary rocks in western and central India of Cretaceous–Tertiary age. *See* Chapter 16.

Deltoid crest A large process at the head of the humerus. *See* Chapter 10.

Dense Haversian bone A type of Haversian bone in which the canals and their rims are very closely packed. *See* Chapter 12.

Dental battery A cluster of closely packed cheek teeth in the upper and lower jaws, whose shearing or grinding motion is used to masticate plant matter. *See* Chapter 6.

Denticles Small bumps or protuberances generally associated with teeth. *See* Chapters 6 and 7.

Derived In an evolutionary context, pertaining to characters that uniquely apply to a particular group and thus are regarded as having been "invented" by that group during the course of its evolutionary history. *See* Chapter 3.

Detritus Loose particulate rock, mineral, or organic matter; debris. *See* Chapter 16.

Developmental biology The study of how organisms develop from a fertilized egg to an adult. *See* Chapter 14.

Diagnostic In phylogeny, a feature that uniquely pertains to a group of organisms. Diagnostic features permit the identification of groups of organisms because, uniquely, all members of that group possess the feature (and ideally all other groups do not). *See* Chapter 3.

Diapsida The large clade of amniotes that includes the common ancestor of lepidosauromorphs and archosaurs, and all its descendants. *See* Chapter 4.

Diastem(a) A gap. *See* Part II.

Digitigrade In anatomy, a position assumed by the foot when the animal is standing, in which the ball of the foot is held high off the ground and the weight rests on the ends of the toes. Opposite of plantigrade. *See* Chapter 8.

Dinosauria A clade of ornithodiran archosaurs. *See* Chapter 4.

Disarticulated Dismembered. *See* Chapter 1.

Disaster biota Organisms that colonize a landscape immediately after an ecological disaster. Disaster biotas tend to have three characteristics: small size, high rates of speciation, and generalist life strategies. *See* Chapter 16.

Distal In anatomy, in the direction away from the central part (or core) of the animal. *See* Chapters 3 and 12.

Diversity The variety of organisms; the number of *kinds* of organisms. *See* Chapter 13.

DNA hybridization A molecular biological technique that measures the difference between two comparable strands of DNA. *See* Chapter 11.

Down A bushy, fluffy, type of feather in which barbules and vanes are not well developed, used for insulation. *See* Chapter 10.

Dynamic similarity A conversion factor that "equalizes" the stride rates of vertebrates of different sizes and proportions, so that speed of locomotion can be calculated. *See* Chapter 12.

E

Ecological diversity The proportion of an ecosystem that is occupied by a particular lifestyle, such as feeding type or mode of locomotion. For a simple example, one might study an ecosystem by dividing it into herbivores, carnivores, and omnivores. *See* Chapter 15.

Ecospace Niches that are available in an ecosystem. Simple categories of ecospace, for example, include "carnivore," "herbivore," and "scavenger." More refined categories could include "grazing" versus "browsing" herbivores, and large and small carnivores. *See* Chapter 16.

Ectothermic Regulating temperature (and thus metabolic rate) using an external source of energy (heat). The opposite of endothermic. *See* Chapter 12.

Electron A negatively charged subatomic particle. Electrons reside in clouds around the nucleus of an atom. *See* Chapter 2.

Element In chemistry, an atom distinguished by the number of protons in its nucleus; in anatomy, discrete part of the skeleton, that is, an individual bone. *See* Chapters 2 and 4.

Enantiornithes A group of sparrow-like, relatively common Mesozoic avialians. *See* Chapter 11.

Encephalization That condition in which an organism bears a head structure that is distinct from the rest of the body and that contains a brain. *See* Chapter 12.

Encephalization Quotient (EQ) An estimate based on brain size and body weight, designed to determine the intelligence of an extinct organism, relative to a living organism whose intelligence can be ascertained. *See* Chapter 12.

Endemic An organism or fauna is said to be endemic to a region when it is restricted to that region. *See* Chapter 13.

Endemism The property of being endemic. *See* Chapter 13.

Endosymbionts Organisms that live within another organism in a mutually beneficial relationship. *See* Chapter 8.

Endothermic Regulating temperature (and thus, metabolic rate) using an internal source of energy. The opposite of ectothermic. *See* Chapter 12.

Enlightenment The Enlightenment was a seventeenth- and eighteenth-century political, social, and philosophical movement that, among other qualities, emphasized the primacy of logic, reason, and observation for understanding the natural world. *See* Chapter 14.

Epicontinental sea Relatively shallow (at most, a few hundred meters) marine water covering a continent (synonym: epeiric sea). *See* Chapter 2.

Era A very large block of geological time (hundreds of millions of years long), composed of Periods. *See* Chapter 2.

Erect stance In anatomy, the condition in which the legs lie parasagittal to (alongside) the body and do not extend laterally from it. *See* Chapter 4.

Esophagus A tube located between the neck vertebrae and the trachea, leading from just behind the mouth down to the stomach. In humans it is about 15 cm long; in sauropods it could have reached 10–15 m in length. *See* Chapter 8.

Estivate In zoology, to spend summers in a state of torpor. *See* Chapter 7.

Euornithopoda A monophyletic group containing the more derived members of Ornithopoda. *See* Chapter 7.

Eurypoda The ornithischian clade Stegosauria + Ankylosauria. *See* Chapter 5.

Eustatic Global. *See* Chapter 2.

Evo-Devo (evolution and development) The coupling of the sciences of evolutionary biology and developmental biology to attempt to understand how new features in organisms have evolved. *See* Chapter 14.

Evolution In biology, descent with modification. *See* Appendix 3.1.

Extraterrestrial From outer space. *See* Chapter 16.

F

Fauna A group of animals presumed to live together within a region.

Femur The upper bone in the hindlimb (thigh bone). *See* Chapter 4.

Fibula The smaller of the two lower leg bones in the hindlimb; the bone that lies alongside the shin bone (tibia). *See* Chapters 4 and 10.

Fit Adjectival form of fitness. Fitness is the degree to which natural selection acts upon an organism to assure the presence of its genetic makeup in the succeeding (descendent) generation. *See* Chapter 3.

Flight feather Elongate feather with well-developed, asymmetrical vanes; usually associated with flight. *See* Chapter 10.

Flood basalt Episodic lava flow from fissures in the Earth's crust.

Flux A measure of change; rate of discharge times volume.

Folsom projectile point A style of tool-making (likely a spearhead) known from the Great Plains of the USA, first found near Folsom, New Mexico, and dating from 9500–8000 BCE (Before the Common Era). *See* Chapter 14.

Footplate On the pubis, an anterior–posterior enlargement of the distal end. Generally associated with theropods. *See* Chapter 10.

Footprint Trace fossil left by the feet of vertebrates. *See* Chapter 1.

Foramen magnum The opening at the base of the braincase through which the spinal cord travels to connect to the brain. *See* Chapters 4 and 5.

Foraminifera (sing. foraminifer) Single-celled, shell-bearing organisms that live in the oceans. *See* Chapter 16.

Fossil Technically, anything buried; generally refers to the buried remains of organisms. *See* Chapter 1.

Fossilization The process of becoming a fossil. *See* Chapter 1.

Fossorial Burrowing. *See* Chapter 7.

Frill In ceratopsians, a sheet of bone extending dorsally and rearward from the back of the skull, made up of the parietal and squamosal bones. *See* Chapter 6.

Furcula Fused clavicles (collarbones); the "wish-bone" in birds and certain non-avian theropods. *See* Chapter 10.

G

Gamete(s) A mature sex cell; either male or female. *See* Chapter 13.

Gastralia Belly ribs. *See* Chapter 10.

Gastrolith Smoothly polished stone in the stomach, used for grinding plant matter. *See* Chapter 5.

Genasauria The ornithischian clade of Thyreophora + Cerapoda. *See* Part II.

General In phylogenetic reconstruction, referring to a character that is non-diagnostic of a group; in this context, synonymous with primitive. *See* Chapter 3.

Generalist In ecology, unspecialized. Basically willing and able to eat anything to stay alive. In terms of diet, at least, humans are generalists; a meat-eater, like a great white shark, would be considered a specialist. *See* Chapter 16.

Generic Adjective from the word genus, the second-smallest grouping in the Linnaean classification (a genus [pl. genera] is composed of species, the smallest formal category in the Linnaean classification). *See* Chapter 4.

Genotype The total genetic makeup of an organism. *See* Chapter 3.

Geochemistry Chemistry particularly as related to geological problems. *See* Chapter 14.

Geochronology The science of determining the age (in years) of the Earth. Adjectival form of the word geochronological. Practitioners of geochronology are, not surprisingly, geochronologists. *See* Chapter 2.

Ghost lineage Lineage of organisms for which there is no physical record (but whose existence can be inferred). *See* Chapter 13.

Gigantothermy Modified mass homeothermy, which mixes large size with low metabolic rates and control of circulation to peripheral tissues. *See* Chapter 12.

Girdle(s) Something that encircles; in this case attachments for the limbs that partially encircle the trunk. *See* Chapter 4.

Gizzard A muscular chamber just in front of the glandular part of the stomach. *See* Chapter 8.

Glycogen A complex, carbohydrate-based molecule used by the body for energy storage. *See* Chapter 5.

Gnathostome A vertebrate with a jaw (formal term: Gnathostomata). *See* Chapter 4.

Gondwana A southern supercontinent comprising present-day Australia, Africa, South America, and Antarctica. *See* Chapter 2.

Gregarious Highly socialized; regularly moving in flocks or herds. *See* Chapters 5 to 8.

Gut Stomach, intestine, and bowels. *See* Part II.

Gymnosperms A group of seed-bearing, non-flowering plants, including pines and cypress. Gymnosperms are not a monophyletic grouping, unless angiosperms are also included within the group. *See* Chapters 5 and 13.

H

Half-life The amount of time that it takes for 50% of a volume of unstable isotope to decay. *See* Chapter 2.

Hard part In paleontology, all hard tissues, including bones, teeth, beaks, and claws. Hard parts tend to be preserved more readily than soft tissues. *See* Chapter 1.

Haversian canal In bone histology, a canal composed of secondary bone. *See* Chapter 12.

Hemal Referring to blood. In bone anatomy, hemal arch. A vertebral process straddling the ventral side of the vertebral column and

pointing ventrally (oriented opposite to the neural arch). *See* Chapter 10.

Hierarchy As applied here, the ordering of objects, organisms, and categories by rank. The military and the clergy are both excellent examples of hierarchies; in these, rank is a reflection of power and, one hopes, accomplishment. Another hierarchical system is money, which is ordered by value. *See* Chapter 3.

Histology The study of tissues. *See* Chapters 12 and 14.

Homeotherm Organism whose core temperature remains constant. *See* Chapter 12.

Homologous Two features are homologous when they can be traced back to a single structure in a common ancestor. *See* Chapter 3.

Horn core The horns in many animals, cows and sheep being two familiar examples, are constructed of a bony central part, the core, covered by a layer, or sheath, made of keratin. Ceratopsian dinosaurs had this type of horn. *See* Figure 6.21.

Hornlets Small horns. *See* Chapter 9.

Humerus The upper arm bone. *See* Chapter 4 (especially Figure 4.5).

Hyoid bone(s) Paired elongate bones in the throat that form a support for the tongue. *See* Chapter 5.

Hypothesis of relationship A hypothesis about how closely or distantly organisms are related. *See* Chapter 3.

I

Ichnofossil Impression, burrow, track, or other modification of the substrate by organisms. *See* Chapter 1.

Ichthyosaurs Dolphin-like marine reptiles of the Mesozoic. *See* Figure 16.11.

Ilium The uppermost of three bones that make up the pelvis. *See* Chapter 4.

Impact ejecta The material thrown up when an asteroid strikes the Earth. *See* Chapter 15.

Impact winter The idea that a temporary period of climatic cooling caused by sunlight blockage by ejecta would follow an impact.

Based on climatic data from large volcanic eruptions such as Krakatoa. *See* Chapter 16.

Interspecific Among different species. *See* Chapter 6.

Intraspecific Within the same species. *See* Chapters 5 and 6.

Iridium (Ir) A non-toxic, platinum-group metal, rare at the Earth's surface. *See* Chapter 16.

Iridium anomaly High concentrations of iridium in a single place; the name comes from the idea that since Ir levels at the Earth's surface are generally low, high concentrations would be considered anomalous. *See* Chapter 16.

Ischium The most posterior of three bones that make up the pelvis. *See* Chapter 4.

Isotopes In chemistry, elements that have the same atomic number but different mass numbers. *See* Chapter 2.

J

Jacket In paleontology, a rigid, protective covering placed around a fossil, so that it can be moved safely out of the field. Commonly made up of strips of burlap soaked in plaster. *See* Figure 1.10.

Jurassic A Period lasting from 200 Ma to 146 Ma.

K

K-strategy The evolutionary strategy of having few offspring, which are cared for by the parents. The symbol K stands for the carrying capacity of the environment. *See* Chapter 7.

K/T (boundary) Common abbreviation for that moment in time, 65.5 Ma, which marks the boundary between the Cretaceous and the Tertiary. *See* Chapter 16.

Keel A flange or sheet of bone, as in the keeled sternum of birds; named for its resemblance to the keel on a sailboat. *See* Chapter 10.

Keratin A protein that forms the basis of nails, horns, hooves, feathers, and hair. *See* Part II.

Kinetic With reference to skull anatomy, movement between bones of the skull. *See* Chapter 7.

L

Lactic acid An organic acid produced as a byproduct of muscle exertion; $CH_3CH(OH)COOH$. *See* Chapter 12.

LAG *See* **Lines of arrested growth**. *See also* Chapter 12.

Lambeosaurines The hollow-crested hadrosaurid dinosaurs.

Land bridge A corridor of land between two continents that allows the passage of organisms from one continent to the other. Such corridors have occasionally existed, for example, between South America and North America, as well as between North America and Asia. *See* Chapter 13.

Laurasia A northern supercontinent. *See* Chapter 2.

Lepidosauromorpha One of the two major clades of diapsid reptiles; the other clade is Archosauromorpha. *See* Chapter 4.

Lines of arrested growth (LAGs) Lines that are inferred to represent times of non-growth, visible in the cross-section of bones. *See* Chapter 12.

Lithostratigraphy The general study of all rock relationships. *See* Chapter 2.

Lower temporal fenestra The lower opening of the skull just behind the eye; *see* **Temporal fenestra**. *See also* Chapter 4.

M

Mandible The lower jaw. *See* Chapter 4.

Mandibular foramen A fenestra in the posterior portion of the mandible. *See* Part II.

Marginocephalia The clade of dinosaurs that includes the most recent common ancestor of pachycephalosaurs and ceratopsians and all of its descendants. *See* Chapter 5.

Mass extinctions Global and geologically rapid extinctions of many kinds, and large numbers, of species. *See* Chapter 15.

Matrix In paleontology, the rock that surrounds fossil bone. *See* Chapter 1.

Megaflora The visible remains of plants, especially leaves. *See* Chapter 16.

Mesotarsal A linear type of ankle in which hinge motion in a fore–aft direction occurs between the upper ankle bones (the astragalus and calcaneum) and the rest of the foot. *See* Chapter 4.

MDT *See* **Minimal divergence time.** *See also* Chapter 13.

Mesozoic An Era lasting from 251 Ma to 65.5 Ma. *See* Chapter 2.

Metabolism The sum of the physical and chemical processes in an organism. *See* Chapter 12.

Metacarpal Bone in the palm of the hand. *See* Chapter 4.

Metapodial A general name for metacarpals and metatarsals. *See* Chapter 4.

Metatarsal Bone in the sole of the foot. *See* Chapter 4.

Micropaleontology The study of microscopic organisms, such as marine plankton. A person who specializes in micropaleontology is a micropaleontologist. *See* Chapter 16.

Microtektite A small, droplet-shaped blob of silica-rich glass thought to have crystallized from impact ejecta. *See* Chapter 16.

Minimal divergence time (MDT) The minimal amount of time missing between the two descendent species and their common ancestor; calculated by comparing phylogeny and age of fossils. *See* Chapter 13.

Minimum number of individuals (MNI) A technique for estimating how many individual organisms are represented in a locality. If we find two left thigh bones (among other elements), then we know that at least two different individuals are represented. If we found six theropod teeth, however, we would not know whether they came from one or more than one (up to six) different individuals; the *least* number of individuals that could have produced those six teeth is one. *See* Chapter 16.

Mold Ichnofossil that consists of the impression of an original fossil. *See* Chapter 1.

Molecular clock The use of the rates of mutation in certain molecules to determine how long ago two organisms diverged from a common ancestor. *See* Chapters 11 and 14.

Molecular evolution The idea that molecules can evolve at particular rates. *See* Chapter 11.

Monophyletic group A group of organisms that has a single ancestor and contains all of the descendants of this unique ancestor (synonymous with clade and "natural group"). *See* Chapter 3.

Monotypic Of one type; generally, composed of one species. *See* Chapter 7.

Morphology The study of shape. *See* Chapter 3.

Mosasaurs Late Cretaceous marine-adapted lizards. *See* Chapter 15.

N

Naris (pl. nares) Opening in the skull for the nostrils. *See* Chapter 4.

Natural selection The process by which certain members of a population are more effectively able to assure the representation of their genes in the succeeding generation (e.g., the descendent generation). *See* Chapter 3.

Neoceratosauria The theropod clade *Coelophysis, Ceratosaurus,* and near relatives. *See* Chapter 9.

Neural arch A piece of bone that straddles the spinal cord; generally with a central process that rises dorsally. *See* Chapters 4 (especially Figure 4.5) and 10.

Neutron Electrically neutral subatomic particle that resides in the nucleus of the atom. *See* Chapter 2.

Node A bifurcation or two-way split point in a phylogenetic diagram (cladogram). *See* Chapter 3.

Nodosauridae Along with Ankylosauridae, one of the clades making up Ankylosauria. *See* Chapter 5.

Non-avian dinosaurs All dinosaurs *except* birds. *See* Chapters 1, 9, and 10.

Non-diagnostic In phylogeny, a feature that does not uniquely pertain to a group of organisms. Non-diagnostic features do not permit the identification of a particular group of organisms because members outside of that group also possess the feature. *See* Chapter 3.

Notochord An internal rod of cellular material that, primitively at least, ran longitudinally down the backs of all chordates. May be thought of as a precursor to the vertebral column. *See* Chapter 4.

Nuchal ligament An elastic ligament running dorsally in the neck from the back of the head to a posterior cervical vertebra. In sauropods, it likely helped to support the head. *See* Chapter 8.

Nucleus Central core of the atom. *See* Chapter 2.

Nutrient cycling As used here, the movement of nutrients from the shallow surface waters to the bottom of the ocean. *See* Chapter 16.

Nutrient turnover The amount of nutrients that pass through the system; in the case of bone development, the amount of metabolic activity associated with bone growth. *See* Chapter 12.

O

Obligate biped Tetrapod that must walk or run on its hind legs. *See* Chapters 4, 9, and 15.

Occipital condyle A knob of bone at the back of the skull with which the vertebral column articulates. *See* Chapter 4.

Occiput The back of the skull. *See* Chapter 6.

Occlusion Contact between upper and lower teeth; necessary for chewing. Teeth that contact each other between the upper and lower jaws are said to occlude. *See* Part II.

Olfactory bulbs An enlarged part of the brain that deals with the sense of smell. *See* Chapter 5.

Ontogeny Biological development of the individual; the growth trajectory from embryo to adult. *See* Chapters 6 and 11.

Opisthopubic The condition in which at least part of the pubis has rotated backward to lie close to, and parallel with, the ischium. *See* Part II.

Orbit Eye socket. *See* Chapter 4.

Ornithischia One of the two monophyletic groups comprising Dinosauria. *See* Chapter 4.

Ornithodira The common ancestor of pterosaurs and dinosaurs, and all its descendants. *See* Chapter 4.

Ornithopoda Biped and quadrupedal herbivorous ornithischian dinosaurs. *See* Chapter 7.

Ornithothoraces A group of Mesozoic birds including Aves. *See* Chapter 11.

Ornithurae Hesperornithiformes, Ichthyornithiformes, and Aves. *See* Chapter 11.

Ornithuromorpha The lineage of ornithothoracean birds, including Aves. *See* Chapter 11.

Osteichthyes Bony fishes that include ray-finned and lobe-finned gnathostomes. *See* Chapter 4.

Osteoderm Bone within the skin; may be small nodule, plate, or a pavement of bony dermal armor. *See* Chapter 5.

Oxidation Bonding of oxygen. *See* Chapter 12.

P

Pachycephalosauria Dome-headed ornithischians of North America and Asia who, together with Ceratopsia, make up Marginocephalia. *See* Chapter 6.

Palate The part of the skull that separates the nasal cavity (for breathing) from the oral cavity (for eating); usually strengthened by a paired series of bones. *See* Chapter 4.

Paleobiology The discipline of paleontology, arising in the 1970s, that actively sought to understand the biology of fossil organisms. *See* Chapter 14.

Paleobotany The study of ancient plants. A person who studies ancient plants is called a paleobotanist. *See* Chapter 13.

Paleoclimate Ancient climate. *See* Chapter 2.

Paleoenvironment Ancient environment.

Paleontology The study of ancient life; distinguished from anthropology, which is the study of humans, and archaeology, which is the study of past civilizations.

A paleontologist is someone who studies paleontology.

Paleozoic An Era lasting from 543 Ma to 251 Ma. *See* Chapter 2.

Palpebral A rod-like bone that crosses the upper part of the eye socket. *See* Part II.

Palynoflora Spores and pollen. *See* Chapter 16.

Pangaea The mother of all supercontinents, formed from the union of all present-day continents. *See* Chapter 2.

Parasagittal stance Stance in which the legs are held under the body. *See* Chapter 4.

Parascapular spine An enlarged spine over the shoulder. *See* Chapter 5.

Parsimony A principle that states that the simplest explanation that explains the greatest number of observations is preferred to more complex explanations. *See* Chapter 3.

Patellar groove A groove at the distal end of the femur to accommodate the patella (knee cap). *See* Chapter 11.

Pectoral girdle The bones of the shoulder; the attachment site of the forelimbs. *See* Chapter 4.

Pectoralis (muscle) The muscle that drives the wing's powerstroke in bird flight. *See* Chapter 10.

Pedestal In paleontology, a pillar of matrix underneath the fossil. *See* Figure 1.10.

Pelvic girdle The bones of the hips; the attachment site of the hindlimbs. *See* Chapter 4.

Perforate acetabulum A hole in the hip socket; a diagnostic character of Dinosauria. *See* Figure 4.5.

Period Subdivision of an Era, consisting of tens of millions of years. *See* Chapter 2 (especially Figure 2.4).

Permineralization The geological process in which the spaces in fossil bones become filled with a mineral. *See* Chapter 1.

Permo-Triassic Relating to events at the Permian/Triassic boundary (251 Ma). *See* Chapter 16.

Phalanx (pl. phalanges) Small bone of the fingers and toes that allows flexibility. *See* Chapter 4.

Phanerozoic That interval of time from 543 Ma to the present; it also refers to the time in Earth's history during which shelled organisms have existed. *See* Chapter 2.

Phenotype The physical appearance and features of an organism. *See* Chapter 3.

Photosynthesis The process by which organisms use energy from the sun to produce complex molecules for nutrition.

Phylogenetic Pertaining to phylogeny. *See* Chapter 3.

Phylogenetic systematics The method of determining organismic relationships that uses parsimony to select among competing hierarchical distributions of shared, derived characters (that is, cladograms). *See* Chapter 3.

Phylogeny The study of the fundamental genealogical connections among organisms. *See* Chapter 3.

Phylum A grouping of organisms whose makeup is supposed to connote a very significant level of organization shared by all of its members.

Phytosaur Long-snouted, aquatic, fish-eating member of Crurotarsi. *See* Figure 13.4.

Planktonic (or planktic) Living in the water column. *See* Chapter 16.

Plantigrade A foot position in which the bottom of the foot (the tissue below the metatarsals) lies flat on the ground. Opposite of digitigrade. *See* Chapter 8.

Plesiosaurs Long-necked fish-eating reptiles with large flippers that inhabited Mesozoic seas. *See* Figure 16.11.

Pleurocoel A well-marked excavation on the sides of a vertebra. *See* Chapter 8.

Pleurokinesis Mobility of the upper jaw. *See* Chapter 7.

Pneumatic Having air sacs or sinuses; the state of having such is called pneumaticity. *See* Chapter 8.

Pneumatic foramina Openings for air sacs to enter the internal bone cavities. *See* Chapters 8 and 10.

Poikilotherm Organism whose core temperature fluctuates. *See* Chapter 12.

Precocial The condition in which the young are rather adult-like in their behavior. *See* Chapter 7.

Predator : prey biomass ratios The ratio of the total estimated weight of predators to the total estimated weight of their prey in a particular ecosystem. *See* Chapter 12.

Predentary The bone that caps the front of the lower jaws in all ornithischians. *See* Part II.

Preparation (prep) lab Where preparation takes place. *See* Figure 1.11.

Prepare To clean a fossil; to get it ready for viewing by freeing it from its surrounding matrix. *See* Chapter 1.

Prepubic process A flange of the pubis that points toward the head of the animal. *See* Chapters 9 and 10.

Primary bone Bone tissue that was deposited or laid down first. *See* Chapter 12.

Primary productivity The sum total of organic matter synthesized by organisms from inorganic materials and sunlight. *See* Chapter 16.

Primitive *See* Ancestral. *See also* Chapter 3.

Process In relation to anatomy, part of a bone that is commonly ridge-, knob-, or blade-shaped and sticks out from the main body of the bone. *See* Chapter 4.

Productivity The amount of biological activity in an ecosystem. *See* Chapter 1.

Prosauropoda Monophyletic group of Late Triassic and Early Jurassic saurischian dinosaurs; the world's first high-browsing herbivores. Once thought to be ancestral to Sauropoda, now believed to share a common ancestor with sauropods. *See* Chapter 9, especially Figure 9.21, for diagnostic characters.

Prospect(ing) To hunt for fossils. *See* Chapter 1.

Proton Electronically charged (+1) subatomic particle that resides in the nucleus of the atom. *See* Chapter 2.

Proximal In anatomy, in the direction toward the central part (or core) of the animal.

Pterosauria Flying ornithodiran archosaurs, closely related to (but not included within) dinosaurs. Uniquely, the wing was supported by an extraordinarily elongate digit IV. *See* Chapter 4.

Pubis One of the three bones that make up the pelvic girdle. *See* Chapter 4.

Pull of the Recent The inescapable fact that as we get closer and closer to the Recent, fossil biotas become better and better known. *See* Chapter 13.

Pygostyle A small, compact, pointed structure made of fused tail bones in birds. *See* Chapter 10.

R

***r*-strategy** The evolutionary strategy where organisms have lots of offspring and no parental care. The letter r is obtained from the symbol *r* for growth rate. *See* Chapter 7.

Radiometric The dating method to determine unstable isotopic age estimations. *See* Chapter 2.

Radius One of the two lower arm bones; the other is the ulna. *See* Chapter 4.

Recent The time interval that encompasses 13 000 years ago to the present. *See* Chapters 2 and 13.

Recurved Curved backward, like a scimitar. *See* Chapter 9.

Regression Retreating of seas due to lowering of sea level. *See* Chapter 16.

Relationship The phylogenetic closeness of two organisms; that is, the genealogical nearness or distance of their most recent common ancestor. *See* Chapter 3.

Relative dating The type of geological dating that, although not providing ages in years before present, provides ages relative to other strata or assemblages of organisms. *See* Chapter 2.

Remodel In bone histology, to resorb or dissolve primary bone and deposit secondary bone. *See* Chapter 12.

Renewable resources Resources that can be synthesized or manufactured, either naturally or artificially. *See* Chapter 1.

Replace To exchange the original mineral with another mineral. *See* Chapter 1.

Reptilia The old Linnaean category for turtles, lizards, snakes, and crocodiles. Reptilia as formulated by Linnaeus and as commonly used is not monophyletic; only the addition of birds to these four groups constitutes a monophyletic group. *See* Chapter 10.

Respiratory turbinate A thin, convoluted or complexly folded sheet of bone located in the nasal cavities of living endothermic vertebrates. *See* Chapter 12.

Rhamphotheca Cornified covering on the upper and lower jaws (for example, a beak). *See* Part II and Chapter 5.

Robust (1) In the context of hypothesis testing, a hypothesis is said to be robust when it has survived repeated tests; that is, despite meaningful attempts, it has failed to be falsified. (2) In anatomy, strong and stout.

Rock An aggregate of minerals.

Rostral Referring generally to the rostrum, or snout region of the skull; in ceratopsians, a unique, diagnostic bone at the tip of the snout on the skull. *See* Chapter 6 (especially Figure 6.19).

Rostral bone A unique bone on the front of the snout of ceratopsians, giving the upper jaws of these dinosaurs a parrot-like profile. *See* Chapter 6.

S

Sacrum The part of the backbone where the hip bones attach. *See* Chapter 4.

Sarcopterygii Lobe-finned fish. *See* Chapter 4.

Saurischia One of the two monophyletic groups that Dinosauria comprises; the other is Ornithischia. *See* Chapter 4 and Part III.

Sauropoda Monophyletic group of long-necked, long-tailed, quadrupedal herbivorous saurischians. *See* Chapter 8, especially Figure 8.20, for diagnostic characters.

Scapula (pl. scapulae) The shoulder blade. *See* Chapters 4 (especially Figure 4.5) and 10.

Sclerotic ring A ring of bony plates that support the eyeball within the skull. *See* Figure 4.6 and Chapters 7 and 10.

Scute(s) A bony plate embedded in the skin; synonym: osteoderm. *See* Part II.

Seasonality Highly marked seasons. *See* Chapter 2.

Secondarily evolve To re-evolve a feature. *See* Chapters 4 and 15.

Secondary bone Bone deposited in the form of Haversian canals. *See* Chapter 12.

Secondary palate A shelf of bone, above the palate, over which air can be directed so that air is not mixed with the food during chewing. All mammals and crocodiles have secondary palates; some turtles do as well. *See* Chapter 5.

Sedimentary rock A rock that generally represents the lithification, or hardening, of sediment. *See* Chapter 1.

Sedimentologist Someone who studies sedimentary rocks and processes. *See* Chapter 1.

Seed A capsule that contains gametes, as well as nutrients. These are generally encased in a protective pod. *See* Chapter 13.

Semi-lunate carpal A distinctive, half-moon-shaped bone in the wrist. *See* Chapter 9.

Sexual dimorphism Size, shape, and behavioral differences between sexes. *See* Chapter 5.

Sexual selection Selection not between all of the individuals within a species, but between members of a single sex. *See* Chapter 6.

Shaft (1) The hollow main vane of a feather. (2) The title and eponymous lead male character in the first and most famous of the 1970s "blacksploitation" films. *See* Chapter 10.

Shocked quartz Quartz that has been placed under such pressure that the crystal lattice becomes compressed and distorted; correctly termed "impact metamorphism." *See* Chapter 16.

Sigmoidal Having an "S" shape. *See* Chapter 10.

Sinus A cavity. *See* Chapter 9.

Skeleton The supporting part of any organism. In vertebrates, the skeleton is internal and consists of tissue hardened by mineral deposits (sodium apatite). Such tissue is called "bone." *See* Chapter 4.

Skull That part of the vertebrate skeleton that houses the brain, special sense organs, nasal cavity, and oral cavity. *See* Figure 4.6.

Skull roof The bones that cover the top of the braincase. *See* Chapter 4 (especially Figures 4.6 and 4.9).

Soft tissue In vertebrates, all of the body parts except bones, teeth, beaks, and claws. *See* Chapter 1.

Specialization In biology, the idea that an organism is adapted for a particular circumstance (for example, diet, climate, ecosystem, a particular host [if it is a parasite], or any other aspect of its existence). Such an organism is termed a specialist. *See* Chapter 16.

Speciate Evolve new species; diversify. *See* Chapter 16.

Species-specific Appying only to a particular species (the one under discussion). *See* Chapter 5.

Specific (1) Diagnostic of a monophyletic group; uniquely evolved. (2) Adjectival form of the word "species," the smallest formal category in the Linnaean classification. Operationally (among living organisms), if two creatures breed and produce viable offspring, they are generally said to be the same species. *See* Chapter 4.

Sphenopsid A primitive type of vascular plant. *See* Figure 13.8.

Sprawling stance Stance in which the upper parts of the arms and legs splay out approximately horizontally from the body. *See* Figure 4.16.

Stable isotope An isotope that does not spontaneously decay. *See* Chapter 12.

Stapes The middle-ear bone that transmits sound (vibrations) from the tympanic membrane to a hole in the side of the braincase (allowing auditory nerves of the brain to sense vibration). *See* Chapter 4.

Stegosauria Thyreophorans (Ornithischia) with paired rows of bony osteoderms running along the back. *See* Chapter 5.

Sternum The breastbone. *See* Chapters 4 and 10.

Strata Layers of rock. *See* Figure 2.2.

Stratigraphy The study of the relationships of strata and the fossils they contain. *See* Chapter 2.

Subatomic Smaller than atom-sized. *See* Chapter 2.

Substitution In molecular biology, the notion that one of the bases in the pairs that compose a DNA molecule can be substituted for another base or base-pair. *See* Chapter 11.

Superposition The geological principle in which the oldest rocks are found at the bottom of a stack of strata and the youngest rocks are found at the top. *See* Chapter 2.

Supracoracoideus (muscle) The muscle that drives the wing's recovery stroke in bird flight. *See* Chapter 10.

Survivorship The pattern of survival measured against extinction. *See* Chapter 16.

Synapsida The large clade of amniotes, including mammals, diagnosed by a single temporal opening. *See* Chapter 4.

Synsacrum A single, locked unit consisting of the sacral vertebrae. *See* Chapter 10.

T

Tarsal Ankle bone. *See* Chapter 4.

Tarsometatarsus The name for the three metatarsals fused together with some of the ankle bones. *See* Chapter 10.

Taxon (pl. taxa) A group of organisms, designated by a name, of any rank within the biotic hierarchy. *See* Chapter 4.

Tectonic Referring to tectonics, that branch of geology dealing with the development and form of the surficial and near-surficial parts of the Earth. *See* Chapter 2.

Temnospondyl Belonging to the group Temnospondyli, large, carnivorous largely Paleozoic amphibians, whose behavior must have been something like that of a crocodile. *See* Chapters 12 and 13 (especially Figure 13.6).

Temporal (1) Referring to time. (2) In anatomy, the side region of the skull, above the jaw. *See* Chapters 2 and 4.

Temporal fenestra (pl. fenestrae) Openings in the temporal region of the skull. *See* Chapter 4.

Terrestrial Referring to land; that is, not marine. *See* Chapter 1.

Testable hypothesis A hypothesis that makes predictions that can be compared and assessed by observations in the natural world. *See* Chapter 3.

Tetanurae Literally, "stiff tails"; the clade of theropods with tails stiffened as a result of overlapping caudal zygapophyses. *See* Chapter 9.

Tethyan Referring to the ancient water mass that eventually became today's Atlantic Ocean. *See* Chapter 2.

Tetrapoda A monophyletic group of vertebrates primitively bearing four limbs. *See* Chapter 4.

Thecodontia A paraphyletic taxon that at one time was used to unite the separate ancestors of crocodilians, pterosaurs, dinosaurs, and birds. *See* Chapters 4 and 10.

Therapsida The clade of synapsids that includes mammals, some of their close relatives, and all of their most recent common ancestors. *See* Chapters 12, 13 (especially Figure 13.3), and 14.

Thermoregulation Control of body temperature. *See* Chapter 5.

Thorax In vertebrates, the part of the body between the neck and abdomen. *See* Chapter 8.

Thyreophora Armor-bearing ornithischians; stegosaurs, ankylosaurs, and their close relatives. *See* Part II and Chapter 5.

Tibia One of the two lower bones in the tetrapod hindlimb; the other is the fibula. *See* Chapters 4 and 10.

Trace fossil Impressions in sediment left by an organism. *See* Chapter 1.

Trachea The windpipe. *See* Chapter 8.

Trackway Group of aligned footprints left as an organism walks. *See* Chapter 1.

Transgression Advancing of seas due to rising sea level. *See* Chapter 15.

Triassic A Period lasting from 251 Ma to 200 Ma. *See* Chapter 2.

Triosseal foramen The hole formed by the coracoid, furcula, and scapula, through which the tendon for the supracoracoideus connects to the humerus for the wing recovery stroke in birds. *See* Chapter 10.

Tubercles Small bumps or protuberances. *See* Chapter 9.

Turn (a fossil) To separate the fossil from the surrounding rock at the base of the pedestal and to rotate it 180°. *See* Figure 1.10.

Tympanic membrane Eardrum. *See* Chapter 4.

U

Ulna One of the two lower arm bones; the other is the radius. *See* Chapter 4.

Unaltered When original mineralogy is unchanged. *See* Chapter 1.

Ungual phalange An outermost bone of the fingers and toes. *See* Chapter 4.

Unstable isotope An isotope that spontaneously decays from an energy configuration that is not stable to one that is more stable. *See* Chapter 2.

Upper temporal fenestra The opening in the skull roof above the lower temporal fenestra; *see* **Temporal fenestra.** *See* also Chapter 4.

V

Vane The sheet of feather material that extends away from the shaft. *See* Chapter 10.

Vascular Pertaining to vessels that conduct fluids. In animals, a region that is vascular has a lot of blood vessels; a vascular plant is a plant that has tubes (xylem and phloem) that conduct water and nutrients. *See* Chapter 13.

Vertebrae The repeated structures that compose the backbone and that, along with the limbs, support the rest of the body. *See* Chapter 4.

Vertebrata The group of all animals containing vertebrae. *See* Chapter 4.

W

Weathering The physical or chemical breaking down of earthly materials (for example, minerals, rocks, and bones). *See* Chapter 1.

Wedge The evolutionary pattern of waxing and waning dominance among groups of organisms. *See* Chapter 14 (especially Figure 14.13).

Z

Zygapophysis A fore-and-aft projection from the neural arches (of vertebrae). *See* Chapter 9.

FIGURE CREDITS

Special Acknowledgement

We give special thanks to the Natural History Museum (NHM) Vienna for allowing us to use photographs from the NHM archive as well as new photographs taken especially for this new edition. We particularly thank NHM Director Christian Köberl and Staff Photographer Alice Schumacher for all their help and generosity. Likewise, special thanks are due to the Austria-American Education Commission (Fulbright), who generously supported Fastovsky during the time that he was a Fulbright Scholar in Austria, writing this book.

Frontispiece © Natural History Museum, London

  1.1 Photograph courtesy of A. K. Behrensmeyer

  1.6 Photograph courtesy of D. J. Varricchio

  1.8 National Geographic Society

  1.9 Photograph courtesy of D. E. Fastovsky

1.10 Photographs courtesy of D.J. Nichols

1.12 Photograph courtesy of the Library, American Museum of Natural History

  2.2 Photograph courtesy of D. E. Fastovsky

  2.4 In Bergreen, W. A., Kent, D. V., Aubry, M.-P. and Hardenbol, J. (eds.) Geochronology, Time Scales and Global Stratigraphic Correlation, SEPM Special Publication no. 54, pp. 95–126.

  3.3 From the Big Book of Hell © Matt Groening, All Rights Reserved, Reprinted by permission of Pantheon Books, a division of Random House, Inc. New York

4.10 From Romer, A. A. and Price, L. I. 1940. Review of the Pelycosauria. Geological Society of America Special Paper no. 28

  5.9 Photograph courtesy of D. E. Fastovsky

5.10 Photograph courtesy of D. E. Fastovsky

5.14 Image courtesy of Scott Hartman; from Richard J. Butler, Roger M. H. Smith, David B. Norman, A primitive ornithischian dinosaur from the Late Triassic of South Africa, and the early evolution and diversification of Ornithischia, Proc. R. Soc. B 2007 274 2041-2046; DOI: 10.1098/rspb.2007.0367. Published 22 August 2007

5.15a Courtesy of the Natural History Museum Vienna

  6.5 Courtesy of the Natural History Museum Vienna

  6.6 Courtesy of the Library, American Museum of Natural History

  6.9 Courtesy of the Natural History Museum Vienna

6.12 Courtesy of the Natural History Museum Vienna

6.19 Courtesy of the Natural History Museum Vienna

6.20 Courtesy of the Natural History Museum Vienna

6.21 Photograph courtesy of D. E. Fastovsky

6.22 Courtesy of L. Witmer; inset - THE ANATOMICAL RECORD 292:1266–1296.

6.25 Museum of Natural History, Ulan Batoor

6.26 Photograph courtesy of Paul Sereno

6.31 Image courtesy of Julia Clark

6.32 Courtesy of the Library, American Museum of Natural History

  7.7 Modified after Stephen L. Brusatte, Dinosaur Paleobiology, Wiley-Blackwell, 2012, Fig 4.9

7.9 Photograph taken at Oxford University Museum of Natural History, 2006. Licensed under the Creative Commons Attribution - Share Alike 3.0 Unported License. (https://commons.wikimedia .org/wiki/File:Compsognathus_longipes_cast2.jpg)

7.10 Dr. Xu Xing, Institute of Vertebrate Paleontology and Paleoanthropology, Chinese Academy of Sciences; (d) Zhang Hailong

7.18 Dr. Xu Xing, Institute of Vertebrate Paleontology and Paleoanthropology, Chinese Academy of Sciences

7.19 Modifed after Stephen L. Brusatte, Dinosaur Paleobiology, Wiley-Blackwell, 2012.

7.20 Dr. Xu Xing, Institute of Vertebrate Paleontology and Paleoanthropology, Chinese Academy of Sciences; (d) Zhang Hailong

7.21 Specimen courtesy of X. Xu

7.22a. Dr. Xu Xing, Institute of Vertebrate Paleontology and Paleoanthropology, Chinese Academy of Sciences; (d) Zhang Hailong

7.22b. Dr. Xu Xing, Institute of Vertebrate Paleontology and Paleoanthropology, Chinese Academy of Sciences; (d) Zhang Hailong

7.23 The American Association for the Advancement of Science

7.24 Dr. Xu Xing, Institute of Vertebrate Paleontology and Paleoanthropology, Chinese Academy of Sciences. Redrawn from Nature 521, 70–73 (07 May 2015), doi:10.1038/nature14423

8.2 Photograph courtesy of J. Ostrom; images copyright Yale Peabody Museum

8.3 Courtesy of the Natural History Museum Vienna

8.4 Photographs courtesy of J. Ostrom; images copyright Yale Peabody Museum

8.5 Modified from Stephen L. Brusatte, Graeme T. Lloyd, Steve C. Wang, Mark A. Norell, Gradual Assembly of Avian Body Plan Culminated in Rapid Rates of Evolution across the Dinosaur-Bird Transition, Current Biology, Volume 24, Issue 20, 20 October 2014, Pages 2386-2392, ISSN 0960-9822, http://dx.doi.org/10.1016/j.cub.2014.08.034.

8.9 From Erickson, G. M., Rauhut, O. W. M., Zhou, Z. *et al.* 2009. Was dinosaurian physiology inherited by birds? Reconciling slow growth in *Archaeopteryx*. *PLoS ONE* 4(10): e7390. doi:10.1371/journal. pone.0007390

8.12 Courtesy of the Natural History Museum Vienna

8.15 Skeletal reconstruction by O. C. Marsh

8.16 Skeletal reconstruction by O. C. Marsh

8.17 Photograph courtesy of Julia Clark

9.12 Courtesy of the Natural History Museum Vienna

9.14 Courtesy of the Natural History Museum Vienna

9.16 Photograph courtesy of Martin Lockley

9.21 Adapted from Stephen L. Brusatte, *Dinosaur Paleobiology*, Wiley-Blackwell, 2012, p106

B.9.1.3 Bates, K. T., Manning, P. L., Hodgetts, D. and Sellers, W. I. 2009. Estimating mass properties of dinosaurs using laser imaging and 3D computer modelling. *PLoS ONE* 4(2): e4532. doi:10.1371/ journal.pone.0004532

B9.2.1a Michael A. Morales, American Museum of Natural History

B9.2.1b Library, American Museum of Natural History

III.1 Photograph by D. E. Fastovsky, taken at Department für Lithosphärenforschung, University of Vienna

III.2 Courtesy of the Natural History Museum Vienna

III.4 Dr. Xu Xing, Institute of Vertebrate Paleontology and Paleoanthropology, Chinese Academy of Sciences; (b) Dr. David Hone, Institute of Vertebrate Paleontology and Paleoanthropology, Chinese Academy of Sciences

10.12 Photograph courtesy of the Institut und Museum für Geologie und Paläontologie, Universität Tübingen

10.13 With permission of the Royal Ontario Museum © ROM

10.19 CT scan images courtesy of L. Witmer

10.23 (a) Photograph courtesy of the Library, American Museum of Natural History; (b) Museum of Natural History, Ulaanbaatar

10.24 From Arbour VM (2009), Estimating Impact Forces of Tail Club Strikes by Ankylosaurid Dinosaurs. PLoS ONE 4(8): e6738. doi:10.1371/journal.pone.0006738

11.7 Photograph courtesy of Michael A. Morales

11.11 Photograph courtesy of Hans Dieter Sues

11.17 Courtesy of the Natural History Museum Vienna

11.20 Courtesy of the Natural History Museum Vienna

11.26a Qingjin Meng, Jinyuan Liu, David J. Varricchio, Timothy Huang & Chunling Gao, Palaeontology: Parental care in an ornithischian dinosaur, *Nature* 431, 145-146 (9 September 2004), doi:10.1038/431145a

11.26b Photographs courtesy of Dr. Kh. Tsogtbaatar, Director, Museum of Paleontology, Mongolian Paleontological Center

11.27 "Goodwin and Horner, 2006 Proc. R. Soc. B (2006) 273, 2757–2761 doi:10.1098/rspb.2006.3643."

B11.1.1 Sereno, P. C., J. A. Wilson, L. M. Witmer, J. A. Whitlock, A. Maga, O. Ide, and T. A. Rowe. 2007. Structural extremes in a Cretaceous dinosaur. *PLoS ONE* 2(11): e1230. doi:10.1371/journal.pone.0001230.

11.32 (a–c) Photograph courtesy of J. M. Clark (d) Courtesy of the Natural History Museum Vienna

11.35 Courtesy of Wendy Sloboda

12.4 Courtesy of the Natural History Museum Vienna

12.8 Courtesy of the Natural History Museum Vienna

12.9 Courtesy of the American Museum of Natural History

12.12 Courtesy of D. B. Weishampel

13.2 Drawing courtesy of D. R. Carrier, Paleobiology, 13: 327

13.5 Redrawn from Dingus and Rowe, 1997

13.6 Fricke, H., and Rogers, R., 2000, GEOLOGY, v. 28, p. 799 - 802, Figure 1

13.8 Photomicrographs courtesy of A. Chinsamy-Turan, University of Capetown

13.9 Photomicrographs courtesy of A. Chinsamy-Turan, University of Capetown

13.10 Estimates for Syntarus and Massopondylus from the work of A. Chinsamy; estimates for Maiasaura from the work of J. Peterson and J. Horner

13.11 Redrawn from Erickson, G. M. 2005. Assessing dinosaur growth patterns: a microscopic revolution. *Trends in Ecology and Evolution*, 20, 677–84

13.12 Redrawn from Erickson, G. M. 2005. Assessing dinosaur growth patterns: a microscopic revolution. *Trends in Ecology and Evolution*, 20, 677–84

13.13 Figure from *Discover* magazine

13.15 Data from Barrick, R. E. Stoskopf, M. K., and Showers, W. J. 1997. Oxygen isotopes in dinosaur bone. In Farlow, J. O. and Brett-Summan, M. K. (eds.), *The Complete Dinosaur*. Indiana University Press, Bloomington, IN, pp 474–90

14.2 Data from Fastovsky *et al.*, 2004

14.11 Data from Fastovsky *et al.*, 2004

15.1 Natural History Museum, London

15.2 Natural History Museum, London

B15.1.1 Courtesy of the Library, American Museum of Natural History

15.5 Photographs courtesy of S. McCarthy and M. Gilbert, 1994, Crystal Palace Foundation

15.6 Illustrated London News (Jan 1854)

B15.3.1 Courtesy of the American Museum of Natural History

15.7 Courtesy of D. B. Weishampel

B15.5.1 Sternberg, C. H., 1985, *Hunting Dinosaurs in the Bad Lands of the Red Deer River*, Alberta Canada, NeWest Press, Edmonton, p. 235

INDEX OF SUBJECTS

INDEX OF GENERA